An Introduction to Textile Coloration

Titles in the Society of Dyers and Colourists – John Wiley Series

An Introduction to Textile Coloration: Principles and Practice
Roger H. Wardman

Physico-chemical Aspects of Textile Coloration
Stephen M. Burkinshaw

Standard Colorimetry: Definitions, Algorithms and Software
Claudio Oleari

The Coloration of Wool and Other Keratin Fibres
David M. Lewis and John A. Rippon (Eds)

An Introduction to Textile Coloration

Principles and Practice

Roger H. Wardman

Formerly Head of School of Textiles and Design
Heriot-Watt University
Edinburgh, UK

WILEY

Registered Office(s)
John Wiley & Sons, Inc., 111 River Street, Hoboken, NJ 07030, USA
John Wiley & Sons Ltd, The Atrium, Southern Gate, Chichester, West Sussex, PO19 8SQ, UK

Editorial Office
The Atrium, Southern Gate, Chichester, West Sussex, PO19 8SQ, UK

For details of our global editorial offices, customer services, and more information about Wiley products visit us at www.wiley.com.

Wiley also publishes its books in a variety of electronic formats and by print-on-demand. Some content that appears in standard print versions of this book may not be available in other formats.

Library of Congress Cataloging-in-Publication data applied for

Paperback: 9781119121565

Cover Design: Wiley
Cover images: Photograph kindly supplied by Theis GmbH & Co. KG, Germany;
(left image) © SHUTTER TOP/Shutterstock;
(right image) © gerenme/gettyimages

Set in 10/12.5pt Times by SPi Global, Pondicherry, India
Printed and bound in Malaysia by Vivar Printing Sdn Bhd

10 9 8 7 6 5 4 3 2 1

Contents

Society of Dyers and Colourists

Society of Dyers and Colourists (SDC) is the world's leading independent educational charity dedicated to advancing the science and technology of colour. Our mission is to educate the changing world in the science of colour.

SDC was established in 1884 and became a registered educational charity in 1962. SDC was granted a Royal Charter in 1963 and is the only organisation in the world that can award the Chartered Colourist status, which remains the pinnacle of achievement for coloration professionals.

We are a global organisation. With our head office and trading company based in Bradford, United Kingdom, we have members worldwide and regions in the United Kingdom, China, Hong Kong, India and Pakistan.

Membership: To become a member of the leading educational charity dedicated to colour, please email members@sdc.org.uk for details.

Coloration Qualifications: SDC's accredited qualifications are recognised worldwide. Please email edu@sdc.org.uk for further information.

Colour Index: The unique and definitive classification system for dyes and pigments used globally by manufacturers, researchers and users of dyes and pigments (www.colour-index.com).

Publications: SDC is a global provider of content, helping people to become more effective in the workplace and in their careers by educating them about colour. This includes textbooks covering a range of dyeing and finishing topics with an ongoing programme of new titles. In addition, we publish *Coloration Technology*, the world's leading peer-reviewed journal dealing with the application of colour, providing access to the latest coloration research globally.

For further information please email: info@sdc.org.uk or visit www.sdc.org.uk.

Preface

In 1993 the Society of Dyers and Colourists published the book *Colour for Textiles: A User's Handbook* by Dr Wilfred Ingamells, which covered the basic science and technology of textile coloration. Dr Ingamells wrote the book at a level at which those without a detailed scientific background could understand the fundamental principles underlying dyeing and printing processes and it proved to be a very successful publication.

During the 25 years since the publication of the book, there have been a considerable number of developments in the dyeing and printing industry, not least of which have been the challenges imposed by the drive towards environmentally friendly processes and restrictions on the use of certain chemicals. In response, the Publications Committee of the Society of Dyers and Colourists (SDC) considered it necessary to produce an updated version of the book, and I, together with Dr Matthew Clark, agreed to take on the task. Unfortunately Dr Clark had to withdraw from the project at an early stage, but did complete Chapter 5 on textile printing, and I gratefully acknowledge his contribution.

One of the aims of rewriting Dr Ingamells' book was to create a supporting textbook for the course 'Textile Coloration Certificate', which had been developed by Dr Clark and Mr Filarowski at the SDC and introduced in 2011. This course is aimed at personnel working in textile dyeing or printing companies, but who do not have a strong scientific background, so that they may attain a good understanding of the chemical principles involved in the processes with which they are involved. Accordingly, given the slightly different aim of this book, there is slightly more technical detail than in Dr Ingamells' book.

It is the intention that candidates who successfully complete this course will be very well prepared to continue their studies to enter the Society's examinations for Associateship, an honours degree level qualification. It is hoped therefore that this book additionally will provide a sound basis for students preparing for the ASDC qualification, though of course it is expected that they will also consult books available on the specialised topics related to textile coloration. At the introductory level of this book, it has not been possible to cover the dyeing of all the variants of textile fibre types. For example, there are many different types of polyester fibres and numerous types of fibre blends, and it is unrealistic to detail the processes involved in all these cases.

It was necessary for me to make some decisions about nomenclature in writing the book. In the teaching of organic chemistry in schools, the names of chemicals established by the rules of the International Union of Pure and Applied Chemistry (IUPAC) are used. In industry the original (trivial) names are still widely used, so to avoid confusion I have used these names also. Thus, for example, I have used 'ethanoic (acetic) acid', instead of just 'ethanoic acid'. Another issue to address was that of commercial names for products. I have tried to avoid giving commercial names as much as possible, so, for example, all dye structures are labelled with their Colour Index numbers.

Finally, in preparing the book, I am grateful to Dr Ingamells for allowing me to use, where I considered it appropriate, parts of his text and diagrams. I gratefully acknowledge the help of the library staff at the Scottish Borders campus of Heriot-Watt University, especially Mr Peter Sandison, Mr Jamie MacIntyre and Mrs Alison Morrison. I am particularly indebted to Mr Andrew Filarowski at the Society of Dyers and Colourists for so carefully going through the manuscripts of Chapters 1–6 and suggesting countless, yet very pertinent, modifications. His advice, support and links with the dyeing companies have been invaluable in the preparation of the book. Dr Jim Nobbs, formerly of Leeds University, very carefully scrutinised Chapter 7 and similarly provided many useful comments. Finally, I would like to thank Mr Alan Ross of High Street Textile Testing Services Ltd for carrying out very thoroughly a similar task on Chapter 8 and for making many useful corrections.

ROGER H. WARDMAN

1
General Chemistry Related to Textiles

1.1 Introduction

This chapter provides a background to the chemical principles involved in coloration processes, which will be beneficial to those with little working knowledge of dyeing chemistry. Chemistry has been classically divided into three branches: inorganic chemistry, organic chemistry and physical chemistry. Inorganic chemistry is the study of elements and their compounds. However carbon is so unique in the breadth of the compounds it forms (chiefly with hydrogen, oxygen, nitrogen and, to a lesser extent, sulphur) that it has its own branch – organic chemistry. Physical chemistry is concerned with the influence of process conditions such as temperature, pressure, concentration and electrical potential on aspects of chemical reactions, such as how fast they proceed and the extent to which they occur.

There are no clear distinctions between the three branches. For example, organometallic compounds are important substances that combine organic and inorganic chemistry, and the principles of physical chemistry apply to these two branches as well. Fundamental to all these branches of chemistry is an understanding of the structure of matter, so the chapter begins with this important aspect.

1.2 Atomic Structure

Modern chemistry is based on the belief that all matter is built from a combination of exceedingly minute particles (*atoms*) of the various chemical elements. Many different elements are found in nature, each possessing characteristic properties; the atoms of any one element are all chemically identical. An element is a substance made up of only one type of atom, for example, carbon is only made up of carbon atoms, and sodium is only made up of sodium atoms. Atoms combine together to form *molecules* of chemical compounds. A molecule is the smallest particle of a chemical element or compound that has the chemical properties of that element or compound.

A single atom consists of a very dense central core or *nucleus*, which contains numbers of positively charged particles called *protons* and uncharged particles, called *neutrons*. Protons and neutrons have equal mass and together they account for the atom's mass. A number of very small negatively charged particles, called *electrons*, circulate around the nucleus in fixed orbits or 'shells', each orbit corresponding to a certain level of energy: the bigger the shell (the further away from the nucleus it is), the greater the energy. These shells are labelled $n = 1$, 2, 3, etc., counting outwards from the nucleus, and each can hold a certain maximum number

An Introduction to Textile Coloration: Principles and Practice, First Edition. Roger H. Wardman.
© 2018 John Wiley & Sons Ltd. Published 2018 by John Wiley & Sons Ltd.

of electrons, given by $2n^2$. The movement of an electron from one energy level to another causes the absorption or emission of a definite amount of energy. Atoms are electrically neutral, so the number of electrons in an atom is exactly the same as the number of protons in its nucleus. The total number of electrons within an atom of a particular element is called the *atomic number* of the element. This is the same as the number of protons in its nucleus. It is the arrangement of the electrons around the nucleus of an atom that determines the chemical properties of an element, especially the electrons in the outermost shells.

It is possible that some of the atoms of an element have a different number of neutrons in their nucleus, but their numbers of protons and electrons are still the same. These atoms are called *isotopes*, and although they have the same chemical properties as the other atoms, their atomic masses are different. Also recent research into atomic structure has shown that the three subatomic particles are themselves made up of other smaller particles such as quarks, but for this book it is sufficient to only consider atoms in terms of protons, neutrons and electrons.

The simplest atom is that of hydrogen, which has a nucleus consisting of just one proton with one electron orbiting around it and has an atomic number of 1. In *deuterium*, an isotope of hydrogen, there is one neutron and one proton in its nucleus. So its atomic mass is 2, but its atomic number is still only 1. There are roughly 6400 atoms of 'normal' hydrogen for every atom of deuterium. Another example is chlorine, which has two stable isotopes – one with 18 neutrons and the other with 20 neutrons in the nucleus. Because each has 17 protons, their atomic weights (the combined weights of protons and neutrons) are 35 and 37, respectively. These two forms are labelled ^{35}Cl and ^{37}Cl. Approximately 75.8% of naturally occurring chlorine is ^{35}Cl and 24.2% is ^{37}Cl, and this is the reason why the periodic table of the elements shows the atomic weight of chlorine to be 35.45.

Within a shell there are *orbitals*, each of which can hold a maximum of two electrons. Within an orbital, the two electrons are distinguished by the fact that they are spinning around their own axis, but in opposite directions. In illustrating this diagrammatically the electrons in an orbital are often shown as upward and downward arrows ↑↓, for example, as in Figure 1.3. The orbital nearest the nucleus is called an s orbital, followed by p, d and f orbitals, which are occupied in the larger atoms. These orbital types have different shapes. The s orbitals are spherical, whilst the p orbitals have two lobes and are dumbbell shaped. The three p orbitals are all perpendicular to each other, in *x*, *y*, *z* directions around the nucleus, so are often labelled p_x, p_y and p_z (Figure 1.1). There are five d and seven f orbitals and these have more complex shapes.

The first shell ($n=1$) can accommodate only two electrons (according to the $2n^2$ rule) and there is just the s orbital. The next element, that of atomic number 2 (helium), has two electrons, both occupying the s orbital. In lithium (atomic number 3), its first shell contains two s electrons, but because that is now full, the third electron goes into the s orbital of the next shell. This second shell ($n=2$) now fills up, and after the s orbital is full, further electrons go into the p orbitals, as shown in Table 1.1. The p orbitals can hold a maximum of six electrons and after they are full the third shell ($n=3$) begins to fill.

Table 1.1 shows that once the three p orbitals of the third shell ($n=3$) are full in argon, the electron of the next element, potassium, goes into the fourth shell, instead of continuing to fill the third shell, which can hold a maximum of 18 electrons. However, after calcium, further electrons go into the third shell, into its d orbital, of which there are five, thus

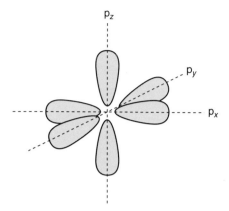

Figure 1.1 The three p orbitals.

Table 1.1 Filling of the shells by electrons in the first 30 elements.

Atomic number	Element	Orbit, n			
		1	2	3	4
1	Hydrogen	1s			
2	Helium	2s			
3	Lithium	2s	1s		
4	Beryllium	2s	2s		
5	Boron	2s	2s 1p		
6	Carbon	2s	2s 2p		
7	Nitrogen	2s	2s 3p		
8	Oxygen	2s	2s 4p		
9	Fluorine	2s	2s 5p		
10	Neon	2s	2s 6p		
11	Sodium	2s	2s 6p	1s	
12	Magnesium	2s	2s 6p	2s	
13	Aluminium	2s	2s 6p	2s 1p	
14	Silicon	2s	2s 6p	2s 2p	
15	Phosphorus	2s	2s 6p	2s 3p	
16	Sulphur	2s	2s 6p	2s 4p	
17	Chlorine	2s	2s 6p	2s 5p	
18	Argon	2s	2s 6p	2s 6p	
19	Potassium	2s	2s 6p	2s 6p	1s
20	Calcium	2s	2s 6p	2s 6p	2s
21	Scandium	2s	2s 6p	2s 6p 1d	2s
⋮					
30	Zinc	2s	2s 6p	2s 6p 10d	2s

holding a total of 10 electrons. After the d orbitals are all filled, at zinc, further electrons then fill up the 4p orbitals from gallium to krypton. Thereafter electrons go on to occupy the 5th orbit in a similar order, starting with the 5s orbital (rubidium and strontium).

1.3   Periodic Table of the Elements

During the nineteenth century, as new elements were discovered, chemists attempted to classify them according to their properties, such as metals and non-metals, or on the basis of their atomic weights. One of the most important methods of classification was Newlands' law of octaves, which he developed in 1865. Newlands considered that elements with similar chemical characteristics differed by either seven or a multiple of seven and created a table comprising rows of the known elements, in sevens. Going down the columns of his table gave, for example, hydrogen (H), fluorine (F) and chlorine (Cl) with similar chemical properties in the first column, then lithium (Li), sodium (Na) and potassium (K) in the second and so on. However Newlands' table was not entirely correct, but his ideas were taken further by Mendeleev who developed what has become known as the *periodic table of elements*. Mendeleev, in focussing on arranging the elements into families with the same valencies (see Section 1.4), produced a more accurate table and left spaces in it for elements he considered had yet to be discovered. In the years since many new elements have indeed been discovered and a complete version of the periodic table is shown in Figure 1.2.

The periodic table lists the elements in columns called *groups* and rows called *periods*. Moving across the table from left to right, in any given period the atomic number increases incrementally, meaning the size of the atom increases. Moving down the table from top to bottom, in any given group the elements have the same number of electrons in their outer shells. In group 1 all the elements – hydrogen (H), lithium (Li), sodium (Na) and so on – have one electron in their outer shell and are all very chemically reactive and readily form ionic bonds (see Section 1.4.1). In group 2, the elements beryllium (Be), magnesium (Mg), calcium (Ca) and so on all have two electrons in their outer shell and, whilst chemically reactive, are not quite as reactive as the group 1 elements, but still form ionic bonds. The elements in group 17 (called the halogens) all have one electron short of a complete outer shell and readily form ionic bonds with elements of groups 1 and 2. The elements in group 16 are two electrons short of a complete outer shell and again readily form ionic bonds with elements of groups 1 and 2. The elements in groups 13, 14 and 15 do not gain (or lose) electrons easily, so they tend to form covalent bonds with other elements (see Section 1.4.2). The elements in group 18 have completely full outer shells and are unreactive. These elements – helium (He), neon (Ne), argon (Ar) and so on – are called inert gases.

The elements in groups 3–12 are the transition elements. Those in period 4 – scandium (Sc) to zinc (Zn) – involve the filling of the inner d orbitals with electrons, as described in Section 1.2. Those elements in period 5 – yttrium (Y) to cadmium (Cd) – involve the filling of f orbitals. In the higher periods (periods 6 and 7), many of the elements are unstable and gradually break down through radioactive decay.

The periodic table in Figure 1.2 shows a solid black line, labelled the Zintl border. This line represents the boundary between metals and non-metals: the elements to the left and below the line are metals, and those above it and to the right are non-metals.

Periodic table of elements — Figure 1.2

Group	1	2	3	4	5	6	7	8	9	10	11	12	13	14	15	16	17	18
Period 1	1 H 1.0079																	2 He 4.0026
2	3 Li 6.941	4 Be 9.0122											5 B 10.811	6 C 12.0107	7 N 14.0067	8 O 15.9994	9 F 18.9984	10 Ne 20.179
3	11 Na 22.9898	12 Mg 24.305											13 Al 26.9815	14 Si 28.0855	15 P 30.9738	16 S 32.066	17 Cl 35.453	18 Ar 39.948
4	19 K 39.0983	20 Ca 40.078	21 Sc 44.9559	22 Ti 47.867	23 V 50.9415	24 Cr 51.996	25 Mn 54.9380	26 Fe 55.845	27 Co 58.933	28 Ni 58.693	29 Cu 63.546	30 Zn 65.409	31 Ga 69.723	32 Ge 72.64	33 As 74.9216	34 Se 78.96	35 Br 79.904	36 Kr 83.798
5	37 Rb 85.4678	38 Sr 87.62	39 Y 88.9059	40 Zr 91.224	41 Nb 92.9064	42 Mo 95.94	43 Tc 98.9062	44 Ru 101.07	45 Rh 102.9055	46 Pd 106.42	47 Ag 107.8682	48 Cd 112.41	49 In 114.818	50 Sn 118.710	51 Sb 121.760	52 Te 127.60	53 I 126.9045	54 Xe 131.29
6	55 Cs 132.9054	56 Ba 137.327	57–71 lanthanoids	72 Hf 178.49	73 Ta 180.9479	74 W 183.84	75 Re 186.207	76 Os 190.2	77 Ir 192.22	78 Pt 195.08	79 Au 196.9665	80 Hg 200.59	81 Tl 204.3833	82 Pb 207.2	83 Bi 208.9804	84 Po (209)	85 At (210)	86 Rn (222)
7	87 Fr (223)	88 Ra (226.0254)	89–103 actinoids	104 Rf (261.1088)	105 Db (262.1141)	106 Sg (266.1219)	107 Bh (264.12)	108 Hs (277)	109 Mt (268.1388)	110 Ds (271)	111 Rg (272)	112 Cn copernicium		114 Fl flerovium		116 Lv livermorium		

Lanthanoids:

57 La lanthanum 138.9	58 Ce 140.12	59 Pr 140.9077	60 Nd 144.24	61 Pm (147)	62 Sm 150.36	63 Eu 151.96	64 Gd 157.25	65 Tb 158.9254	66 Dy 162.50	67 Ho 164.9304	68 Er 167.26	69 Tm 168.9342	70 Yb 173.04	71 Lu 174.967

Actinoids:

89 Ac actinium	90 Th 232.0381	91 Pa 231.0359	92 U 238.0289	93 Np 237.0482	94 Pu (244)	95 Am (243)	96 Cm (247)	97 Bk (247)	98 Cf (251)	99 Es (252)	100 Fm (257)	101 Md (260)	102 No (259)	103 Lr (262)

Zintl border

Figure 1.2 Periodic table of elements.

Based on information from IUPAC, the International Union of Pure and Applied Chemistry (version dated 1st May 2013). For updates to this table, see http://www.iupac.org/reports/periodic_table.

1.4 Valency and Bonding

The number of electrons in the outermost shell considerably influences the chemical reactivity of the elements. For example, those elements with just one electron in their outermost shell (hydrogen, sodium, lithium, potassium) are very reactive, whilst those with eight electrons are very unreactive (the 'inert' gases helium, neon, argon, etc.), having what is termed *stable octets*. Although something of an oversimplification, it is convenient to assume that when bonds form between atoms of different elements, the atoms achieve an electronic configuration of a stable octet in their outermost shell. The achievement of a stable octet can be brought about either by atoms giving or receiving electrons or by the sharing of electrons.

1.4.1 Giving or Receiving of Electrons: Formation of Ionic Bonds

A good example of ionic bond formation is that of sodium chloride, formed by the reaction between sodium and chlorine atoms. As can be seen in Table 1.1, the sodium atom has an electronic configuration of 2.8.1, which means there are two (s) electrons in the first shell, eight (two s and six p) electrons in the second shell and a single (s) electron in its outermost shell. The chlorine atom has a configuration of 2.8.7 with seven (two s and five p) electrons in its outermost shell. Their electronic configurations are represented in Figure 1.3.

In the reaction between the two, one electron (the outermost) is transferred from the sodium atom to the chlorine atom. Sodium is then left with the configuration 2.8 and chlorine with 2.8.8, thus both having stable octets (Scheme 1.1).

Sodium
1s   2s $2p_x$ $2p_y$ $2p_z$ 3s

| ↑↓ | ↑↓ | ↑↓ | ↑↓ | ↑↓ | ↑ |

Chlorine
1s   2s $2p_x$ $2p_y$ $2p_z$ 3s $3p_x$ $3p_y$ $3p_z$

| ↑↓ | ↑↓ | ↑↓ | ↑↓ | ↑↓ | ↑↓ | ↑↓ | ↑↓ | ↑ |

Figure 1.3 Electronic arrangements of the sodium and chlorine atoms (the ↑↓ arrows indicate electrons with opposite spins).

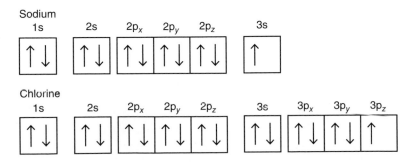

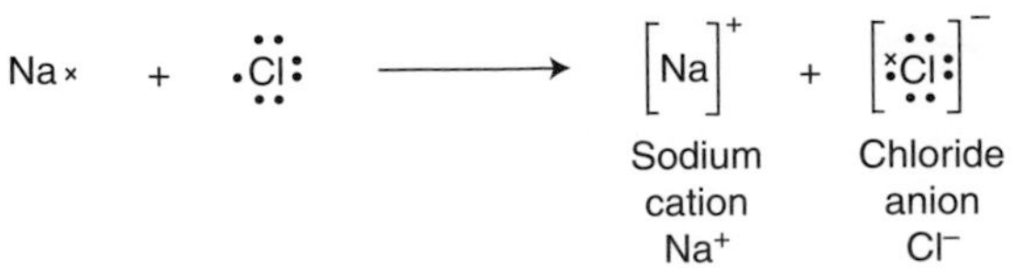

Scheme 1.1 Reaction between sodium and chlorine.

(The electron on the sodium atom represented by x is identical with those on the chlorine atom represented by •: they are only given different notations to show where the electrons come from.)

Since atoms started out electrically neutral, the loss of one electron in sodium leaves the atom positively charged (it is now a positive *ion*, or *cation*) and the gain of one electron leaves the chlorine atom negatively charged (a negative ion or *anion*). Crystals of sodium chloride are therefore made up of equal numbers of sodium and chloride ions. Since opposite charges attract each other, there is strong electrostatic attraction between the two kinds of ions, which makes sodium chloride a very stable compound. This type of bonding is called ionic bonding and is typical of compounds called *electrolytes*. When simple electrolytes of this type dissolve in water, they split up (*dissociate*) into their separate ions. It is for this reason that they allow an electric current to pass through water.

Sodium and chlorine react in equal numbers because they each have a *valency* of one, the sodium atom needing to lose one electron and the chlorine atom needing to gain one electron. In the case of atoms of an element with two electrons in their outermost shell, it is necessary for them to lose these two electrons to achieve a stable octet, so they have a valency of two. For example, a calcium atom needs to react with two atoms of chlorine, with the result that the ionic compound formed, calcium chloride, has the formula $CaCl_2$ (Scheme 1.2):

$$Ca\overset{x}{\underset{x}{}} \quad + \quad 2\, .\overset{..}{\underset{..}{Cl}}: \quad \longrightarrow \quad \left[Ca\right]^{2+} \quad + \quad 2\left[\overset{..}{\underset{..}{\overset{x}{.}Cl}}:\right]^{-}$$

Scheme 1.2 Reaction between calcium and chlorine atoms.

It is reasonable to expect that atoms of elements with three or four electrons in their outermost shells will need to react with three or four atoms of chlorine, respectively. However, in these cases, the removal of so many electrons is less easy and such elements tend to form covalent bonds (Section 1.4.2) where electrons are shared instead.

There are many simple electrolytes and the two most commonly used in dyeing and printing are sodium chloride (NaCl) and sodium sulphate (Na_2SO_4), the latter being known as Glauber's salt. The reason why there are two ions of sodium to one sulphate ion in Glauber's salt is that the sulphate ion is a complex ion, but requires two electrons to achieve stability: it has a valency of 2. A sodium atom has one electron in its outer shell (a valency of 1) and so two sodium atoms are required to satisfy this valency requirement. However, because a calcium atom has two electrons in its outer shell (it also has a valency of 2), it can react with the sulphate on a 1:1 basis, so calcium sulphate has the formula $CaSO_4$.

Water-soluble dyes are also electrolytes, but in this case the coloured part of the molecule is very large and usually an anion, whilst the cation, usually a sodium ion, is very small by comparison. In fact water-soluble dye molecules are all synthesised to contain at least one group of atoms known to confer water solubility on the dye molecule through the formation of ions. Very often this is the sulphonic acid (or sulphonate) group, $-SO_3H$, or the carboxylic acid group, $-COOH$, both of which form sodium salts that dissociate in water. In each case, their valency is one, so they form salts with sodium ions in a 1:1 ratio. The dissociation of a dye molecule with a sulphonate group is shown in Scheme 1.3, in which D represents the coloured part of the dye molecule.

$$
\begin{array}{c}
\quad\quad O \\
\quad\quad \| \\
D - S - ONa \\
\quad\quad \| \\
\quad\quad O
\end{array}
\longrightarrow
\begin{array}{c}
\quad\quad O \\
\quad\quad \| \\
D - S - O^- \quad + \quad Na^+ \\
\quad\quad \| \\
\quad\quad O
\end{array}
$$

Scheme 1.3 Dissociation of the sulphonate group.

1.4.2 Sharing of Electrons: Formation of Covalent Bonds

A covalent bond differs from the bonding in an ionic compound in that there is no transfer of electrons from one atom to another. Instead two atoms share two electrons, each atom providing one electron of the pair. Scheme 1.4 represents a covalent bond in the simple inorganic molecule of chlorine (Cl_2) as an example. As in Schemes 1.1 and 1.2, although electrons are given different symbols on the two atoms, this is just to show where they come from; in practice there is no difference between them.

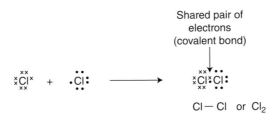

Scheme 1.4 Formation of the chlorine molecule.

By sharing a pair of electrons each of the chlorine atoms achieves a stable octet of electrons in their outer shells.

Elements whose atoms have four electrons in their outermost shells need to either gain or lose four electrons if they are to achieve a stable octet by forming ionic bonds. To do this requires too much energy and so instead they react with other atoms by forming covalent bonds instead, through the sharing of electron pairs. Such elements have a valency of 4. Typical of such an element is carbon, and the structures of all organic molecules, including dye molecules, are based on carbon atoms linked by covalent bonds.

The simplest organic compound is methane, CH_4 (Scheme 1.5). In methane there are four covalent bonds from the carbon atom, one to each hydrogen atom, arranged in the form of a symmetrical tetrahedron with the carbon atom in the middle (Figure 1.4).

The carbon atom has the electronic structure shown in Figure 1.5. There are only two unpaired electrons (in the p_x and p_y orbitals), so before the four covalent bonds can be formed, one of the 2s electrons must be promoted to the vacant p_z orbital, giving the electronic arrangement shown in Figure 1.6.

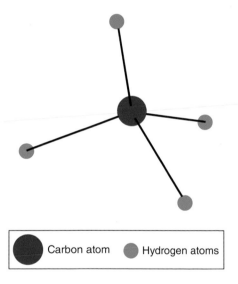

Scheme 1.5 Formation of methane (outer electrons only are shown); x electron from carbon, • electron from hydrogen.

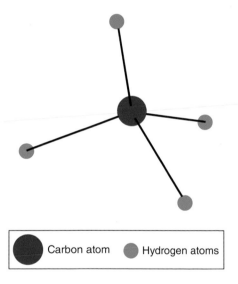

Carbon atom   Hydrogen atoms

Figure 1.4 Tetrahedral molecular structure of the methane (CH_4) molecule.

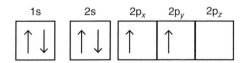

Figure 1.5 Electronic structure of the carbon atom.

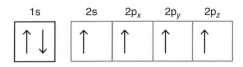

Figure 1.6 Electronic structure of the sp^3-hybridised carbon atom.

There are now four unpaired electrons, giving carbon a valency of four. The formation of the four orbitals, which now each contain one electron, is called *hybridisation*. The four sp^3 hybrid orbitals are equivalent and oriented tetrahedrally from the nucleus of the carbon atom. In methane the four bonds with hydrogen are formed by the overlap of the four sp^3 hybrid orbitals with the s orbitals of the hydrogen atoms.

On bonding with four hydrogen atoms, the carbon atom has eight electrons in its outermost shell and each hydrogen atom has two. The electronic requirements of both carbon and hydrogen atoms are satisfied and the compound is very stable. Such bonds do not dissociate in water; indeed most covalent compounds are insoluble in water and do not conduct electricity. Covalent bonds are the most stable of all chemical bonds and they cannot be broken easily. As indicated in Figure 1.4, the bonding is directional and covalent compounds therefore exist as molecules with a definite shape.

Organic chemistry is essentially the chemistry of carbon and the compounds it forms with atoms of other elements, most notably hydrogen, but also oxygen and nitrogen. Nitrogen with five electrons in its outermost shell and oxygen with six readily form covalent bonds, for example, with hydrogen to form ammonia and water, respectively (Scheme 1.6).

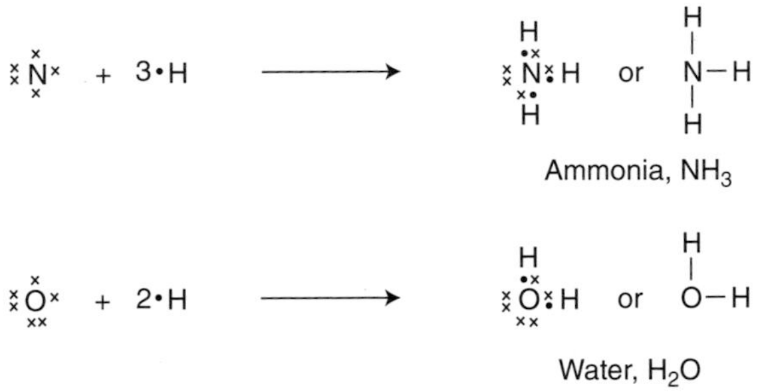

Scheme 1.6 Covalent bonding by nitrogen and oxygen atoms with hydrogen atoms.

Nitrogen and oxygen atoms (and also sulphur atoms) can also form covalent bonds readily with carbon atoms, so there is a vast range of organic compounds involving these elements. Dye and organic pigment molecules are comprised mainly of atoms of these elements covalently bound to each other. For all of these compounds, carbon has a valency of 4, so all C atoms can form four single bonds. The atoms of nitrogen, with a valency of 3, have three single bonds; of oxygen (valency 2) two single bonds, and of hydrogen (valency 1) one single bond. The classes of organic compounds are dealt with later in this chapter.

1.4.3 Secondary Forces of Attraction

Ionic and covalent bonds are referred to as primary forces. Most application classes of dyes are attracted to fibres by ionic bonds (e.g. acid dyes on wool and nylon, basic dyes on acrylics) and by covalent bonds (reactive dyes). In addition to these types of attractive forces between molecules, there exist other types called secondary forces because they are usually weaker than the primary forces. Although weaker, they have an important influence on the physical properties of organic compounds and indeed serve to enhance the attraction between dye molecules and fibres. The main types of secondary forces of attraction are dipolar forces, hydrogen bonding, π–H forces (pronounced 'pi–H') and dispersion forces. Collectively, these forces are often referred to as van der Waals forces.

Dipolar Forces

These forces occur between molecules that are *polar* in character. A molecule is polar if there exists some charge separation across it where some parts of the molecule are partially positive in character ($\delta+$) and other parts are partially negative ($\delta-$). Polarity in a molecule occurs when there is unequal sharing of the electrons of the covalent bond between two different atoms, because one has greater electronegativity than the other, the latter tending to pull the electron pair towards it from the other less electronegative atom. The polar molecules then attract each other, the slightly positive part of one molecule being attracted to the slightly negative end of the other (Figure 1.7).

Hydrogen Bonding

Hydrogen bonding can be regarded as a special case of dipole–dipole attraction. It is so named because it involves hydrogen and because hydrogen is such a small simple atom, when it is bound to another atom that is electronegative, such as oxygen, the shared electron pair forming the covalent bond between them is pulled so much more towards the oxygen that a strong charge separation occurs, creating polarity. The hydrogen atom then serves as a bridge, linking two other electronegative atoms. The most common example of a molecule that readily forms hydrogen bonds is water (Figure 1.8). Aside from forming between

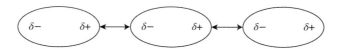

Figure 1.7 Dipole–dipole attraction between molecules.

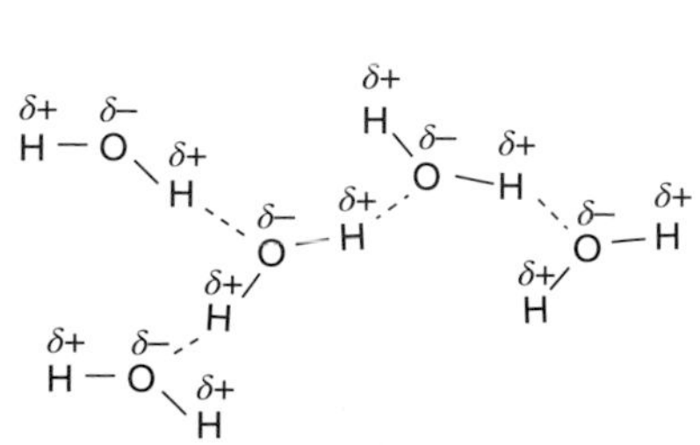

Figure 1.8 Hydrogen bonding in water (— covalent bond, ---- hydrogen bond).

neighbouring molecules (called intermolecular hydrogen bonds), these bonds can also occur between different functional groups of more complex organic molecules such as dyes and pigment molecules. This is called intramolecular hydrogen bonding and is often responsible for the good technical performance, especially the lightfastness, of many dyes and pigments.

π–H Bonding

This is a variation on hydrogen bonding in that it involves the interaction between the π-electron (pronounced 'pi' electron) system of an aromatic ring (see Section 1.8.1.2) that typically occurs in dye and pigment molecules and hydrogen donor groups such as the hydroxy (—OH) group. In the —OH group the O atom is slightly negatively charged and the H atom slightly positively charged. The π-electrons are attracted to the slightly positively charged hydrogen atom: this is called π–H bonding and is illustrated in Figure 1.9.

Dispersion Forces

These are very weak forces that exist between non-polar molecules, in which there is no charge separation in the molecule. They are explained by the fact that electrons in the orbits of the atoms that make up a molecule are in constant motion, so that at any instant small dipoles will be created. These transient dipoles will either induce oppositely oriented dipoles in neighbouring molecules or get into phase with the transient dipoles in them, creating an attractive force. Although these forces are very weak, they are considerable in number, so cumulatively can have an important effect.

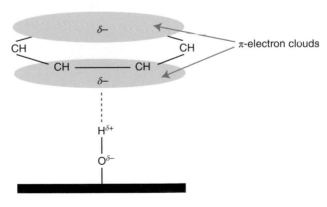

Figure 1.9   π–H bonding.

1.5 Chemical Reactions

1.5.1 Types of Chemical Reaction

Chemical reactions generally fall into three categories: synthesis, decomposition and displacement. In a *synthesis reaction*, two reactants combine to form a single product (Scheme 1.7).

$$A + B \rightarrow AB$$

Scheme 1.7 General formula representing a synthesis reaction.

The arrow $\rightarrow$ indicates that the reaction proceeds from left to right and is called the *forward reaction*. An example of a synthesis reaction is the formation of ammonia (NH_3) in the Haber process from nitrogen (N_2) and hydrogen (H_2), the equation for which is given in Scheme 1.8.

$$N_2 + 3H_2 \rightarrow 2NH_3$$

Scheme 1.8 Formation of ammonia from nitrogen and hydrogen.

When writing chemical equations it is important that they balance, that is, the number of atoms of each element is the same on each side of the arrow. Thus, in Scheme 1.8 the equation shows that two molecules of ammonia are formed and the equation balances because on each side of the arrow there are two atoms of nitrogen and six atoms of hydrogen. The equation is written in this way because it so happens that gases such as nitrogen, hydrogen and oxygen exist as diatomic molecules (N_2, H_2 and O_2 respectively).

In a *decomposition* reaction a compound breaks down into two or more products, represented in Scheme 1.9.

$$AB \rightarrow A + B$$

Scheme 1.9 General formula representing a decomposition reaction.

An example of this type of reaction is the decomposition of calcium carbonate into calcium oxide and carbon dioxide (Scheme 1.10).

$$CaCO_3 \rightarrow CaO + CO_2\uparrow$$

Scheme 1.10 Decomposition of calcium carbonate.

The $\uparrow$ arrow next to the CO_2 indicates that carbon dioxide is evolved as a gas.
Displacement reactions can be single displacement or double displacement (Scheme 1.11).

$$A + BX \rightarrow AX + B \qquad\qquad AX + BY \rightarrow AY + BX$$

Single displacement   Double displacement

$$\text{For example:} \quad Zn + H_2SO_4 \rightarrow ZnSO_4 + H_2\uparrow \qquad NaCl + AgNO_3 \rightarrow AgCl\downarrow + NaNO_3$$

Scheme 1.11   Single and double displacement reactions.

The $\downarrow$ arrow next to AgCl indicates that silver chloride is precipitated from solution.

The $\rightarrow$ arrow in all of the above reactions indicates that they go to completion; for example, in the above single displacement reaction between zinc and sulphuric acid, the zinc reacts with the acid until it is all used up. However many reactions do not go to completion; for example, the reaction between hydrogen and nitrogen to form ammonia reaches an equilibrium, because ammonia itself can decompose back into hydrogen and nitrogen. This state of affairs is represented by a $\rightleftharpoons$ sign. The reaction between nitrogen and hydrogen in the Haber process for synthesising ammonia shown above should be written as

$$N_2 + 3H_2 \rightleftharpoons 2NH_3$$

There are two aspects of chemical reactions that are important for practical reasons, especially in the case of industrial processes:

(1)   The rate at which reactions take place (reaction kinetics)
(2)   The driving forces of reactions and the extent to which they occur (thermodynamics)

These two aspects will be described in sections 1.5.2 to 1.5.5.

1.5.2   Rates of Chemical Reactions and Chemical Equilibria

For an industrial process to be commercially viable, a chemical reaction must proceed at a rate that produces the product in a reasonably quick time. The *law of mass action* states:

> *The rate of a chemical reaction is directly proportional to the active masses of the reactants.*

For a reaction in which a reactant R gives a product P (Scheme 1.12), the active mass of the reactant is indicated by [R]. As the reaction progresses, [R] decreases. The rate of reaction, given by $-d\,[\mathrm{R}]/dt$, is the decrease in its concentration with respect to time, the negative sign indicating the decreasing rate of reaction. By law of mass action,

$$-\frac{d\left[\mathrm{R}\right]}{dt} \alpha \left[\mathrm{R}\right], \quad \text{or} \quad -\frac{d\left[\mathrm{R}\right]}{dt} = k\left[\mathrm{R}\right], \tag{1.1}$$

where k is the rate constant for the reaction.

$$\mathrm{R} \rightarrow \mathrm{P}$$

Scheme 1.12   General formula representing a general reaction in which a reactant changes to a product.

For the general reversible reaction shown in Scheme 1.13,

$$A + B \rightleftharpoons C + D$$

Scheme 1.13 General formula representing a reversible reaction.

The position of the equilibrium depends on the relative rates of the forward and reverse reactions and is established when they are equal. By law of mass action the rate (V_f) of the forward reaction between A and B is given by

$$V_f \propto [A] \times [B], \quad \text{or} \quad V_f = k_f [A] \times [B], \tag{1.2}$$

where [A] and [B] are the concentrations (active masses) of the reactants A and B and k_f is the rate constant. The corresponding equations for the rate of the backward reaction between C and D (V_b) are

$$V_b \propto [C] \times [D], \quad \text{or} \quad V_b = k_b [C] \times [D] \tag{1.3}$$

As reactants A and B react together, their concentrations gradually decrease, so V_f decreases. At the same time products C and D are formed, and as more of them are formed, their rate of reaction gradually increases (V_b increases). There then comes a point when $V_f = V_b$: this is the point at which equilibrium is reached. Then

$$k_f [A] \times [B] = k_b [C] \times [D] \tag{1.4}$$

or

$$\frac{k_f}{k_b} = K \frac{[C] \times [D]}{[A] \times [B]} \tag{1.5}$$

The term K is called *equilibrium constant*. It is the ratio of the forward and backward reaction rates, but its value indicates where the position of equilibrium lies, that is, if it is in favour of the products or the reactants. Thus, if its value is large, then the yield of the products is high, and conversely if its value is small, the yield of products is low.

Whilst Equations 1.2, 1.3, 1.4 and 1.5 apply to a general chemical reaction between two reactants, the same logic can be applied to the adsorption of dyes by fibres. This is a physical process whereby dye transfers from aqueous solution into the solid fibre, but because dye molecules can desorb from the fibre, an equilibrium is established. This concept is considered further in Chapter 6.

The position of equilibrium of a reversible chemical reaction varies with conditions such as pressure (in the case of reactions between gases) or temperature. The way in which the position of equilibrium is affected is given by application of *Le Chatelier's principle*:

For a reaction at equilibrium, if one of the conditions changes, for example the temperature, the equilibrium shifts in a direction that nullifies the effect of the change.

If this principle is applied to the Haber process for synthesising ammonia, for example,

$$N_2 + 3H_2 \rightleftharpoons 2NH_3 - 92\,kJ/mol$$

heat is evolved in the forward reaction (the reaction is exothermic), so if heat is removed by lowering the temperature of the reaction, the equilibrium will compensate by producing more heat, that is, the equilibrium shifts in favour of the forward reaction and more ammonia is produced. Therefore it might be expected that the best yield of ammonia is produced at lower temperatures and indeed this is the case, but at lower temperatures, the rate at which ammonia is formed is very slow, so higher temperatures are preferred (about 400–450 °C) but at high pressure (about 200 atm). A very high pressure favours the forward reaction because 1 vol of nitrogen reacts with 3 vols of hydrogen (total 4 vols) to give 2 vols of ammonia. Thus ammonia occupies only half the volume of the reactants, and so its formation nullifies the effect of the high pressure.

Dyeing processes occurring in solution are unaffected by changes in atmospheric pressure but they are exothermic, so in principle, at equilibrium more dyes will be adsorbed by a fibre at lower temperatures. As with the Haber process though, the time taken to reach equilibrium at lower temperatures makes the dyeing process too long, so higher temperatures are preferred when diffusion of dye into the fibre is much quicker. Also, at higher dyeing temperatures, the fabric is wetted out more thoroughly.

1.5.3   Effect of Temperature on Rate of Reaction

The rate of reactions usually increases with temperature and the relationship between them is given by the *Arrhenius equation*:

$$k = Ae^{-E/RT} \tag{1.6}$$

where

k is the rate constant
A is the Arrhenius constant
e is the base of natural logarithms (for explanation of natural logarithms, see Section 1.5.5.4)
T is the absolute temperature
E is the *energy of activation*

This equation shows that the rate of reaction increases as the temperature increases. Typically the rate of a chemical reaction doubles for a 10 °C rise in temperature. Figure 1.10 shows the energy distribution of molecules at a low temperature and also at a higher temperature. It can be seen that at the higher temperature the energy of the molecules increases and that more molecules have energies in excess of the activation energy, E.

In any chemical reaction the reactants must possess sufficient energy so that when the molecules collide, a reaction between them occurs. The minimum energy required for a reaction to take place is the energy of activation, E, so that if reacting molecules possess less than this minimum energy collide, no reaction will occur. This is known as the *collision theory of reaction rates*.

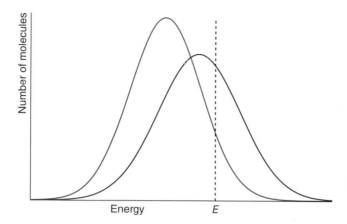

Figure 1.10 Effect of temperature on the energy of molecules: blue line low temperature; red line high temperature. *E* is the activation energy.

As the temperature increases, a greater number of molecules will possess more than this minimum energy and so achieve the necessary activated state for reaction to occur. The factor $e^{-E/RT}$ in the Arrhenius equation gives the proportion of collisions at a given temperature for which the colliding molecules have at least this minimum energy. Some reactions take place through the formation of an intermediate or *activated state* (the activated state in Figure 1.11), which exists for only a very short period of time, but its formation requires the additional energy given by *E*. This is the *activated state theory* of reaction rates.

1.5.4   Catalysts

A catalyst is a substance that increases the rate of a chemical reaction without changing itself chemically during the reaction. Although there are different mechanisms by which catalysts function, there are general characteristics that can be identified:

(1)   The catalyst is chemically unchanged at the end of the reaction, and the same amount is present at the end of the reaction as at the beginning. But in some cases there can be a change in the physical form of the catalyst.
(2)   Only a small amount of catalyst is required to considerably accelerate a reaction.
(3)   In the case of reversible reactions, the catalyst does not alter the final position of equilibrium, but it only accelerates the rate at which equilibrium is established.
(4)   The catalyst does not actually initiate a chemical reaction.

In the presence of a catalyst the same overall chemical process occurs, but by an alternative pathway, one that involves a smaller energy of activation. This means that under a given set of conditions, more of the reacting molecules will possess the minimum energy required for reaction to take place, so the process will take place more quickly. In Figure 1.11 this is the equivalent of the dotted line showing the activation energy being lower when the reaction is carried out in the presence of a catalyst.

(a)

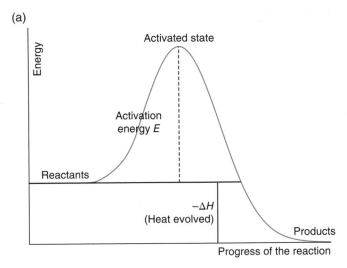

(b)

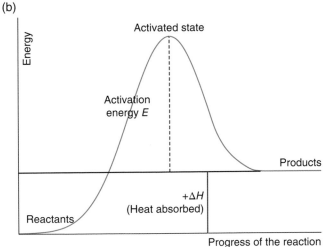

Figure 1.11   Energy pathway of a chemical reaction. (a) An exothermic reaction and (b) an endothermic reaction.

Catalysts can be classed as either *homogeneous* or *heterogeneous*. Homogeneous catalysts have the same phase as the reactants, for example, both may be gases (not all that common) or both may be liquids. An example of this is the use of concentrated sulphuric acid as a catalyst in the formation of esters (see Section 1.8.3.4). Heterogeneous catalysts are in a different phase from the reactants, usually a solid catalyst being used with gaseous or liquid reactants. Typical of this situation is the solid iron catalyst used in the reaction between the hydrogen and nitrogen gases in the Haber process.

Enzymes are an important class of catalyst, especially in textile fibre treatments, such as de-sizing (see Section 2.5.1.3) and bio-finishing, as well as in detergents generally. Unlike most other types of catalysts, they are highly specific in their action and are very sensitive to conditions of temperature and pH.

1.5.5 Thermodynamics of Reactions

Thermodynamic studies are concerned with the energetics of chemical reactions. They can be used to predict the outcome of reactions and calculate equilibrium constants of reversible reactions. Thermodynamic predictions are based only on the differences between the initial and final states, that is, once the system has reached its final equilibrium position. However they do not give any indication about the rate at which reactions occur, so although they may predict a reaction to give a good yield of a product, it may be that the rate at which the product forms takes such a long period of time that it is commercially unviable. Nevertheless, thermodynamic studies applied to dyeing processes can yield much useful information about the mechanism with which dye adsorption occurs and the influence of process conditions on (equilibrium) dye uptake.

The Sections 1.5.5.1–1.5.5.5 give a brief explanation of the laws of thermodynamics and their applications. Later, in Chapter 6, the relevance of thermodynamic studies to dyeing processes will be explained.

1.5.5.1 The First Law of Thermodynamics

The first law of thermodynamics is concerned with energy and the change of energy as systems move from one state to another, for example, when a chemical reaction takes place. It is also known as the *law of conservation of energy*, which states that energy cannot be created or destroyed.

In chemical thermodynamics the heat energy content of a system is called *enthalpy*, which has the symbol H. When a reaction takes place, the system may gain or lose heat energy, so it is the change in enthalpy, ΔH, that is of interest. ΔH is called the *heat of reaction*. It is positive if a chemical reaction is endothermic (absorbs heat) and negative if it is exothermic (liberates heat), as shown in Figure 1.11. A negative value for ΔH will favour a chemical reaction taking place.

The value of ΔH for a reaction can be determined from what are called the *standard enthalpies of formation* of the reactants and products involved. The standard enthalpy of formation, ΔH_f^o, of a compound is the change in enthalpy when it is formed from its elements in their most stable form at normal atmospheric pressure and a temperature of 298 K. (K is the temperature in *Kelvin*, measured on the absolute temperature scale. The units of Kelvin and centigrade temperature interval are identical. A temperature in centigrade is equal to the temperature in Kelvin minus 273.15.)

For example, when carbon reacts with oxygen to form carbon dioxide, the equation is represented as shown in Scheme 1.14.

$$C(graphite) + O_2(g) \rightarrow CO_2(g) \qquad \Delta H_f^o = -393.5 \text{ kJ/mol}$$

Scheme 1.14 Formation of carbon dioxide from carbon and oxygen.

The equation shows the carbon in its most stable form at normal atmospheric pressure and 298 K is solid graphite, whilst oxygen and carbon dioxide are gases. Values of the standard enthalpies of formation of compounds are available in tables in specialised textbooks. The values of ΔH_f^o for all elements are zero.

The standard enthalpy change for a chemical reaction, ΔH°, is given by

$$\Delta H^{\circ} = \Delta H_f^{\circ}\left(\text{products}\right) - \Delta H_f^{\circ}\left(\text{reactants}\right) \qquad (1.7)$$

For example, when methane burns in oxygen to give carbon dioxide and water (Scheme 1.15), the standard enthalpies of the formation of the reactants and products are as follows:

$$CH_4(g) + 2O_2(g) \rightarrow CO_2(g) + 2H_2O(l)$$

Scheme 1.15 Burning of methane in oxygen.

Products	Reactants
$CO_2(g)$ –393.5	$CH_4(g)$ –74.81
$H_2O(l)$ –285.8	$O_2(g)$ 0

Applying Equation 1.7 to obtain ΔH° for the reaction $= [-393.5 + 2(-285.8)] - (-74.8 + 0)$

$$= -965.1 + 74.8 \Rightarrow -890.3\,\text{kJ/mol}$$

1.5.5.2 *The Second Law of Thermodynamics*

One of the driving forces that make naturally occurring processes such as chemical reactions occur is the minimisation of energy. It might be expected therefore that if a reaction is exothermic (enthalpy is decreased due to the liberated heat), it will occur spontaneously. This is usually the case if ΔH is large, but not necessarily if it is small. Conversely there are some endothermic reactions (enthalpy increases) that are also spontaneous. This apparent contradiction implies that there must be some other factor involved in governing whether or not a reaction proceeds and this factor is the *entropy*, which is given the symbol S. According to the second law of thermodynamics, the total entropy, that is, of a system and surroundings, increases during all naturally occurring processes.

Entropy is a measure of the randomness, or disorder, of a system. The tendency for entropy to increase is the second driving force of reactions. The concept of entropy can be difficult to understand but one way of imagining it is when liquid molecules vaporise into gaseous form. In doing so molecules move from a condition in which they are highly ordered to one where they have much more freedom to move. The opposite situation occurs during a dyeing process because the dye molecules move from a state of relatively high randomness when they are in solution to one where they are much more constrained once they have become adsorbed in the solid fibre. As with enthalpy changes, ΔH, it is the change in entropy, ΔS, as a chemical reaction takes place that is of interest. Depending on the reaction (or dyeing process), ΔS may be positive, negative or zero.

During a reversible process at a constant temperature, T, a small change in entropy dS is related to an infinitesimal amount of heat absorbed, q_{rev}, by the equation

$$dS = \frac{dq_{rev}}{T} \qquad (1.8)$$

and for a finite change between the initial and final states

$$\Delta S = \frac{q_{rev}}{T} \qquad (1.9)$$

Although the notion of a reversible change is required to obtain ΔS, it does not matter what path the process takes in transitioning from the initial to the final state or whether it is thermodynamically reversible or not. A positive value for ΔS will favour a reaction occurring.

1.5.5.3 The Third Law of Thermodynamics

The third law of thermodynamics states that all truly perfect crystals at absolute zero temperature have zero entropy. By using heat capacity data it is possible to calculate the entropy of a compound in its standard state, S^o. Values of S^o are available in tables in specialised textbooks.

1.5.5.4 Free Energy

In attempting to predict whether a chemical reaction will occur or not, it has been established above that the loss of enthalpy (i.e. a negative value of ΔH) and a gain in entropy (i.e. a positive value of ΔS) will each have a favourable influence. Conversely, if both are unfavourable (i.e. ΔH is positive and ΔS is negative), the reaction will not occur. However, if one is favourable and the other is not, a measurement is required that comprises two components of enthalpy and entropy changes. This measure is the *free energy*, G, also known as the *Gibbs free energy*, which is defined by the equation

$$G = H - T \cdot S \qquad (1.10)$$

As with enthalpy and entropy, thermodynamics is only concerned with the change in free energy, ΔG, when a process takes place. For a reaction occurring at constant temperature T,

$$\Delta G = \Delta H - T \cdot \Delta S \qquad (1.11)$$

If values for T, ΔH and ΔS are applied to Equation 1.11, the value of ΔG obtained indicates whether a reaction will take place. If ΔG has a negative value, the reaction will occur; if it is positive, the reverse reaction occurs; and if it is zero, equilibrium is attained. However, it should be remembered that the magnitude of ΔG is not related to the rate at which the reaction takes place; it only indicates the tendency for it to occur. In practice the reaction may be very slow or very fast – thermodynamics gives no indication of the rate.

It is convenient to determine the change in free energy under standard conditions (1 atm and 298 K), and this can be achieved using the equation

$$\Delta G^\circ = \Delta H^\circ - T \Delta S^\circ \qquad (1.12)$$

Often, it is necessary to know the value of ΔG° at some other temperature. This can be done, but it involves knowledge of the heat capacities of the products and reactants, so that the enthalpy changes involved in heating (or cooling) the reactants and products and of the entropy values at the new temperature can be calculated.

Another important equation in thermodynamics is the relation between the Gibbs free energy and the equilibrium constant of a reaction:

$$\Delta G^\circ = -R \cdot T \cdot \ln K \qquad (1.13)$$

where

R is the gas constant,
T is the absolute temperature,

ln is the symbol for natural logarithm, that is, logarithm to the base 'e', which has a value of 2.718. In the equation $a = b^n$, n is the logarithm of a. Common logarithms are to the base 10, so in $a = 10^n$ n is the logarithm of a and is written '$\log_{10} a$' or often just '$\log a$'. In the case of natural logarithms $a = 2.718^n$, n is now the natural logarithm of a and is written 'ln a'. The relationship between natural and common logarithms is: $\ln a = 2.303 \log a$.

K is the equilibrium constant (see Equation 1.5).

Interpretation of this equation shows that for a reaction that has reached equilibrium, if $\Delta G^\circ < 0$, then $K > 1$, and vice versa. If $\Delta G^\circ = 0$, then $K = 1$ and the position of equilibrium is such that there are equal amounts of products and reactants.

1.5.5.5   Interpreting Thermodynamic Data

In Section 1.5.2 reference was made to Le Chatelier's principle and thermodynamic values can be used to provide quantitative evidence in support of it. If Equations 1.12 and 1.13 are combined,

$$-R \cdot T \cdot \ln K = \Delta H^\circ - T \Delta S^\circ \qquad (1.14)$$

so

$$\ln K = -\frac{\Delta H^\circ}{RT} + \frac{\Delta S^\circ}{R} \qquad (1.15)$$

Entropy changes, ΔS°, vary little over small changes in temperature, so $\Delta S^\circ / R$ can be regarded as constant, and

$$\ln K = -\frac{\Delta H^\circ}{R} \cdot \frac{1}{T} + C \qquad (1.16)$$

Differentiating Equation 1.16 gives

$$\frac{d(\ln K)}{dT} = \frac{\Delta H^\circ}{RT^2} \tag{1.17}$$

This equation, known as the Van't Hoff equation, indicates that if ΔH° is negative, then K will decrease with an increase in temperature, and vice versa. This is significant in dyeing processes for which ΔH° is negative, indicating that for dyeing processes at equilibrium, less dye will be adsorbed by the fibre at higher temperatures. This behaviour will be considered further in Chapter 6.

Equation 1.16 can be developed to form

$$\ln \frac{K_1}{K_2} = -\frac{\Delta H^\circ}{R}\left(\frac{1}{T_1} - \frac{1}{T_2}\right) \tag{1.18}$$

This equation provides a useful method for determining the value of ΔH° of a dyeing process if the values of K_1 and K_2 at temperatures T_1 and T_2, respectively, are determined experimentally.

1.6 Acids, Bases and Salts

1.6.1 Acids and Bases

Acids are substances that dissociate in solution to liberate hydrogen ions, H^+. For example, hydrogen ions are formed when hydrogen chloride dissolves in water, forming hydrochloric acid:

$$HCl \xrightarrow{aq} H^+ + Cl^-$$

Acids are characterised by their sour taste and corrosive action. The most common inorganic acids are hydrochloric acid (HCl), sulphuric acid (H_2SO_4) and nitric acid (HNO_3). These acids are described as being 'strong' acids because they dissociate almost completely in water. There exist also acids, usually organic acids, that are described as being 'weak' acids because they only dissociate partially in water. These acids contain the carboxyl group —COOH, typical examples being methanoic acid (formic acid, HCOOH) and ethanoic acid (acetic acid, CH_3COOH).

Bases are substances that create hydroxyl ions (OH^-) in water. Typical of the bases are alkalis, which are the hydroxides of metals. A widely used alkali in preparation and dyeing processes is sodium hydroxide (caustic soda), NaOH. As with acids, bases can be described as being 'strong' or 'weak', for example, sodium hydroxide is a 'strong' alkali. Weaker alkalis are ammonium hydroxide (NH_4OH) and organic bases such as methylamine (CH_3NH_2) or ethylamine ($C_2H_5NH_2$).

1.6.2 The pH Scale

The pH scale is a convenient way of expressing the strength of solutions of acids or bases. A solution with a pH value of 7 is neutral. pH values of less than 7 represent acidity – the lower the value, the more acidic the solution. Values greater than 7 represent alkalinity (basicity) – the higher the value, the more alkaline the solution. The full scale ranges from 0 to 14.

These figures are not arbitrary. They are derived from the concentration of hydrogen ions in the solution, written [H⁺]. Consider first the case of water, H_2O. Even liquid water undergoes dissociation to some extent: about one water molecule in every 10 million molecules dissociates:

$$H_2O \rightarrow H^+ + OH^-$$

Each positively charged hydrogen ion is accompanied by a negatively charged hydroxide ion, so that the numbers of positive and negative charges in the solution are balanced, and

$$\left[H^+ \right] = \left[OH^- \right] = 10^{-7}$$

When [H⁺] and [OH⁻] are multiplied together, a constant value of 10^{-14} is obtained, that is:

$$\left[H^+ \right] \times \left[OH^- \right] = 10^{-14}$$

An acidic solution contains more hydrogen ions than hydroxide ions. Even in solutions of weak acids, the number of hydrogen ions is very large, so the pH value expresses the concentration of hydrogen ions on a logarithmic scale:

$$pH = -\log \left[H^+ \right] \quad \text{or} \quad pH = \log \left(\frac{1}{\left[H^+ \right]} \right) \tag{1.19}$$

For pure water the pH = $-\log(10^{-7})$, which is 7.

For acidic solutions [H⁺] is greater than 10^{-7}. For example, in a solution of 0.01 M hydrochloric acid, the [H⁺] is 10^{-2}, so the pH value will be 2.

In alkaline solutions the concentration of hydrogen ions is less than in neutral water, and the concentration of hydroxide ions is higher. For example, in a solution of 0.01 M sodium hydroxide, NaOH, the [OH⁻] will be 10^{-2}, so the [H⁺] will be 10^{-12}, in which case the solution has pH = 12.

1.6.3 Salts and Salt Hydrolysis

When an acid is neutralised by a base, salt and water are formed. The hydrogen atoms of the acid are replaced by atoms of a metal. Typical examples are the reaction between hydrochloric acid and sodium hydroxide:

$$HCl + NaOH \rightarrow NaCl + H_2O$$

and between sulphuric acid and sodium hydroxide:

$$H_2SO_4 + 2NaOH \rightarrow Na_2SO_4 + 2H_2O$$

Note that in the second case, sulphuric acid has two hydrogen atoms that must be replaced, so two molecules of sodium hydroxide are required. The salts formed in each case, sodium chloride and sodium sulphate, are named according to the metal and the acid from which they were derived. These two salts are important as auxiliaries in dyeing processes. Sodium chloride is used in the application of anionic dyes (such as direct dyes, reactive dyes and vat dyes) to cotton, and sodium sulphate is used in the application of acid dyes to wool and nylon.

When salts are dissolved in water, they do not necessarily produce solutions with neutral pH; it just depends on the strengths of the acid and base from which they were formed. The salts in the two examples above were both produced from a strong base (sodium hydroxide, NaOH) and a strong acid (hydrochloric acid, HCl, and sulphuric acid, H_2SO_4). In solution both of these salts give neutral solutions.

However, a salt made from a strong base but a weak acid will give an alkaline solution (pH > 7). An example of this is sodium ethanoate (sodium acetate): the acid from which the salt is derived is ethanoic acid (acetic acid), which is a weak acid, but the base, sodium hydroxide, is, as noted above, a strong base.

Conversely, a salt made from a weak base and a strong acid will give an acidic solution (pH < 7). In this case an example is ammonium sulphate, $(NH_4)_2SO_4$, for which the weak base is ammonium hydroxide, NH_4OH, and the strong acid is sulphuric acid.

The last combination, that of salts produced from a weak base and a weak acid, such as ammonium ethanoate (ammonium acetate), CH_3COONH_4, is likely to be neutral, but may be slightly acidic or slightly alkaline, depending on the relative strengths of the acid and base.

Other important salts in dyeing processes are those of orthophosphoric acid, H_3PO_4. This acid is tribasic and three sodium salts of it are possible: sodium dihydrogen orthophosphate, NaH_2PO_4; disodium hydrogen orthophosphate, also called 'sodium phosphate', Na_2HPO_4; and trisodium orthophosphate, Na_3PO_4. Orthophosphoric acid is a weak acid and trisodium orthophosphate gives an alkaline solution. Disodium hydrogen orthophosphate is neutral in solution and sodium dihydrogen orthophosphate gives a slightly acidic solution.

This behaviour of salts is termed *salt hydrolysis*. Salts that give acidic or alkaline solutions are used in dyeing processes where conditions of pH that are not neutral are required. In some dyeing processes it is highly desirable that a certain pH is maintained throughout, and to ensure that this happens, a *buffer solution* is used. This is the subject of the next section.

1.6.4   Buffer Solutions

A buffer solution is capable of maintaining a constant pH value, even if a small quantity of acid or alkali is added to it. Such a solution usually consists of two components: a mixture of a weak acid and its salt, such as ethanoic (acetic) acid/sodium ethanoate (acetate), or of a weak base and its salt, such as ammonium hydroxide/ammonium sulphate. Buffer solutions of any desired pH can be made by appropriate formulation of the two components, though they do have what is called a *buffer capacity*, that is, there is a limit to the amount of acid or alkali that they can absorb without their pH changing.

1.7 Redox Reactions

Oxidation–reduction reactions (*redox reactions*) are important in textile coloration because they are an essential part of the process of the application of vat and sulphur dyes. In rather oversimplified terms, when a compound is oxidised, it gains oxygen; when it is reduced, it loses oxygen.

Reduction of a substance can also be thought of as gaining hydrogen atoms and oxidation as losing hydrogen atoms. For example, when hydrogen reacts with oxygen to form water, hydrogen becomes oxidised and oxygen is reduced. In a redox reaction there is always a compound acting as a *reducing agent* (hydrogen in this example). The reducing agent becomes oxidised during the reaction by the compound that is being reduced, which is acting as an *oxidising agent* (in this case oxygen).

In the water molecule, each hydrogen atom shares the only electron it possesses by pairing with one of the six electrons of the oxygen atom to form a covalent bond (see Scheme 1.6). Thus the hydrogen atom has lost one electron to become oxidised and the oxygen atom is reduced by gaining electrons. This is a more general way of expressing the phenomena of oxidation and reduction:

- Oxidation entails the loss of electrons by the oxidised compound.
- Reduction entails a net gain of electrons by the reduced compound.

Vat dyes such as indigo and compounds derived from anthraquinone are applied after the temporary reduction of two carbonyl groups (C=O) in a conjugated chain (see Section 1.8.1); this converts the dye into a colourless water-soluble form. The conversion is carried out using a strong reducing agent such as sodium dithionite and sodium hydroxide. In the reaction (Scheme 1.16) the two oxygen atoms become reduced to —O⁻.

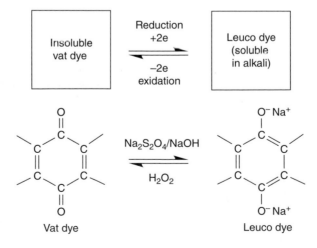

Scheme 1.16 Reduction of vat dyes.

The reduced (soluble) form is called the *leuco* form and is applied from an alkaline solution. Once adsorbed on the fibre, the dye can be re-oxidised back to the insoluble carbonyl form by air or by the use of an oxidising agent.

Sulphur dyes are also applied using a redox reaction mechanism (Scheme 1.17), in which sodium sulphide is used as the reducing agent.

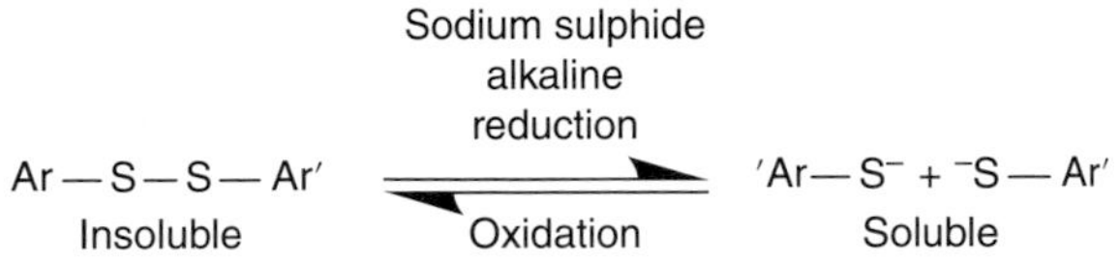

$$Ar-S-S-Ar' \xrightarrow[\text{Oxidation}]{\substack{\text{Sodium sulphide} \\ \text{alkaline} \\ \text{reduction}}} {}'Ar-S^- + {}^-S-Ar'$$

Insoluble   Soluble

Scheme 1.17   Redox reactions of sulphur dyes.

1.8   Organic Chemistry

Organic chemistry is the study of compounds containing carbon. There are so many compounds containing carbon, well over one million, that a branch of chemistry different from inorganic chemistry (which deals with the chemistry of all the other elements) is justified. Carbon atoms have the ability to combine with themselves to form chains and these chains can have different lengths, be straight or branched. In addition, the ends of chains may join up at their ends to form closed rings.

Organic compounds that comprise carbon and hydrogen atoms are referred to as *hydrocarbons*. However, as noted in Section 1.1, carbon can also form compounds involving atoms of other elements, such as oxygen and nitrogen, which greatly extends the range and characteristics of organic compounds. Some organic molecules are highly complex in structure, containing small groups of atoms, known as functional groups. Examples of functional groups are alcohols (–OH), amines ($-NH_2$) and carboxylic acids (–COOH), which will be described in later sections. The properties of an organic compound depend on the properties of the individual functional groups that make up the molecule.

In simple terms, there are two classes of organic compounds: *aliphatic* and *aromatic*. Aromatic compounds are ring structures based on benzene and are so called because when they were first discovered they were considered to have pleasant aromas. The rest are aliphatic compounds, but many of these also have distinctive aromas.

The rules for naming organic compounds are now very formalised: as organic chemistry developed and new compounds discovered, trivial names were given to them, some of which are still used today. In the textile manufacturing, dyeing and finishing industries, some of these trivial names are still used, so it is necessary to be somewhat 'bilingual' and know the two names often used for the same chemical.

1.8.1   The Hydrocarbons

Hydrocarbons (compounds made up of just hydrogen and carbon atoms) can occur as straight (and branched) chain structures or as ring structures. The chain compounds are called *aliphatic* hydrocarbons, whilst the ring structures are called *aromatic* compounds.

1.8.1.1 Aliphatic Hydrocarbons

There are three families, called *homologous series*, of aliphatic hydrocarbons:

(1)   Alkanes (paraffins)
(2)   Alkenes (olefins)
(3)   Alkynes (acetylenes)

Alkanes have the general formula C_nH_{2n+2}, where *n* is the number of carbon atoms in the molecule. The simplest compound is methane, CH_4 (**1.1**), where $n = 1$. The next member of the series is ethane, C_2H_6 (**1.2**), followed by propane, C_3H_8 (**1.3**), and then butane, C_4H_{10} (**1.4**). These alkanes have the following structures.

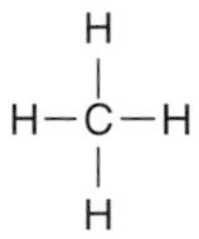

| 1.1 | 1.2 | 1.3 | 1.4 |

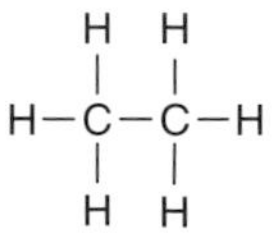

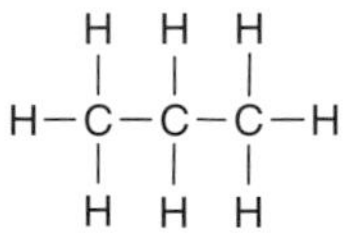

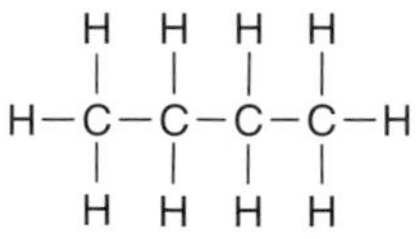

It can be seen that moving from one member to the next upwards in this homologous series adds a CH_2 group. These are all straight-chain hydrocarbons, but it is also possible to write the structure of butane with a branched chain (**1.5**), which is called an *isomer* of butane. This isomer is named 2-methylpropane (or isobutane).

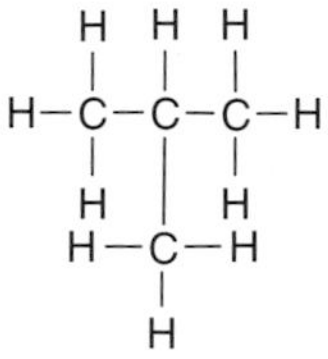

1.5

Going up the homologous series further, there are greater possibilities for branched structures to be drawn.

An important structural feature of carbon chains is their zigzag form, illustrated in Figure 1.12.

As illustrated in Figure 1.4, the four bonds of a carbon atom are oriented tetrahedrally, and because the angles between single bonds are fixed, a chain of carbon atoms cannot lie flat in a straight line. Instead it is forced into a zigzag arrangement. This is important because it governs the three-dimensional shapes of organic molecules, which may in part determine their physical and chemical properties.

Alkanes are chemically unreactive, but burn readily in air to give carbon dioxide, CO_2, and water, H_2O. On ascending the series, the physical properties of the members gradually change, so the first four are all gases, those between pentane (C_5H_{12}) and heptadecane ($C_{17}H_{36}$) are all liquids, and those from $C_{18}H_{38}$ upwards are solids (waxes).

All the atoms of the compounds in the alkane series are joined by single covalent bonds and they are referred to as *saturated* compounds. This distinguishes them from similar

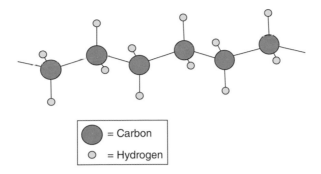

Figure 1.12 Zigzag form of hydrocarbon chains.

compounds in which some double bonds exist between the constituent atoms. Such compounds are known as *unsaturated* compounds.

Alkenes are the simplest homologous series of unsaturated hydrocarbons and have the general formula C_nH_{2n}. The first member of the series is ethene (ethylene), C_2H_4, (**1.6**), which contains two carbon atoms linked by a double bond.

$$
\begin{array}{cc}
\text{H} & \text{H} \\
| & | \\
\text{C} & = \text{C} \\
| & | \\
\text{H} & \text{H}
\end{array}
$$

1.6

The presence of the double bond does not double the strength of the bonding between the two carbon atoms. The second bond is much less chemically stable than the first and is therefore more reactive. Each carbon atom undergoes hybridisation (see Section 1.4.2), but in this case hybridisation of an s orbital and two p orbitals occurs, giving three sp^2 hybrid orbitals (Figure 1.13).

These orbitals overlap with an sp^2 hybrid orbital of the other carbon atom and the s orbitals of two hydrogen atoms to form three sigma (σ) bonds. The remaining p orbital interacts with that on the other carbon atom, giving a pi (π) bond. The σ bonds between the carbon atoms and the hydrogen atoms are all in the same plane and the p orbitals are at right angles to this plane (Figure 1.14).

The σ bonds have true covalent character, but because the π bond is strained, it is easily broken. It is this reactive bond that is responsible for the characteristic chemical reactions

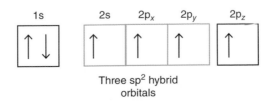

Three sp^2 hybrid
orbitals

Figure 1.13 sp^2 hybridisation.

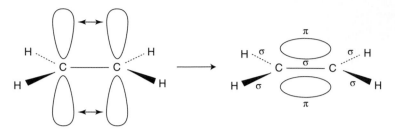

Figure 1.14   σ and π bonds in ethane.

of double-bonded hydrocarbon compounds. The next member of the alkene series is propene, C_3H_6 (**1.7**).

$$
\begin{array}{ccc}
\underset{\overset{|}{H}}{\overset{\overset{H}{|}}{H-C}}-\overset{\overset{H}{|}}{C}=\overset{\overset{H}{|}}{C}-H &
\underset{\overset{|}{H}}{\overset{\overset{H}{|}}{H-C}}-\underset{\overset{|}{H}}{\overset{\overset{H}{|}}{C}}-\overset{\overset{H}{|}}{C}=\overset{\overset{H}{|}}{C}-H &
\underset{\overset{|}{H}}{\overset{\overset{H}{|}}{H-C}}-\overset{\overset{H}{|}}{C}=\overset{\overset{H}{|}}{C}-\underset{\overset{|}{H}}{\overset{\overset{H}{|}}{C}}-H \\
\textbf{1.7} & \textbf{1.8} & \textbf{1.9}
\end{array}
$$

It does not matter where the double bond occurs in propene because the molecule is the same from whichever angle it is viewed. However, in the case of the next highest member of the series, butane (C_4H_8), two isomers are possible. If the double bond is located between the first and second carbon atoms, it is named 1-butene (**1.8**), but if it is located between the second and third carbon atoms, it is named 2-butene (**1.9**).

The reactions of these unsaturated compounds include the polymerisation reactions used in the formation of synthetic fibres such as acrylics and polyolefins (Chapter 2) and the dye–fibre reactions of certain dyes.

Alkynes contain triple bonds between carbon atoms. They have the general formula C_nH_{2n-2} and the simplest member is ethyne (acetylene), C_2H_2 (**1.10**), and the next member up the series is propyne, C_3H_4 (**1.11**). In the formation of alkynes, sp hybridisation occurs, so that between the two carbon atoms, in addition to the two σ bonds, there are two π bonds, both of which are highly strained, making them very chemically reactive gases.

$$
H-C\equiv C-H \qquad H-C\equiv C-\overset{\overset{\displaystyle H}{|}}{\underset{\underset{\displaystyle H}{|}}{C}}-H
$$

$$
\textbf{1.10} \qquad\qquad\qquad \textbf{1.11}
$$

1.8.1.2   Aromatic Hydrocarbons

The molecules of aromatic compounds contain ring structures of a special kind. The structural formulae of aromatic rings may be drawn as if alternating single and double bonds existed between adjacent carbon atoms. Compare, for example, the six-carbon aliphatic cyclic compound cyclohexane, C_6H_{12} (**1.12**), with the six-carbon aromatic compound benzene, C_6H_6, (**1.13**).

1.12   **1.13**

An arrangement of alternating single and double bonds (as in Structure **1.13**) is called a *conjugated system*, and it is very important in the molecular structure of dyes. In 1865 Kekulé proposed the closed ring structure for benzene with the three alternating single and double bonds. For this reason representing the structure of benzene in the way shown in **1.13** is called the Kekulé structure.

The bonds in the benzene ring are of a very special kind, however, and do not form a conjugated system. All six bonds in the ring are exactly alike and none of them undergo the characteristic reactions of double bonds in, say, alkenes; they are in fact much more stable (unreactive) than double bonds. The π electrons are distributed above and below the plane of the ring, forming diffuse clouds of negative charge: they are said to be *delocalised* and the phenomenon is known as *resonance*. The delocalising of electrons makes aromatic compounds different from aliphatic compounds. In recognition of the nature of the benzene ring, its structural formula is sometimes represented as a hexagon with a central circle (**1.14**) rather than with the alternating single and double bonds.

1.14

For simplicity's sake, the benzene ring is often drawn without showing the individual carbon and hydrogen atoms, as in **1.14**, their presence being implied by the notation.

All dye molecules contain aromatic ring structures, though it is more usual to find fused rings, such as naphthalene (**1.15**) (two rings fused together) or anthracene (**1.16**) (three rings fused together).

1.15   **1.16**

Where the rings join, they share two carbon atoms, and thus naphthalene with two rings has 10 carbon atoms, not 12. Similarly, anthracene with three rings has 14 carbon atoms rather than 18. As naphthalene and anthracene contain delocalised electrons *over all the rings,* it is inappropriate to use the delocalised symbol which is used for benzene on its own, for that would indicate, incorrectly, two or three separate delocalised systems. Thus in this book, *Kekule structures* are used.

Table 1.2 Common functional groups in organic compounds.

functional group	formula
Halide	F, Cl, Br, I
Alcohol	—OH
Carboxylic acid	—COOH
Ester	—COO—
Aldehyde	—CHO
Ketone	—CO—
Ether	—O—
Amine	—NH_2
Amide	—$CONH_2$
Cyano	—CN
Nitro	—NO_2
Sulphonic acid	—SO_3H

1.8.2 Functional Groups

One or more of the hydrogen atoms in hydrocarbons can be replaced by other atoms or, more commonly, groups of atoms, which gives rise to different homologous series. The most common functional groups are shown in Table 1.2.

It is most usual for the functional groups shown in Table 1.2 to replace one of the hydrogen atoms in an alkane. When this is the case, the general formula becomes $C_nH_{2n+1}X$, where X is the functional group (—OH, —COOH, etc.). Often for convenience, the hydrocarbon part, C_nH_{2n+1}, called an *alkyl* group, is written R, so the general formula of an alcohol is represented as R—OH and a carboxylic acid as R—COOH. Alkyl groups are named by replacing the final letter *e* of the parent hydrocarbon (i.e. the hydrocarbon with the same number of carbon atoms) by 'yl', so the CH_3— group is the *methyl* group, the C_2H_5— group is the *ethyl* group and so on.

Just as a hydrogen atom of an alkane can be replaced by a functional group, so can one of the hydrogen atoms of an aromatic ring. In the case of benzene, the removal of one of the hydrogen atoms creates the *phenyl* group, C_6H_5—, and if this hydrogen atom is replaced by an alcohol group, the compound phenol, C_6H_5OH, is formed. The chemistry of the organic compounds with functional groups, relevant to fibres and dyes, is described in Sections 1.8.3 and 1.8.4.

1.8.3 Important Functional Groups of Aliphatic Compounds

1.8.3.1 Halides

If one of the hydrogen atoms of an alkane is replaced by a halogen atom, an alkyl halide is formed. A typical example is monochloromethane, CH_3Cl (**1.17**), though fluorine, bromine and iodine derivatives are also known and are used as solvents. Chlorine derivatives tend to be mostly used because of their cheapness, such as di- (**1.18**), tri- (**1.19**) and tetra- (**1.20**) substituted chlorinated compounds of methane.

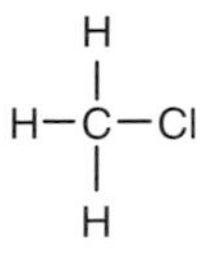

$$
\begin{array}{cccc}
\mathrm{H} & \mathrm{H} & \mathrm{Cl} & \mathrm{Cl} \\
| & | & | & | \\
\mathrm{H-C-Cl} & \mathrm{Cl-C-Cl} & \mathrm{Cl-C-Cl} & \mathrm{Cl-C-Cl} \\
| & | & | & | \\
\mathrm{H} & \mathrm{H} & \mathrm{H} & \mathrm{Cl} \\
\textbf{1.17} & \textbf{1.18} & \textbf{1.19} & \textbf{1.20}
\end{array}
$$

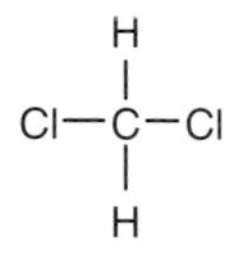

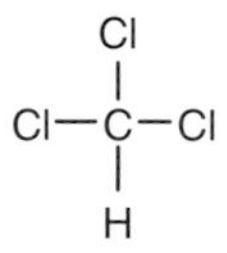

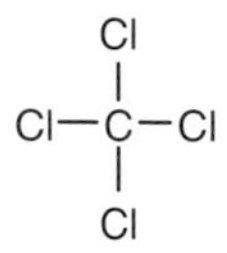

Dichloromethane (**1.18**) is used as a solvent in the dry spinning of cellulose triacetate fibre (see Section 2.7.4). Complex fluorinated hydrocarbon compounds are used in water-, oil- and soil-repellent finishes for fabrics.

1.8.3.2 Alcohols

If aliphatic hydrocarbons have just one of their hydrogen atoms replaced by an —OH group, in which case the general formula is $C_nH_{2n+1}OH$, they are called *monohydric* alcohols. The first two members of this series are methanol, CH_3OH (**1.21**), and ethanol, C_2H_5OH (**1.22**). Alcohols are named by replacing the final letter *e* of the parent hydrocarbon by *ol*. When the structural formulae of these alcohols are written, it is seen that for methanol and ethanol it does not matter which C atom the —OH group is located on – the structures will be the same. However, for the next members of the series, propanol and butanol, the —OH group can be placed at different positions on the carbon chain, giving rise to isomers. Thus propanol has two isomers, propan-1-ol (or *n*-propanol) (**1.23**) and propan-2-ol (or isopropanol) (**1.24**). In the case of butanol, four isomers are possible: butan-1-ol (**1.25**), butan-2-ol (**1.26**), 2-methyl-1-propanol (sometimes named isobutanol) (**1.27**) and 2-methyl-2-propanol (sometimes called *tert*-butanol) (**1.28**). For higher alcohols, such as pentanol, hexanol and so on, even greater numbers of isomers are possible.

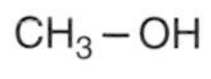

$$
\underset{\textbf{1.21}}{CH_3-OH} \qquad \underset{\textbf{1.22}}{CH_3-CH_2-OH} \qquad \underset{\textbf{1.23}}{CH_3-CH_2-CH_2-OH} \qquad \underset{\textbf{1.25}}{CH_3-CH_2-CH_2-CH_2-OH}
$$

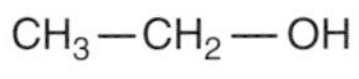

$$
\underset{\textbf{1.24}}{CH_3-\underset{\underset{OH}{|}}{CH}-CH_3} \qquad\qquad \underset{\textbf{1.26}}{CH_3-CH_2-\underset{\underset{OH}{|}}{CH}-CH_3}
$$

$$
\underset{\textbf{1.27}}{\begin{array}{c} CH_3 \\ \diagdown \\ \\ CH_3 \diagup \end{array} CH-CH_2-OH}
$$

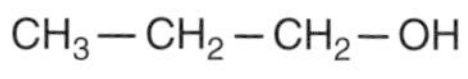

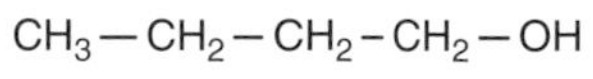

$$
\underset{\textbf{1.28}}{CH_3-\underset{\underset{CH_3}{|}}{\overset{\overset{CH_3}{|}}{C}}-OH}
$$

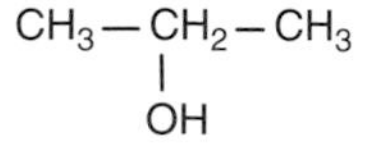

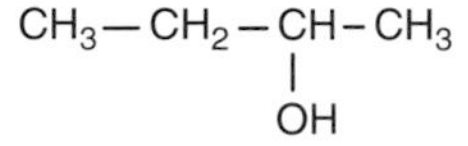

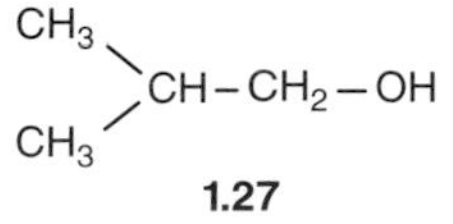

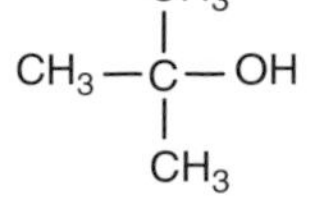

Alcohols containing two —OH groups are called *dihydric* alcohols, the most well known of which is ethane-1,2-diol (ethylene glycol, **1.29**), used in the manufacture of polyester (see Section 2.8.1.3). Propane-1,2,3-triol, better known as glycerol (**1.30**), is a *trihydric* alcohol.

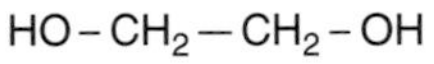

$$HO-CH_2-CH_2-OH \qquad HO-CH_2-CH-CH_2-OH$$
$$\qquad\qquad\qquad\qquad\qquad\qquad | $$
$$\qquad\qquad\qquad\qquad\qquad OH$$

1.29 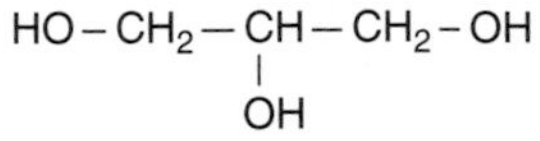   **1.30**

The lower members of the monohydric alcohols are liquids, and on ascending the homologous series, the boiling point rises. Methanol and ethanol boil at 64 and 78 °C, respectively. After decanol though, the higher members are solids. Methanol, ethanol and propanol all mix readily with water, but the greater hydrophobic character of the longer hydrocarbon chains of the higher members dominates and they become less miscible. Nevertheless, alcohols are a useful range of solvents for a variety of organic compounds.

1.8.3.3 *Carboxylic Acids*

Carboxylic acids have the general formula R—COOH, in which the structure of the carboxylic acid group is shown in (**1.31**).

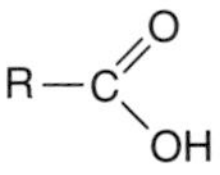

$$R-C\!\!\begin{array}{c} {}^{\displaystyle O} \\ \diagup\diagdown \\ {}_{\displaystyle OH} \end{array}$$

1.31

They are named by replacing the final letter *e* of the parent hydrocarbon by *oic acid*. The first member of the series is methanoic acid (formic acid, HCOOH, **1.32**) because there is one carbon atom, as in methane. The next member is ethanoic acid (acetic acid, CH_3COOH, **1.33**) because it has two carbon atoms, as does ethane. These two acids are used in textile wet processing operations. Ethanoic acid is also used in the manufacture of cellulose acetate fibres (see Section 2.7.4).

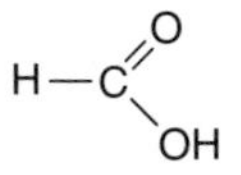

$$H-C\!\!\begin{array}{c} {}^{\displaystyle O} \\ \diagup\diagdown \\ {}_{\displaystyle OH} \end{array} \qquad CH_3-C\!\!\begin{array}{c} {}^{\displaystyle O} \\ \diagup\diagdown \\ {}_{\displaystyle OH} \end{array}$$

1.32 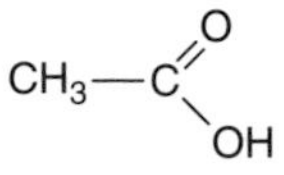   **1.33**

Pure anhydrous ethanoic acid is a colourless liquid. When it freezes at 16.7 °C, it is called *glacial* ethanoic acid because of its ice-like appearance. On ascending the series of carboxylic acids, the solubility in water decreases, and those where $R=C_8H_{17}-$ or longer are waxy solids. These organic acids are weak acids (see Section 1.6.1) because they only dissociate partially in water (Scheme 1.18).

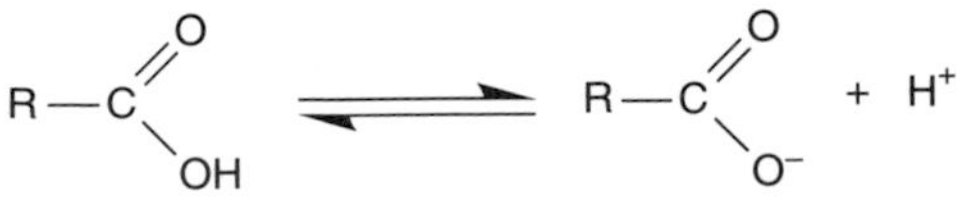

$$R-C\!\!\begin{array}{c} {}^{\displaystyle O} \\ \diagup\diagdown \\ {}_{\displaystyle OH} \end{array} \quad \rightleftharpoons \quad R-C\!\!\begin{array}{c} {}^{\displaystyle O} \\ \diagup\diagdown \\ {}_{\displaystyle O^-} \end{array} \quad + \; H^+$$

Scheme 1.18 Dissociation of carboxylic acids.

Methanoic and ethanoic acids readily form salts with alkalis, for example:

$$CH_3COOH + NaOH \rightarrow CH_3COONa + H_2O$$

The salt formed in this case is sodium acetate.

Carboxylic acids containing two acid groups are called dicarboxylic acids, the simplest being oxalic acid (**1.34**).

COOH
|
COOH

1.34

The dicarboxylic acid 1,6-hexanedioic acid (**1.35**) is used in the manufacture of nylon 6.6 (see Section 2.8.1.1).

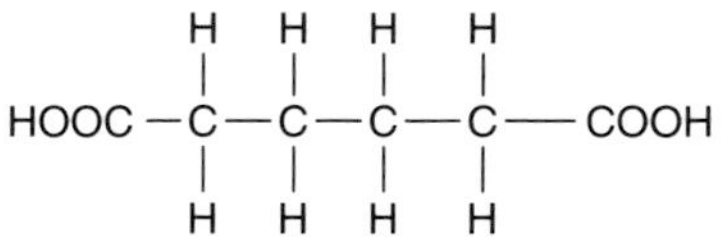

1.35

This acid is also called adipic acid and its formula is often written as HOOC—$(CH_2)_4$—COOH.

1.8.3.4 Esters

Aliphatic esters are produced by the reaction between a carboxylic acid and an alcohol (Scheme 1.19); in the presence of concentrated sulphuric acid, which acts as a catalyst,

Scheme 1.19 Formation of esters.

where the R and R′ groups may be the same or different. The formation of esters is a reversible reaction, and indeed esters can be hydrolysed back to their component carboxylic acid and alcohol.

Esters are named as though they are alkyl 'salts' of the acid, though they are not salts at all. The final letters *oic* of the carboxylic acid from which the ester is formed is replaced by the letters *oate*. For example, if the acid is ethanoic acid and the alcohol ethanol, the ester formed is ethyl ethanoate (**1.36**).

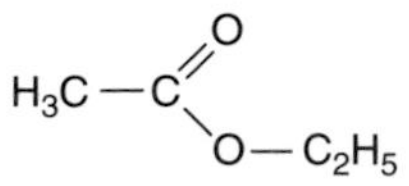

1.36

Lower members of the series are pleasant smelling volatile liquids, whilst higher members are oils, fats and waxes.

Esters are of interest in textile fibre chemistry because of the commercial importance of polyester fibres produced by the reaction between a dicarboxylic acid and a diol (see Section 2.8.1.3). Cellulose acetate fibres (acetate and triacetate) are also esters: in this case it is the —OH groups of cellulose that are esterified.

1.8.3.5   Aldehydes and Ketones

Aldehydes (**1.37**) and ketones (**1.38**) are closely related compounds in that both contain a carbonyl group and differ only in that aldehydes have a hydrogen atom bonded to it whereas ketones have a second alkyl group bonded to it.

$$R-C\overset{O}{\underset{H}{\diagup}} \qquad R-C\overset{O}{\underset{R'}{\diagup}}$$

Aldehyde	Ketone
1.37	**1.38**

Aldehydes are named by replacing the final letter *e* of the parent hydrocarbon by 'al'. The most important aldehydes are methanal (formaldehyde, HCHO) and ethanal (acetaldehyde, CH_3CHO). Ketones are named by stating the names of the two alkyl groups, followed by 'ketone'. Of the ketones, dimethylketone (acetone, CH_3COCH_3) is the most common, being widely used as a solvent. It is used in the manufacture of acetate fibre (see Section 2.7.4). It is also a good solvent for many disperse dyes.

1.8.3.6   Ethers

Ethers have the general formula R—O—R', that is, two alkyl groups bonded to an oxygen atom. They are named in a similar way to the ketones by stating the names of the two alkyl groups, followed by 'ether'. The most common ether is diethyl ether, $C_2H_5OC_2H_5$, often simply called 'ether'. It is a dangerously inflammable liquid and often used as a solvent in organic synthesis.

1.8.3.7   Amines

Amines can be regarded as derivatives of ammonia, NH_3, in which the hydrogen atoms are replaced by alkyl groups. One, two or all three of the hydrogen atoms can be replaced, giving primary (**1.39**), secondary (**1.40**) and tertiary (**1.41**) amines, respectively.

$$R-N\overset{H}{\underset{H}{\diagup}} \qquad R-N\overset{H}{\underset{R'}{\diagup}} \qquad R-N\overset{R''}{\underset{R'}{\diagup}}$$

Primary amine	Secondary amine	Tertiary amine
1.39	**1.40**	**1.41**

The R, R' and R" alkyl groups can be the same or different. Structure **1.39** contains only one alkyl group, and if R=—CH₃, then the amine has the formula CH_3NH_2 and is named methylamine. If R=—C₂H₅, the amine, $C_2H_5NH_2$, is named ethylamine. In the case of secondary amines (**1.40**), in which both R and R' are —CH₃ groups, it is dimethylamine, $(CH_3)_2NH_2$, whilst if R=—CH₃ and R'=—C₂H₅, the compound is named ethylmethylamine.

The tertiary amine (**1.41**) in which R, R' and R" are all —CH$_3$ groups is trimethylamine, (CH$_3$)$_3$NH$_2$. The lower members of primary amines smell strongly of ammonia and have many of the properties of ammonia, giving alkaline solutions in water.

The amine functional group (—NH$_2$) is reactive and reacts with other types of functional groups. Of particular interest in textile chemistry is its reaction with a carboxylic acid to form an amide (Scheme 1.20).

Scheme 1.20 Formation of amides.

The —CONH— group is called the amide group and is an important structural feature of wool keratin and of nylons (polyamides). It has the ability to form hydrogen bonds with water, thereby providing these fibres with moisture absorbency properties.

An amide that is an important auxiliary chemical in dyeing and printing is a compound called urea (**1.42**). Urea is an amide of carbonic acid (**1.43**) and is used in textile printing. It is incorporated into print pastes to assist the solution of acid and direct dyes in the limited amount of water present in the paste. It also acts as a *humectant*, a substance that retains moisture, during steaming of print pastes (see Sections 5.2 and 5.7.2).

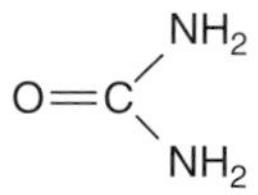

1.42 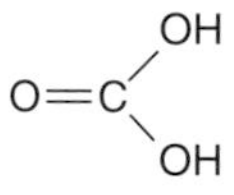   **1.43**

Quaternary ammonium compounds (**1.44**) are another important type of amine. These compounds have the following general formula:

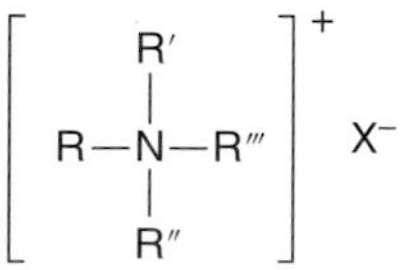

1.44

where X$^-$ is a halide ion (usually Cl$^-$ or Br$^-$) and the four alkyl groups may be the same or different. These compounds are cationic surface-active agents (see Section 1.8.5.2), for example, cetyltrimethylammonium chloride (**1.45**), which is widely used in textile wet processing.

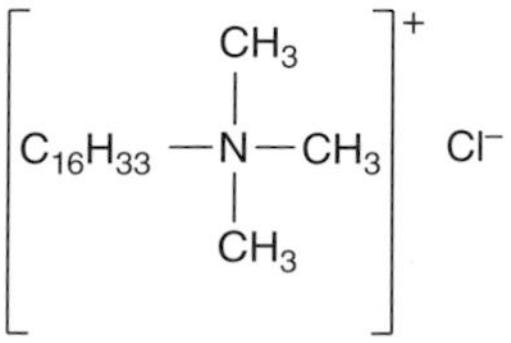

1.45

1.8.3.8   Cyano and Nitro Groups

Cyano (—CN) and nitro (—NO$_2$) groups are both important functional groups in organic synthesis, and both are extensively used in dyes and pigments where they act as aux-ochromes (see Section 3.3). They have an electron-attracting property that influences the colour of the dye or pigment molecules.

1.8.4   Important Functional Groups of Aromatic Compounds

In Section 1.8.3 examples were given of aliphatic organic compounds in which functional groups are bonded to alkyl groups. In a similar manner, one (or more) of the six hydrogen atoms on a benzene ring can be substituted by a functional group, and such compounds are widely used in the synthesis of dyes and pigments.

If just one of the six hydrogen atoms of a benzene ring is replaced by a functional group X, the C$_6$H$_5$— group to which the functional group is bonded is called the *phenyl* group, which is the simplest *aryl* group (**1.46**).

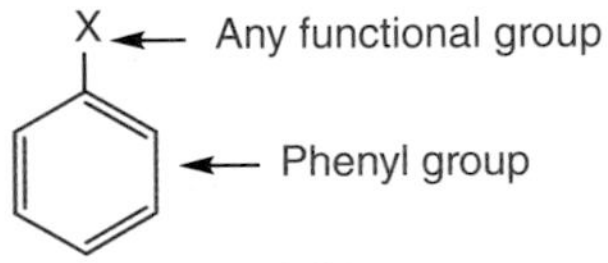

1.46

Aryl groups correspond to the alkyl groups of aliphatic chemistry and are all based on aromatic ring structures. The halogen derivatives are named as might be expected, for example, C$_6$H$_5$Cl (**1.47**) is called phenyl chloride (and also monochlorobenzene). However many aromatic compounds often have less obvious names. Thus, C$_6$H$_5$OH (**1.48**) is called phenol, C$_6$H$_5$NH$_2$ (**1.49**) is called aniline, C$_6$H$_5$COOH (**1.50**) is called benzoic acid, and C$_6$H$_5$CHO (**1.51**) is called benzaldehyde.

In general, the chemical reactivity of the functional groups bonded to aromatic rings is very similar to that of the groups when present in aliphatic compounds.

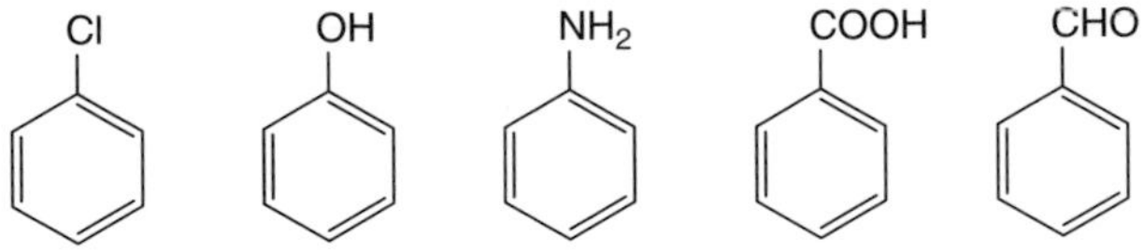

Cl	OH	NH$_2$	COOH	CHO
1.47	**1.48**	**1.49**	**1.50**	**1.51**

Difunctional compounds of benzene are very common and can exist in three isomeric forms depending on the relative positions on the ring of the two functional groups. The carbon atoms of the benzene ring are numbered 1–6. These six carbon atoms are equivalent, so positions 1,2- and 1,6- are equivalent, as are the 1,3- and 1,5-disubstituted positions.

If the two groups are in the 1,2- (or 1,6-) positions, the disubstituted compound is referred to as the *ortho-* (*o*-) compound. If they are in the 1,3- (or 1,5-) positions, they are referred to as the *meta-* (*m*-) compound. The last possible combination is the 1,4-disubstituted compound, referred to as *para-* (*p*-).

The aromatic ring of the disubstituted compounds (C_6H_4-) is now called the phenylene group.

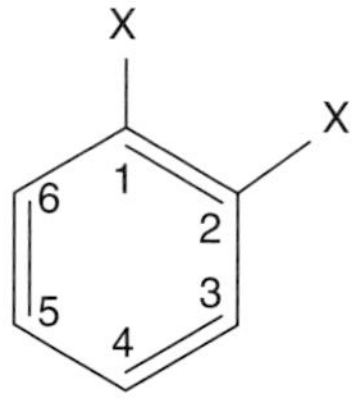

ortho 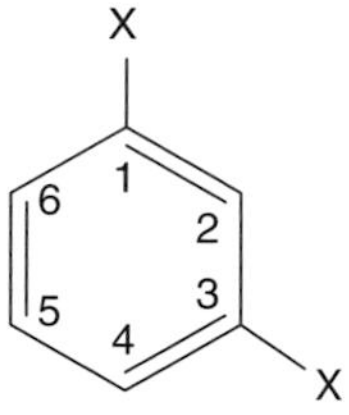   *meta*

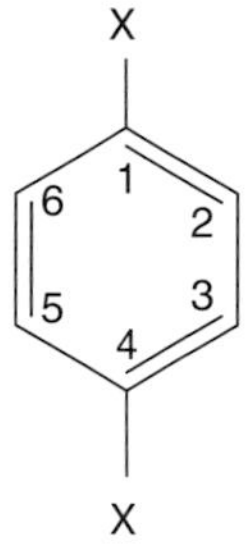

para

A feature of aromatic organic compounds is that the presence of a functional group in a ring has a direct influence on the position taken up by a second functional group. This is because they influence the electron density on the carbon atoms of the ring.

Groups that direct a second functional group to the *meta* position are electron attracting, such as $-CN$, $-CHO$, $-COOH$ and $-NO_2$.

Groups that direct a second functional group to the *ortho* and *para* positions are electron donating, such as $-NH_2$, $-OH$ and $-CH_3$.

Difunctional compounds are much used in the manufacture of dyes and pigments, typical of which is 4-methylaniline, more commonly known as *p*-toluidine (**1.52**). An important difunctional compound in the textile industry is benzene-1,4-dicarboxylic acid, more commonly known as terephthalic acid (**1.53**), which is used to manufacture polyester fibre (see Section 2.8.1.3).

1.52    **1.53**

Trifunctional compounds are also important in the manufacture of dyes and pigments, and an example of such a compound is 4-amino-3-nitrotoluene, or 2-nitro-*p*-toluidine (**1.54**), widely used as a diazo component.

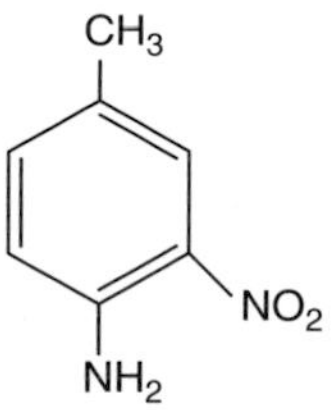

1.54

1.8.5 Important Compounds in Textile Dyeing

1.8.5.1 Sequestering Agents

An important issue to be considered in dyehouses is the purity of its water supply. The presence of metal ions in water can adversely affect a number of operations by reacting with dyes or chemicals used in processing. For example, magnesium (Mg^{2+}) or calcium (Ca^{2+}) ions in hard water can form insoluble complexes with soaps, and if these complexes are deposited on textile substrates, they will create difficulties in subsequent dyeing processes. The presence of transition metal ions such as iron (Fe^{2+}) and copper (Cu^{2+}) can catalyse the decomposition of hydrogen peroxide in bleaching processes, and if these ions are already adsorbed into the substrate, then localised accelerated attack may occur.

Sequestering agents are compounds that are able to form soluble complexes with metal ions in what is termed a *chelation* reaction. For this reason, sequestering agents are also known as *chelating agents*. The soluble complex formed with the metal ions does not interfere with the process (such as bleaching or dyeing) and is washed out at the end of the operation with the exhaust liquors. Sequestering agents have structures that enable them to form closed rings with polyvalent metal ions by the sharing of a lone pair of electrons with them. In this way they 'lock up' the metal ions and prevent them from any further reaction.

The lone pairs of electrons that form the bonds with the metal ions come from electron-donating atoms in the sequestering agent, usually nitrogen (in amines) or oxygen (in phosphates). Typical of the former type is ethylenediaminetetraacetic acid (most usually known as EDTA, **1.55**), which forms complexes with metal ions (**1.56**).

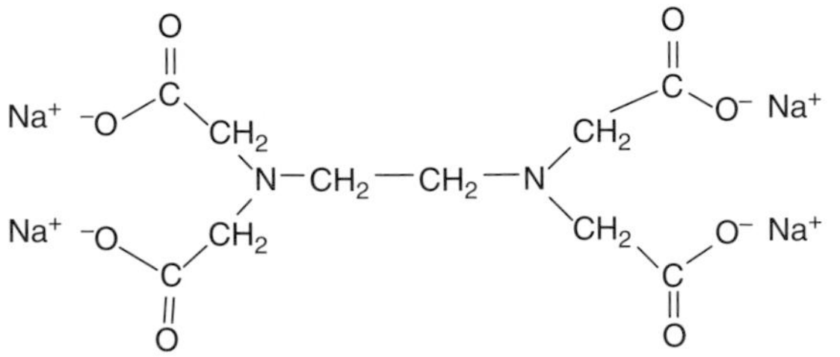

1.55

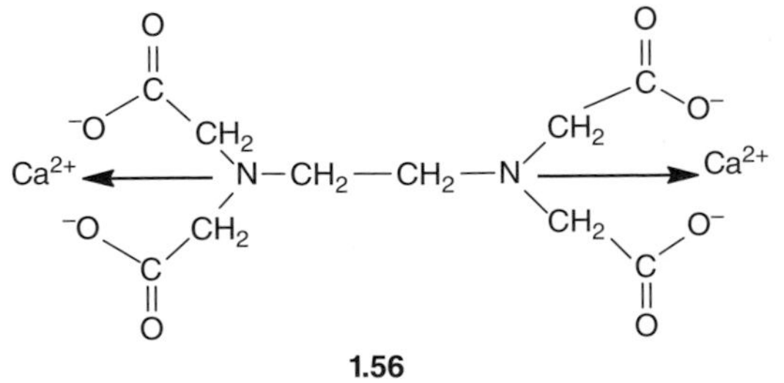

1.56

Typical of the polyphosphate type is sodium hexametaphosphate, $Na_2(Na_4P_6O_{18})$, which forms the water-soluble complex anion $Na_2(Ca_2P_6O_{18})$ by an ion-exchange reaction.

1.8.5.2 Surface-Active Agents (Surfactants)

Surface-active agents are widely used in the manufacture of dyes, in dyeing processes and in preparatory processes for dyeing. Many types of surface-active agents exist and they have different names depending upon their function:

Detergents are used in washing and scouring processes. Some are called *wetting agents* and are incorporated into formulations of dye powders to aid wetting out when the powder is being dissolved in water and also to aid the wetting of fibres prior to dyeing operations.

Levelling and *retarding* agents are also surfactants, these types being used in the application of ionic dyes to fibres.

Dispersing agents are another type of surfactant and are used in the formulation of disperse dyes, whose function is to ensure a stable dispersion of the dyes in the dyebath during the dyeing process, aiding uniform uptake of dye by the fibre.

The molecules of surface-active agents comprise two components, a hydrophilic (water-loving) water-soluble head and a hydrophobic (water-hating) water-insoluble tail. This tail is usually a long-chain hydrocarbon. Surface-active agents may be anionic, cationic, amphoteric or non-ionic in nature.

Anionic surface-active agents are the most commonly used surface-active agents in dyeing systems, in which the hydrophilic head is usually a sulphonate ($-SO_3Na$) or a carboxyl group ($-COONa$). Examples of these types are sodium heptadecanoate (**1.57**) and sodium dodecylbenzenesulphonate (**1.58**), a widely used detergent in domestic washing powders.

$C_{12}H_{25}$ —⟨benzene ring⟩— $COO^- Na^+$ $C_{12}H_{25}$ —⟨benzene ring⟩— $SO_3^- Na^+$

1.57   **1.58**

Cationic types of surface-active agent contain a positively charged group at the head. This positively charged group is usually a quaternary ammonium salt, and an example of such a structure is **1.45** in Section 1.8.3.7.

The amphoteric types contain both positively and negatively charged groups (e.g. betaines such as **1.59**), whilst the non-ionic types do not contain any charged groups at all: these instead contain a number of ethylene oxide groups to impart hydrophilic character, such as **1.60**.

$$C_8H_{17} - \overset{\overset{\displaystyle CH_3}{|}}{\underset{\underset{\displaystyle CH_3}{|}}{N^+}} - CH_2COO^-$$

C_9H_{19} —⟨benzene ring⟩— $(OCH_2CH_2)_xOH$

$x = 10–15$

1.59   **1.60**

Surface-active agents used as dispersing agents are usually polyelectrolytes with complex structures, some of which are not known for certain. They fall into two classes:

(1) Condensation products of methanal (formaldehyde) with arylsulphonic acids, two examples of which are a condensate of naphthalene-2-sulphonic acid and methanal (**1.61**) and a condensation product of phenols with methanal and sodium sulphite (**1.62**)

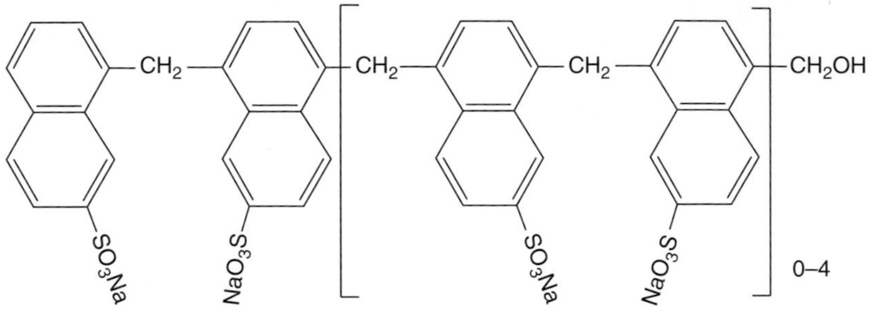

1.61

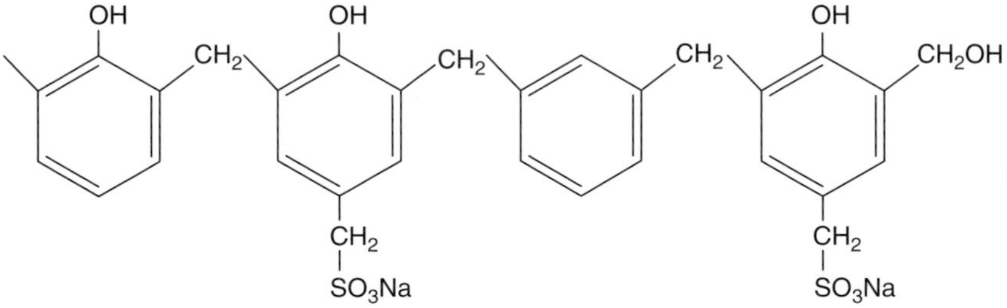

1.62

(2)   Lignin sulphonates, a group of compounds of very complex structure, probably mixtures of compounds, formed by the treatment of wood pulp with sulphite or bisulphite

1.8.5.3   Carriers

The dyeing of polyester fibres is problematic because the rate of diffusion of disperse dyes through the fibre is so low at temperatures where dye uptake is very low, making the process unfeasible. The problem is due to the high degree of crystallinity of polyester fibres (see Section 2.8.1.3) and its low glass transition temperature, T_g (see Section 2.2). Whilst dyeing can be carried out at temperatures above the T_g, that is, at about 130 °C, an alternative is to apply dyes at temperatures at the boil in the presence of a chemical called a *carrier*. The effect of the carrier on the polymer chains of the polyester fibres is similar to that of a raised temperature, allowing more rapid ingress of dye and the development of deep shades. Their smell is objectionable, however, and unless they are removed completely from the fibre, they can lower the lightfastness of the dyeing.

   The types of chemicals used as carriers are non-ionic aromatic organic compounds of fairly low molecular weight (around 150–200). Typical of the carriers used fall into four main chemical groups:

(1)   Phenols, such as *o*-phenylphenol
(2)   Hydrocarbons, such as biphenyl

(3)   Chlorinated hydrocarbons, such as dichlorobenzene
(4)   Esters, such as butyl benzoate

Phenols and chlorinated hydrocarbons are no longer marketed due to concerns about aquatic toxicity. Indeed such is the concern about the use of carriers that they are now little used in polyester dyeing, the high-temperature method being preferred. An exception, however, is their use in the dyeing of polyester in its blends with wool, since wool is easily damaged above boiling temperature, especially at 130 °C.

1.9   The Use of Chemicals by Industry

1.9.1   REACH

A system for controlling the use of chemicals in Europe – Registration, Evaluation, Authorisation and Restriction of Chemicals (REACH) – became law in the United Kingdom in 2007. This legislation places controls on the supply and use of certain chemicals, especially those that can be harmful to human health or the environment. REACH applies to a wide range of chemical substances, including dyes and pigments, as well as chemicals used in dyeing and printing processes.

Under REACH, there are three groups of chemicals that are specifically controlled:

- A candidate list of substances of very high concern (SVHC). Suppliers of a chemical on this list have to give information about it and how to use it safely, for example, a safety data sheet.
- Priority SVHCs from the candidate list. These chemicals require authorisation by the European Chemicals Agency (ECHA) for their supply and use.
- Restricted substances, which are particularly hazardous, have controls on how they can be supplied or used. These chemicals are restricted to protect workers, consumers and the environment. The restriction may limit the concentration of a chemical for a particular use or ban a use entirely.

The ECHA has published lists of chemicals in the three groups. Chemicals that are classed as SVHCs are substances that the ECHA considers to have hazards that might be carcinogenic, mutagenic or toxic for reproduction. Substances that remain in the environment for a long time and are bioaccumulative and toxic are also included.

Since dyeing, printing and finishing companies use chemicals, not just dyes and pigments but also auxiliary chemicals necessary for the processes by which they are applied to textiles, they must ensure they meet the requirements of the REACH legislation. An example of the impact of REACH on the dyeing industry is the classification of chromium as an SVHC, requiring authorisation before it is used. As a result, the use of chrome dyeing in the United Kingdom has declined substantially to the point where it is very little used (see Section 3.5.1.2).

Some countries within the EU, such as Denmark and Sweden, together with France and Germany are urging the EU to move more quickly towards meeting its goal of a toxic-free environment. Outside the EU, some countries have started to implement regulations similar to those of EU REACH. For example, in 2010 *China REACH* came into force, though some countries, notably United States, have criticised the principle of REACH for harming global trade. The Toxic Substances Control Act (TSCA) in the United States regulates the introduction of new chemicals, though it does not categorise chemicals as toxic and non-toxic, and indeed when the law was introduced in 1976, all chemicals were considered to be safe. TSCA has been widely criticised over the years, and in 2016 a reform bill was passed, though it is still unlikely to be as rigorous as REACH in Europe.

A number of countries have adopted a framework to establish a more globalised system of chemicals registration under the Globally Harmonised System of Classification and Labelling of Chemicals (GHS). It was created by the United Nations in 1992 and designed to replace the various classification and labelling standards used in different countries by using consistent criteria on a global level. However it has not yet been fully implemented in many countries.

1.9.2 Effluent Disposal

In Europe, the Urban Waste Water Treatment Directive of the EU requires all member states to implement a control regime for trade effluent discharge. Within the United Kingdom, the Water Industry Act of 1991 meets this requirement. Trade effluent, the waste water produced by industry, can be disposed of to the public sewerage system in the United Kingdom, provided the prior consent of the local water authority is obtained. In granting consent, the water authority states the maximum permitted concentration of pollutant and in some cases states the load of a particular pollutant that can be discharged in a 24 hour period. The majority of water authorities specify the limits on parameters such as pH, biological oxygen demand (BOD), chemical oxygen demand (COD), total suspended solids and temperature, in addition to pollutants such as heavy metals and colour.

As noted in Section 1.9.1, chrome dyeing is in significant decline. This is due to the restrictions not only on the use of dichromate under REACH legislation but also on the discharge of chromium to effluent. The methods developed to minimise the amount of hexavalent chromium being discharged into public sewers, outlined in Section 3.5.1.2, have varying degrees of effectiveness, so local water authorities impose the statutory limits governing the discharge of chromium in effluent. These limits, which vary from country to country and even between regions in a given country, are very strict, typically no more than 0.5 mg/l, often less. Because it is very difficult to analyse separately for trivalent and hexavalent chromium, limits are normally set for total chromium in effluent. In the United Kingdom, local water authorities state, under the conditions of their consent, the maximum permitted levels of chromium to be discharged. A similar situation pertains for the discharge of other hazardous chemicals, such as sulphides that are used in the dyeing of sulphur dyes (see Section 4.2.2.4). Whilst not especially hazardous chemicals, local authorities are also concerned about the discharge of high concentrations of electrolytes, such as sodium chloride and sodium sulphate, used in the dyeing of direct, reactive and acid dyes.

For example, it is recognised that sulphate can damage concrete structures, and a general guideline on sulphate is a limit of 1000 mg/l.

A group created by a number of clothing brands to limit hazardous wastewater discharges, the Zero Discharge of Hazardous Chemicals (ZDHC) Programme, was formed in 2011. Many leading brands have recognised the concern of the public concern and environmental lobby organisations for the need to limit hazardous wastewater discharges and are striving to move towards zero discharge by setting global guidelines. This is an example of manufacturing industry itself, rather than governments, taking action to ensure the textile and dyeing industries adopt more environmentally friendly processes.

2

Textile Fibres

2.1 Introduction

For thousands of years, humans have exploited the attributes of fibres for the provision of clothing, for protection from the environment and to enhance our appearance and reflect our personality. Whilst many people associate textiles with clothing and home furnishings, textiles also have important industrial applications. These types of textile products, which are usually referred to as 'technical textiles', are produced for their technical performance in specialised applications, rather than their aesthetic properties. However, there is no clear distinction between technical textiles and textiles for clothing, because a very important sector is the 'performance apparel' market, which comprises garments produced to meet specific demands, such as protection from hazardous chemicals, flameproof garments or garments for active sportswear.

The production of textiles involves interactions between a wide range of industrial sectors, as the diagram in Figure 2.1 illustrates. The production of natural fibres involves farming, either livestock farming for hair fibres or plantation farming for vegetable fibres, whilst man-made fibres are products of the chemical industry. The conversion from fibres, to yarns, to fabrics involves what is understood to be the textile industry, though the dyeing and finishing of fabrics involve further use of chemicals. These products can be used in the garment industry, for home textiles or in the industrial textile industries. Related sectors supporting these activities include the machinery sector (engineering), retail and distribution and various textile services.

During the nineteenth century, the only fibre types available were natural fibres such as flax, cotton, wool and silk. Indeed in the United Kingdom, it was not even until the early eighteenth century that cotton started to be imported. There is now a very wide range of textile fibres available, but they can be broadly classified into two main groups:

- Natural fibres
- Man-made fibres

Each of these two groups can be subdivided into classes according to their chemical nature. Natural fibres can be divided into two main subgroups:

- Protein fibres, such as wool and hair, which are made of the protein called keratin
- Cellulosic fibres, which are obtained from the stem, leaves or seeds of plants

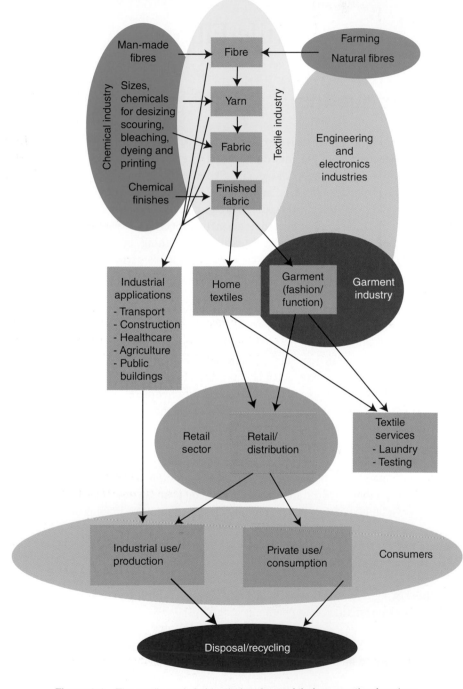

Figure 2.1 The textile and clothing industries and their supporting functions.

Man-made fibres can be subdivided into three groups:

- Regenerated fibres, which are fibres from natural sources, but chemically processed to impart novel characteristics. These fibres are mainly cellulosic in nature, but regenerated protein fibres, alginates and chitin also have niche market areas.
- Synthetic fibres, which are produced from non-renewable resources, principally oil.
- Inorganic fibres, such as glass fibres.

The classes of fibres available are shown in Figure 2.2, which shows the fibre types by what are called their *generic names*.

Man-made textile fibre types are given *generic names*, which are based on common chemical groups, which give fibres their characteristic properties. Common examples are:

- Polyester, which contains the $-COO-$ group
- Polyamide, or nylon, which contains the $-CONH-$ group
- Acrylic, which contains the $-(CH_2-CH \cdot CN)-$ group

These names are used for customs purposes and textile product labelling where it is a legal requirement in both the EU and the United States for all garments to contain a label showing their fibre content. They are assigned in Europe by the Bureau International pour la Standardisation des Fibres Artificielles (BISFA) and in America by the Federal Trade Commission (FTC) and are included in the International Organisation for Standardisation (ISO) standard ISO 2076 Textiles – Man-made fibres – Generic names.

Generic names cover fibres with the same basic technologies and properties from several producers and are completely different from trade names, which are used by individual companies for marketing purposes. Many synthetic fibres are commonly used in garments, and commodity garment labels often just state the fibre type as polyester, nylon, acrylic and so on rather than using a particular manufacturer's trade name. However, for high-performance garments, such as active sportswear, outdoor garments and specialised items such as sleeping bags, the labels often refer to the fibre manufacturer's trade name. Examples are Pertex®, a form of polyamide woven in a particular way, and Meraklon®, a polypropylene fibre, both of which possess excellent wicking properties.

In addition to generic names, the various man-made fibre types have been given codes, which assists in communicating fibre types easily. Some examples are given in Table 2.1. One unexpected code in this system is the code PES assigned to polyester. Many industrialists and academics prefer to use the code PET for this fibre type.

In 2014 the total annual production of textile fibres was approximately 98 million tonnes. The production of the main fibre types is shown in Table 2.2. The most produced fibre type is polyester, the production being mainly in China. Of the natural fibres, cotton is by far the most important. Cellulosic regenerated fibres, which include fibres such as viscose, are an increasingly important group.

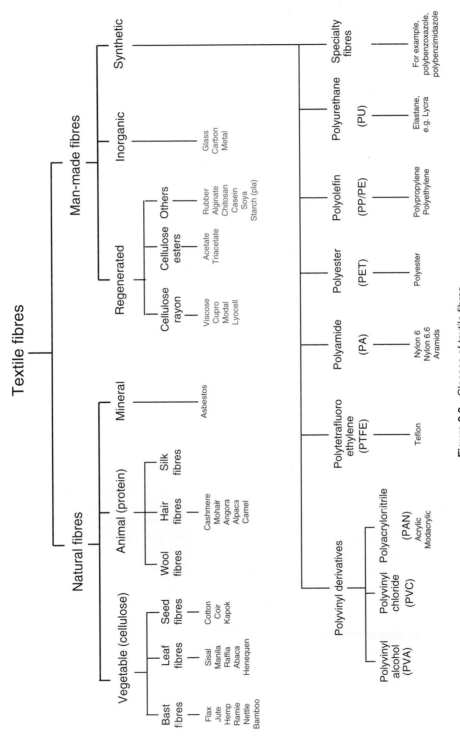

Figure 2.2 Classes of textile fibres.

Table 2.1   Codes for some man-made fibres.

Acrylic	PAN	Polyester	PES
Elastane	EL	Polyethylene	PE
Lyocell	CLY	Polylactide	PLA
Modacrylic	MAC	Polypropylene	PP
Polyamide	PA	Viscose	CV

Table 2.2   Worldwide production of some textile fibre types, thousand metric tonnes, 2014.

synthetic fibres		regenerated fibres		natural fibres	
Polyester	49 140	Cellulosic	5 029	Cotton	26 268
Polyamide	4 552	Lyocell	150	Wool	1 121
Acrylic	1 849			Silk	138
Polypropylene	5 007			Jute	3 085
Spandex	420			Linen	245
Others	133			Others	1 054
Totals	**61 101**		**5 179**		**31 911**

2.2   Nature of Fibre-Forming Polymers

Textile fibres are very long in relation to their width and this characteristic is present at the molecular level also. The molecules that form the structural material of natural fibres are exceptionally long chains of repeating identical units. Each unit is a single simple molecule, but within the chain many such units are chemically linked in a head-to-tail arrangement. The molecule used as the building block of the long-chain molecule is referred to as the *monomer*, and the long-chain molecules themselves are called *polymer* molecules. The word polymer is derived from the Greek words *poly* (many) and *meros* (parts or segments). The chemical properties of the polymer chains are governed by the monomer, so that a monomer with hydrophilic groups will produce a hydrophilic polymer and a hydrophobic monomer, a hydrophobic polymer.

Not all polymers are suitable for forming fibres. To be capable of forming fibres, polymers must meet the following requirements:

- The polymer molecules must be very long in relation to their diameter. When polymers are formed, the molecules are not all of the same size, so there is a distribution of molar masses. This distribution may be a 'normal' distribution (i.e. it is symmetrical), but usually it is skewed somewhat (see Figure 2.3). If the average molar mass is too low, the polymer cannot be extruded into fibres or the extruded fibres are too weak. If it is too high, it will be difficult to dissolve the polymer in a solvent (or it will not melt easily) for extruding because it will be too viscous to extrude through a spinneret to form a fibre.
- Polymer chains that can form intermolecular attractions with parallel chains (e.g. by hydrogen bonding) give fibres of high strength. The polymer chains must be linear, with no branched chains or bulky side groups, since these will prevent close approach of parallel chains and reduce intermolecular attraction.

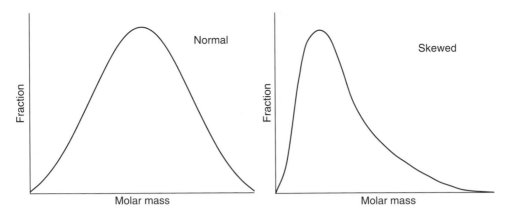

Figure 2.3   Normal and skewed distributions of molecular masses in polymers.

- For synthetic fibres the polymer chains must be flexible. Chain flexibility and intermolecular attractions affect the *melting temperature*, T_m and the *glass transition temperature*, T_g:
  Melting temperature—the temperature at which a substance changes from the solid to the liquid state
  Glass transition temperature—the temperature at which the internal polymer structure changes from a hard, glassy state to a more rubbery, flexible state.

Whilst the polymer molecules that constitute fibres are themselves very long in relation to their width, it is their orientation and ability to form intermolecular bonds that has a large bearing on the resultant fibre properties. Intermolecular attractions between polymer chains influence the strength of a fibre and also its stiffness. To impart acceptable strength to a fibre, it is necessary that neighbouring polymer chains can align closely with each other, so that the forces of attraction hold them together, thereby promoting cohesion and crystallinity.

Within a fibre *crystalline regions* form in areas where there is strong and close alignment of polymer chains. If these crystalline regions are oriented fundamentally in the direction of the axis of the fibre, then the fibre is said to possess a high *degree of orientation*.

Regions also occur within the polymeric structure where there is little alignment of the polymer chains. These are termed *amorphous regions* and are regions where dyes and other molecules can access the fibre structure with relative ease. The polymer structure is represented schematically in Figure 2.4.

Chain flexibility and intermolecular attractions have profound effects on the melting temperature, T_m, and also on their glass transition temperature, T_g. The thermal energy at temperatures above T_g weaken the intermolecular forces between adjacent polymer chains and voids between them in the non-crystalline regions increase in size.

The glass transition temperature of a fibre is important in dyeing because at temperatures above T_g the polymer structure becomes more accessible to dye molecules and on subsequent cooling to below T_g the dye molecules become trapped in the polymer matrix. T_g values tend not to be exhibited by natural fibres because of their complex morphological structures.

The melting temperature is an important consideration for fibres for domestic use. If T_m is low, then the polymer will simply melt or become tacky when ironed after washing. This

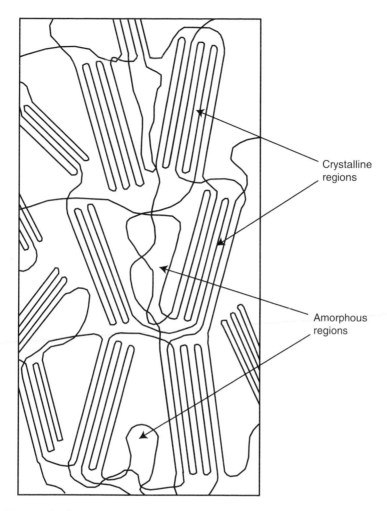

Crystalline
regions

Amorphous
regions

Figure 2.4 Crystalline and amorphous structure of polymer molecules in fibres.

is certainly an issue with synthetic fibres – polypropylene fibres melt at around 170 °C, nylon at 215 °C and nylon 6.6 and polyester at about 260 °C. Natural fibres such as wool, silk, cotton and linen do not melt; rather they decompose under the action of heat, but at temperatures much higher than the melting temperatures of synthetic fibres.

The chemical character of the polymer molecules determines many properties of the resulting fibres:

- Polymer chains containing many $-CH_2-$ groups or benzene rings will have strong hydrophobic (water-hating) character and will be more resistant to attack by chemicals.
- Conversely, polymer chains containing many polar groups, such as $-OH$, $-NH_2$ and $-CN$, will be more hydrophilic (water-liking) because H_2O molecules can form hydrogen bonds with them.
- Ionic groups (e.g. $-COO^{\ominus}$, $-NH_3^{\oplus}$) on the polymer chains (or at the ends of the chains) provide sites for the adsorption of dye ions.

Table 2.3 Important fibre properties.

property	examples
Mechanical	Tenacity, extensibility, elongation at break, elasticity
Thermal	Melt temperature, T_g, thermal decomposition, flammability
Electrical	Electrostatic build-up
Optical	Reflection, refraction, lustre, transparency
Surface	Wetting, friction
Biological	Resistance to microorganisms, biocompatibility, biodegradation
Chemical	Resistance to chemical attack, moisture absorbance, resistance to sunlight

2.3   Properties of Textile Fibres

Textile fibres are used for a wide range of applications, but in order for them to be useful, they must possess adequate properties in various categories. Typical categories that have to be considered in deciding if a fibre is suitable for a particular end use are shown in Table 2.3.

Of the properties listed in Table 2.3, only the mechanical ones are discussed below.

2.4   Mechanical Properties of Textile Fibres

2.4.1   Fibre Length

Most natural textile fibres exist as *staple fibres* and their length varies considerably. For example, the lengths of cotton fibres range from 12 to 36 mm (depending on the quality), whilst those of wool fibres range from 50 to 400 mm. Fibres whose lengths are less than 10 mm are not suitable because they are too short to be spun into yarn.

The characteristics of yarns depend on the staple fibre length. A fluffy, spongy yarn with a soft handle is obtained from shorter fibres, where many loose ends remain disoriented in the yarn. Fibres with longer staple lengths give smoother, finer yarns with a higher lustre and higher strength.

Synthetic fibres are produced in the form of *continuous filaments*, which are long, continuous strands of fibre. They can be used in this form but it is usual for them to be cut into predetermined lengths (i.e. *staple fibres*) to suit the type of yarn needed.

Natural and synthetic fibres are often blended together when making yarns (e.g. wool/nylon, cotton/polyester), giving the benefits of both fibre types. For this purpose the length of the synthetic filament may be cut to match that of the natural fibre, thus making it possible to use the same spinning machinery for both fibres. The only natural fibre that is obtained as a continuous filament is silk, produced from silk worms.

2.4.2   Fibre Fineness

The fineness of fibres also has an important bearing on the properties of yarns and fabrics made from them. There are various ways of representing fineness (the *count*). It can be expressed as the diameter in microns, where 1 micron, $\mu m = 0.001$ mm. Fine fibres have diameters lesser than

18 μm, medium fibres between 18 and 25 μm and coarse fibres greater than 25 μm. In general, the finer the fibres, the better their quality, because they feel softer and are more comfortable next to the skin. Fibres with a diameter of more than 40 μm tend to scratch and stick to the skin because they do not deflect easily. Coarse wool fibres, of the type used for making carpets, typically have such high diameters. Other wool fibres range in diameter between 15 and 40 μm, depending on the breed of animal and the position on the fleece on which it grew. Merino sheep produce the finest fibres, with a diameter of 17–25 μm. The fibres from cashmere goats have diameters of 13–15 μm. Cotton fibres typically have diameters in the range of 10–20 μm.

Man-made fibres can be produced with any desired diameter and unlike natural fibres their diameter is uniform along the length of the fibre. A useful feature of man-made fibres is that they can be produced with very small diameters, the so-called *microfibers*, which have diameters of 5–10 μm. *Nanofibres* are another type of fibre, which is currently the subject of intensive research. These fibres have diameters of around 0.1–0.2 μm.

Another widely used and ISO recommended (ISO 1044) system for expressing the fineness of fibres is the *tex system*. In this system tex is the mass (in grams) of 1 km of fibre, yarn or thread. So, for example, 1 km of a 30 tex yarn will weigh 30 g.

To avoid the use of small fractions of a unit for very fine yarns, a subsidiary unit, the *decitex* (dtex), is used. Decitex is the mass (in grams) of 10 km of fibre, yarn or thread. Thus 10 km of 20 dtex yarn will weigh 20 g. Microfibres, referred to above (and see Section 2.11), typically have values of <1 dtex.

For single fibre filaments the tex unit is impracticable because the values obtained in tex or dtex units are very small fractional values. For this purpose the *millitex* (mtex) unit is used. Millitex is the mass in grams of 1000 km of filament.

A unit that was used extensively in the past (and unfortunately still is in some publications) is denier, which is the mass in grams of 9000 m of yarn. What makes matters worse is that the tenacity of fibres or yarns is still sometimes quoted in units of grams/denier.

Fibres for clothing need to be soft and pliable, so their diameters need to be small. However, if the diameter is halved, there is a fourfold reduction in breaking strength. There is a practical limit to the smallest diameter that can be processed if fibre damage is to be avoided, this being around 10 μm, which is found with the finer Sea Island cottons and with Canton and Japanese silks.

Fibre diameter also has a significant influence on the dyeing properties of the fibre, because the surface area of a given mass of fibres is higher for finer fibres than for coarser ones. The finer fibres are therefore able to take up dye more rapidly. However a disadvantage of finer fibres is that considerably more dye is required to achieve a given depth of shade than is needed for coarser fibres. For example, fibres with a diameter of 25 μm must absorb about twice as much dye per unit mass as fibres of 44 μm diameter to appear the same depth.

2.4.3   Fibre Strength

Mechanical properties have an important bearing on the end uses to which fibres can be put. These properties are obtained from stress–strain curves, measured by the elongation produced as a progressively increasing stress (i.e. force) is applied along the axis of the fibre. Stress–strain diagrams for some natural and synthetic fibres, typical of the grades used in the manufacture of clothing, are shown in Figure 2.5.

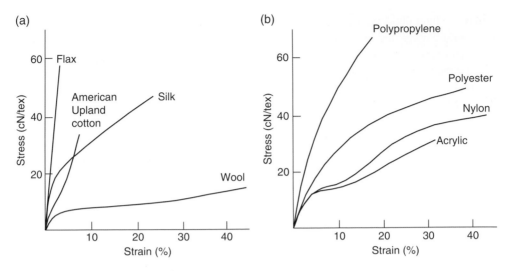

Figure 2.5 Typical stress–strain diagrams of (a) natural and (b) synthetic fibres.

One of the properties that can be obtained from stress–strain curves is *tenacity*, which is the stress (or force, in Newtons) which has been applied to a fibre at its breaking point. Because fibres are fine, only a small force is required to break them, so it is measured in centiNewtons (cN), a force that is 1/100th of a Newton.

Also, the strength of a fibre depends on its fineness (tex), so to compare the tenacity values of different fibre types, the force is divided by the tex. The units of tenacity are therefore usually quoted as cN/tex (or cN tex^{-1}).

It can be seen in Figure 2.5a that flax is the strongest of the natural fibres whilst wool is the weakest. The curves also show that wool extends very readily, even to low stresses, whilst flax has the opposite behaviour and is a much more brittle fibre. Of the synthetic fibres (Figure 2.5b), polypropylene is the strongest, but grades of synthetic fibres can be produced with different tenacities and high-tenacity versions of most are produced for specialised products, such as high-tenacity nylon for ropes. Indeed some forms of polyethylene fibres have tenacities of nearly 400 cN/tex.

Whilst extensibility is an important characteristic of fibres, so too is their *elasticity*. Elasticity is the ability of a fibre, yarn or fabric to return to its original length, shape or size; immediately after the removal of stress and for small amounts of stretch, most fibres recover completely. However, over longer extensions it is unlikely they will recover, thus resulting in some permanent extension.

The mechanical properties of fibres depend very much on the environmental conditions of temperature and humidity. Most fibre types absorb moisture from the atmosphere, though some types are more absorbent than others. This is termed *absorbency*.

Absorbency is expressed as the *moisture regain value*, which is the ratio of the mass of moisture in a material measured under standard conditions to the oven-dry mass, usually expressed as a percentage. It is calculated using the formula

$$\frac{M_{\text{Cond}} - M_{\text{dry}}}{M_{\text{dry}}} \times 100 \tag{2.1}$$

where

M_{cond} = mass of material under standard conditions (65% R.H. and 20 °C)
M_{dry} = mass of oven-dry material

The moisture regain values for some fibre types are given in Table 2.4. In general, natural fibres and regenerated fibres have higher moisture regain values than synthetic fibres. Fibres with high values are better able to dissipate static electricity charges. In warm, dry weather the build-up of static charges in synthetic fibres can make them not only difficult to process but also uncomfortable to wear.

2.5   Chemistry of the Main Fibre Types

2.5.1   Cellulosic Fibres

Cellulosic fibres can be obtained from the stem, leaf or seed of a plant. By far the most important of these is cotton, obtained from the seed. Other important cellulosic fibre types such as flax, jute, ramie, bamboo and hemp are *bast* fibres (obtained from the stem of plants) and are covered in Section 2.5.2.

2.5.1.1   Cotton

Cotton is produced from a plant of the genus *Gossypium*. It requires full sun, high humidity and rainfall for growth. The most widely used variety is *Gossypium hirsutum*, which is Upland cotton, but also most of what is produced as Egyptian cotton. It is a long staple fibre. Sea Island and Egyptian cotton, which has extra long staple (ELS) length, is a variety of *Gossypium barbadense*, also called pima cotton as well as ELS staple cotton. *Gossypium herbaceum* and *Gossypium arboreum* are shorter staple length cottons, common to India and East Asia.

Cotton is produced commercially in many countries, but the main producers are India, China, the United States and Pakistan. After mechanical harvesting, cotton is ginned to

Table 2.4   Moisture regain values of some textile fibres (measured at 20 °C and 65% relative humidity).

fibre	moisture regain (%)
Wool	14–18
Viscose	12–14
Silk	10–11
Cotton	7–8
Nylon 6.6	4.1
Acrylic	1–2
Polypropylene	0

remove solid dirt (dust, sticks, husks, seed coats, etc.) from the seed and then baled, sampled and graded. Fibre samples are graded on the basis of:

- Colour (the whiter the better)
- Staple length
- Micronaire (the fineness and maturity)
- Trash and dust content
- Strength

Since cotton is grown in many countries, inevitably there is a wide range of qualities. The highest qualities are the Sea Island and some Egyptian cottons, with fibre lengths of 25–65 mm. Most cotton is American Upland cotton, with fibre lengths of 13–33 mm. The varieties produced in the Asian countries generally have fibre lengths of only 10–25 mm, which have to be spun into thicker yarns to obtain the necessary strength for weaving.

2.5.1.2   Chemistry of Cotton

The cellulose molecule is derived from the simple sugar β-D-glucose, which can exist as a straight chain (**2.1**) or a closed ring structure (**2.2**):

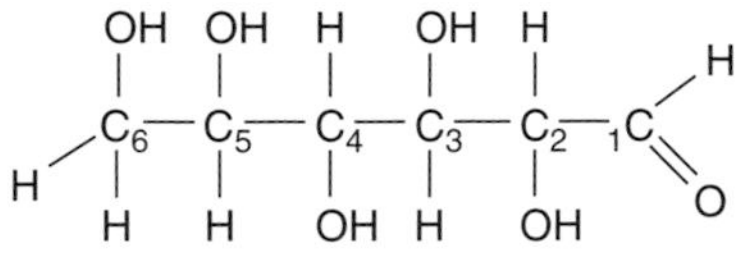

2.1 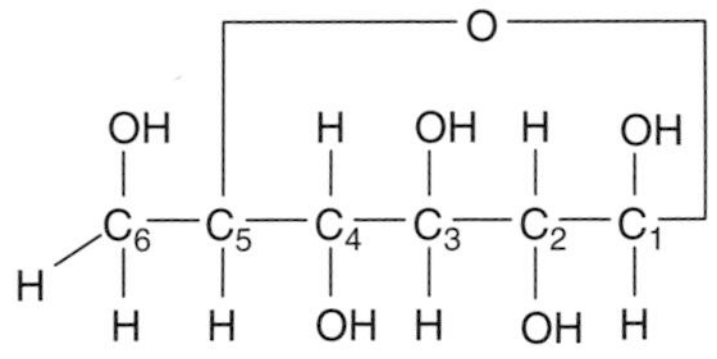   **2.2**

The closed ring structure, β-D-glucopyranose, is usually written as a projection formula (**2.3**):

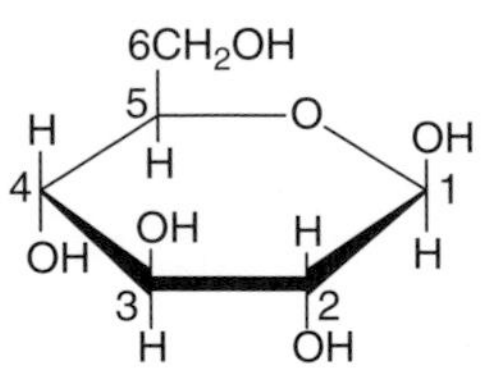

2.3

Two of these molecules combine through the elimination of water from the hydroxyl groups on the first and fourth carbon atoms to form cellobiose, which is the repeating unit of the cellulose polymer. In the formation of cellobiose, the two glucose rings are inverted with respect to each other (Scheme 2.1).

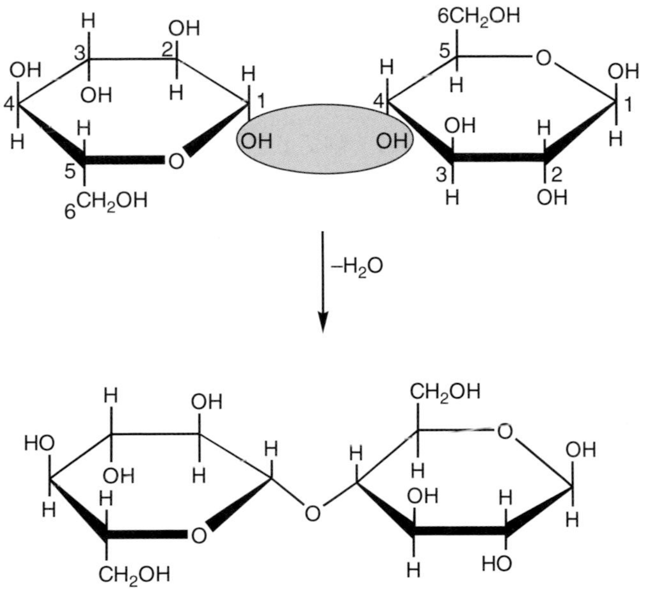

Scheme 2.1 Formation of cellobiose from two glucose units.

By a similar process, there is loss of a water molecule from the —OH groups as cellobiose units combine; cellulose (**2.4**) is formed, the repeating unit of which is then

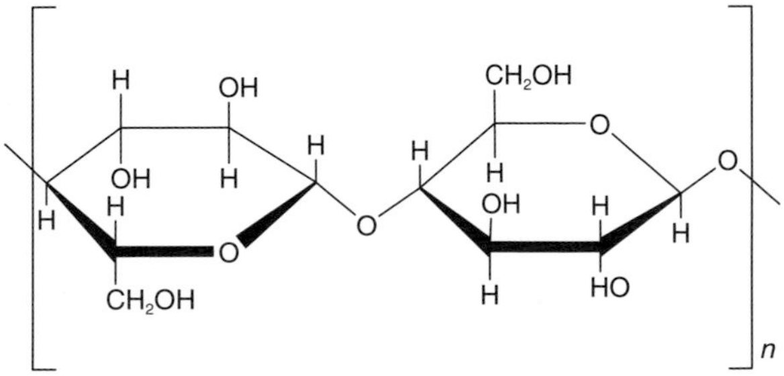

2.4

It is not possible to give an exact figure for the value of n (the *degree of polymerisation* (*DP*)) in the cellulose polymer of the cotton fibre. The value varies with the source of the cotton and even for a particular sample there is a distribution of values. Values generally quoted for the *DP* of cotton are about 3 000, corresponding to a molar mass of over 400 000, though *DP* values in excess of 10 000 have also been quoted. It is clear though that the length of cellulose molecules is extremely long in relation to their width.

The representation of the cellulose polymer molecule by the projection formula in Structure **2.4** is misleading because it implies a flat planar ring structure. The glucopyranose rings are actually 'chair' shaped (**2.5**), and the —OH groups are not oriented perpendicular to the rings either. Instead they lie in the same plane as the ring, and it is this feature of the geometric structure that enables adjacent polymer chains to align closely with each other:

2.5

Each glucopyranose ring in the chain contains three hydroxyl groups and it is their presence that allows cellulose to form hydrogen bonds. The ability to form hydrogen bonds is an important characteristic of the cellulose molecule. The close alignment of adjacent cellulose chains enables hydrogen bonds to form between them, a feature that is responsible for the high degree of crystallinity in the cotton fibre. The cellulose chains can also form hydrogen bonds with water molecules, which gives cotton fibres their moisture absorbent properties. The formation of hydrogen bonds between adjacent cellulose chains has a particularly important effect, for without them a water molecule would become attached to each hydroxyl group in the cellulose chains and the fibre would dissolve. Cellulose will not dissolve in water however and in fact is insoluble in all but a few organic solvents.

2.5.1.3   *Morphology of Cotton*

Cotton fibres have a complex morphological structure due to the way the cotton seed grows. During growth, dark green triangular pods called bolls form and within these bolls are about 20 seeds, which gradually develop (Figure 2.6).

Figure 2.6   Cotton bolls.

There are four main parts to the fibre structure: the cuticle, the primary wall, the secondary wall and the lumen (Figure 2.7a). The outermost layer of the fibre is the thin waxy cuticle, which protects the fibre from its environment. However this layer acts as a barrier to dyes and has to be removed by scouring. Beneath this layer is the primary wall, which is composed of cellulose molecules laid down during growth and spiralling round the longitudinal fibre axis at an angle of about 70° to the main fibre axis. The secondary wall makes up the bulk of the cotton fibre and consists of several layers of cellulose fibrils, each about 20 nm thick. These layers spiral at angles varying from 20 to 30° near the primary wall to about 40° near to the lumen. It is this spiralling of the fibrils along the fibre axis that gives the cotton fibres their inherent high strength. The lumen is the main pore in the centre of the fibre through which a sap, made up of a dilute solution of sugars, proteins and minerals, could pass during the growth stage.

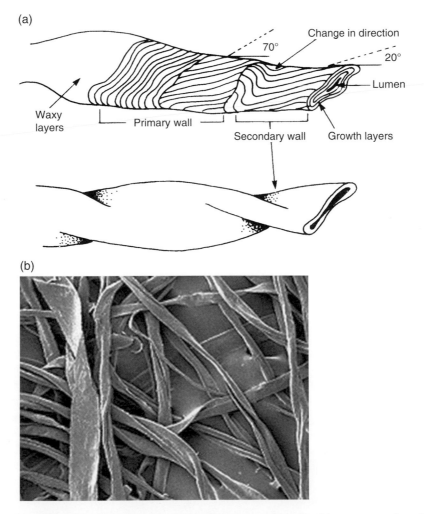

Figure 2.7 Cotton fibre, (a) structure represented diagrammatically, (b) as a scanning electron photograph.

On reaching maturity the boll bursts open, the sap in the seed hairs dries out, leaving residues of mainly proteins and minerals, and the lumen decreases in size. The fibres collapse from a circular cross-sectional shape to a bean shape with a hollow centre, and the fibres become flatter and convoluted along their length (Figure 2.7b). It is these convolutions that prevent parallel fibres from slipping past each other in the direction indicated by the arrow, thus contributing to the strength of the yarns when they are twisted together during spinning.

Cotton is subject to attack by pests and diseases, any of which may prevent the cotton from growing to its full maturity. These fibres are termed 'dead cotton' and they are characterised by having virtually no secondary wall structure. They have a very thin ribbon form and easily entangle into knots, which are called *neps*. It is difficult to remove these neps from mature cotton, and because they tend to reflect light off their top surface, they appear as white spots on dyed cotton fabrics. Another problem is that, in addition to the normal length cotton fibres, cellulosic fibres of much shorter length called cotton *linters* also grow in the bolls. These shorter fibres are unsuitable for spinning, so they are removed from the normal cotton fibres. However cotton linters are a useful additional source of cellulose and are used typically for the production of paper, viscose and acetate rayon fibres.

As cotton is a natural fibre it is understandable that the fibres contain impurities. Also, for ease of weaving, chemicals called 'sizes' are added to reduce friction between the yarns, so before any dyeing operations can be carried out, it is necessary to remove these chemicals. This is achieved by desizing and scouring processes, which also remove the waxy outer layer of the fibres. Additionally the fibres contain a small amount of naturally occurring pigment that imparts a slight yellowish cast to the fibres. This pigment is removed by bleaching, usually with hydrogen peroxide, though care has to be taken to avoid overbleaching when the glucose rings along the cellulose chain are damaged, to give derivatives such as those shown in Scheme 2.2.

After the desizing, scouring and bleaching processes have been efficiently carried out, the final cleaned fibres are almost pure cellulose.

2.5.1.4 Properties of Cotton

In cotton fibres the cellulose molecules exist as fully extended chains aligned sufficiently close in register to enable crystallisation to occur within the polymer matrix. The crystalline material is difficult to deform, and this, together with the already extended configuration of the polymer molecules, makes cotton fibres hard to stretch. All the molecules in the cross section of the fibre take a share of the imposed stress, and consequently cotton fibres will withstand a high tensile force before breaking. However crystalline material is also brittle, and cotton fibres are easily fractured on bending, a property that is associated with a reduced resistance to abrasion. So, cotton fibres have tenacities in the range 25–40 cN/tex, but unusually for textile fibres, they become 10–20% stronger when wet. This feature enables it to withstand aggressive wet processing operations, such as what may occur in dyeing and finishing or laundering. However, cotton fibres only stretch by about 5–10% and their elastic recovery is poor.

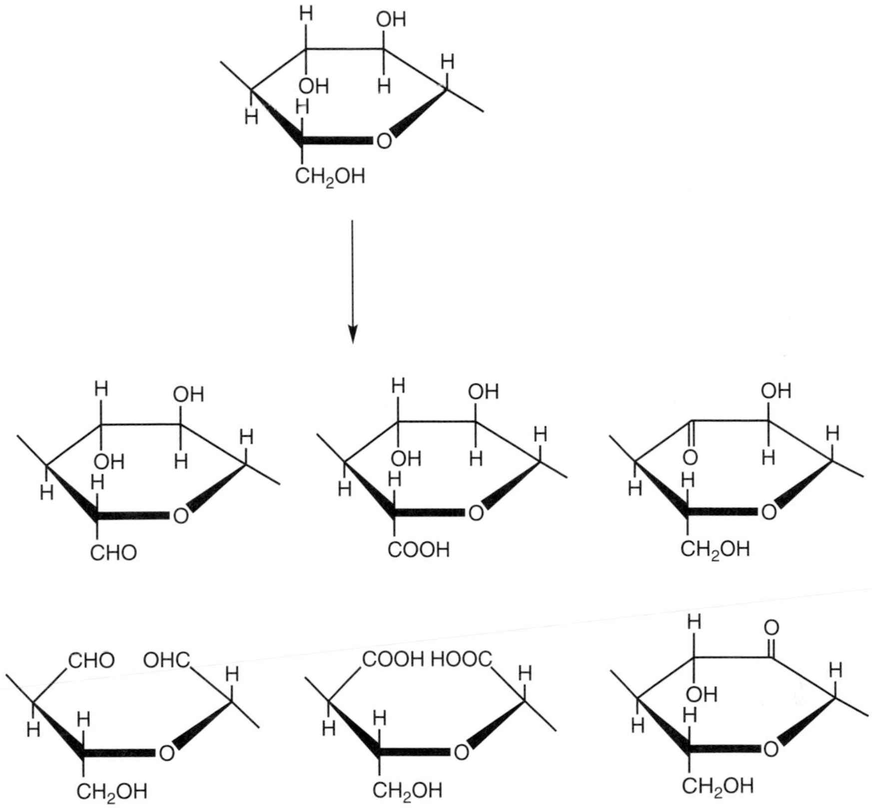

Scheme 2.2 Degradation products due to over-oxidation of the glucose rings of cellulose.

The popularity of cotton results from its softness and absorbency (moisture regain 8.5%), together with its availability at relative low cost. It is used extensively for underwear, jeans, shirting, dresses, knitwear, sportswear, leisurewear, workwear and childrens' wear. It is also used for towelling and interiors, such as curtains and bedding. Cotton is often blended with other fibre types, most commonly polyester, but also nylon and viscose.

Cotton is very stable to alkaline solutions but is rapidly degraded under acid conditions. All wet processing (desizing, scouring, bleaching, dyeing) therefore has to take place in solutions that are higher than pH 7.

The treatment of cotton with solutions of 25–30% caustic soda causes the fibres to swell in diameter and shrink longitudinally. This process is called *mercerisation* and is due to a change in the crystal structure, the fibres becoming denser, more lustrous (if the process is carried out on the fibres under tension) and more dycable. In fact the visual depth of mercerised cotton is higher than for unmercerised cotton, when the same amount of dye is present.

The three —OH groups on each glucose ring of the cellulose molecule not only make the fibre very hydrophilic but also enable the cellulose chains to form strong cross-links with neighbouring chains. These groups are chemically reactive, and use is made of this property in the application of reactive dyes (see Section 4.2.2.4) and crease-resistant finishes.

The interaction between the polymer chains of cellulose is so strong that cotton does not melt on heating; the chains break down and degrade before the temperature rises to a level at which they could be torn apart to form a liquid. Also cotton can be safely washed in boiling water and can be ironed at temperatures up to 200 °C. Nevertheless, because cellulose consists of the elements carbon, oxygen and hydrogen, cotton burns readily in a flame, and for some end uses it is necessary to treat cotton materials with a flameproof agent.

2.5.1.5   Organic Cotton

The cotton shrub is susceptible to attack by many types of insects and by various diseases, and in order to prevent what can be substantial losses, the plants are sprayed with pesticides. On a global basis significant quantities of pesticides are used annually. In response to this huge usage of pesticide, there has been a move to the production of 'organic' cotton. To be certified as organic cotton, the production of the cotton must be approved according to Regulation (EC) 834/2007, USDA National Organic Program (NOP) or any other standard approved by the International Federation of Organic Agriculture Movements (IFOAM) family of standards.

The main criteria are as follows:

- GMO seeds are not allowed.
- The use of fungicides or insecticides is forbidden.
- Herbicides for weed control must be avoided.
- Chemicals for defoliation are avoided.
- Pest control is achieved by means of a balance between pests and their natural predators through healthy soil and by the use of beneficial insects and good biological and culture practices.

As might be expected, yields per hectare are lower for organic cotton than for cotton produced using pesticides, so prices are higher. However there is a demand for organic cotton from increasingly environmentally aware domestic consumers. Independent certification is required to satisfy consumers that the cotton has been produced within a minimum set of standards, and the most demanding label for organic textiles is the Global Organic Textile Standard (GOTS). This standard applies to all natural fibres, not just cotton, though it is concerned more with the processing of the fibres into garments than with their actual growing. The standard imposes a number of severe restrictions on processing conditions, many of which have direct relevance to dyeing and finishing processes. The main features of GOTS that have direct relevance to dyeing are as follows:

- All chemical inputs (process chemicals, dyes and auxiliaries) must be assessed and must meet basic requirements on toxicity and biodegradability: there is a list of approved chemicals to help textile companies (and the dye and chemical suppliers) to meet this requirement.
- All chemical inputs intended to be used to process GOTS goods are subject to approval by a GOTS *Approved Certifier* prior to their usage. Preparations must have been evaluated

and their trade names registered on approved lists prior to their usage by a GOTS Approved Certifier.

- A list of chemicals banned from use is provided. The list includes toxic heavy metals, methanal, functional nanoparticles and aromatic solvents.
- Bleaches must be based on oxygen, so no chlorine-based bleaches may be used.
- The use of azo dyes that release carcinogenic amine compounds is prohibited.
- The effluent of all wet processing units must be purified in a functional waste water treatment plant.

The certification is required for the entire processing chain, from harvesting the fibres to final packaging and labelling. Despite these strict demands and the inevitable price premium that the consumer has to pay, the application of GOTS is increasing rapidly.

2.5.2 Other Cellulosic Fibres

Other cellulosic fibres of any commercial significance are the bast fibres, obtained from the stems of plants. These fibre types include flax, hemp, jute, ramie, bamboo and nettles. Some of these fibres, notably flax and ramie, have been cultivated for thousands of years, with evidence of their use around 5000 BC. The fibres obtained can be very long indeed, for example, flax fibres can be over 1 m in length, though they actually comprise much shorter fibres, called 'ultimate fibres', which are bound together by gums and resins. Also present in the stems are pectins, hemicelluloses and waxes, all of which contribute to the necessary rigidity required of the stem, but which make extraction of the cellulosic fibres difficult. Nevertheless, the fibres, once extracted, are noted for their strength and, especially in the cases of ramie, hemp, bamboo and nettles, their softness.

Different methods are used to extract the fibres from the stems. For example, flax, hemp and jute are extracted by a retting process, whereby the stems are allowed to rot in warm aqueous environments, formerly in bogs or dams, though now in concrete structures. The stems can be just left in fields (dew retting). Afterwards the stems are removed and dried and the woody parts removed by a mechanical crushing process. Ramie fibres are more difficult to extract and a mechanical beating process has to be employed. Most bamboo fibre is obtained by a process similar to that for making viscose (see Section 2.7.2).

2.6 Protein Fibres

Protein fibres are derived mostly from animal hair, the main exception being silk. Of the hair fibres, wool is the most important though other types such as cashmere and mohair are also of commercial significance. The principal protein present in hair fibres is keratin, whilst silk is comprised of the protein fibroin. The main difference between the two proteins is that keratin contains substantial amounts of sulphur-containing amino acids, but fibroin does not contain any such amino acids.

2.6.1   Wool

Wool is the hair fibre of sheep. Hair fibres from other animals are referred to as 'hair' fibres, not wool. Sheep can be reared in most countries, but the most important producers are Australia, the former USSR and New Zealand. The United Kingdom is the world's ninth largest producer at around 2% of world production. There are many breeds of sheep and therefore qualities of wool available. The finest type is Merino wool, with diameters of 17–25 μm and fibre lengths of 60–100 mm. Merino wools are mostly produced in Australia. British wools are coarser than Merino though some cross-bred lambswools are fine enough to be used in clothing. Cross-bred wools are coarser, with diameters of 25–40 μm and length of 75–150 mm. They are also more lustrous. Carpet wools are coarser still, with diameters up to 100 μm.

The term 'lambswool' refers to fibres that are shorter and finer than adult wools and are characterised by being tapered at one end. Only the fibres from lambs' first shearing can be called 'lambswool'; after the first shearing, both ends of the fibre have a blunt, cut edge. The term 'virgin wool' refers to all new wool, not lambswool.

The term 'shoddy' is reclaimed wool, some of which will have been recovered from garments that were never sold. Other shoddy wool may have been worn. The term 'all wool' indicates a mixture of wool fibres that may be a mix of 'virgin' and 'shoddy' wools.

2.6.1.1   Chemistry of Wool

Protein molecules are comprised of different amino acids as the monomer units. The general formula of an amino acid is (**2.6**)

$$H_2N - \overset{\overset{\displaystyle H}{|}}{\underset{\underset{\displaystyle X}{|}}{C}} - COOH$$

2.6

All amino acid monomer building units essentially consist of a central C atom, to which is bonded an amino group ($-NH_2$), a carboxylic acid group ($-COOH$), a hydrogen atom and an X group. Because the four bonds of a carbon atom are arranged in space as a tetrahedron, the keratin molecules formed from them exist in their normal state with their backbone coiled in a helix. This form is known as α-keratin.

The side group X can be one of 18 different chemical groups, so that in all there are 18 different amino acids in keratin. Seven of the groups are hydrocarbons and are non-polar in character, three contain hydroxyl groups and so confer some water-liking character where they occur in the chain, four contain acidic groups and two contain acidic groups. Some examples of these types of amino acids are shown in Table 2.5.

Growth of the keratin chain occurs when the amino group ($-NH_2$) of one amino acid reacts with the carboxylic acid group ($-COOH$) of another. In the reaction (Scheme 2.3) one molecule of water is lost and a peptide linkage ($-CONH-$) is formed.

Table 2.5 Examples of different types of amino acids occurring in protein fibres.

type of amino acid	example of side group, X	
Non-polar (contain hydrocarbon in the side group)	Glycine X=—H	Alanine X=—CH$_3$
		Leucine $X=-CH_2-CH\overset{CH_3}{\underset{CH_3}{\big<}}$
Polar (contain —OH in the side group)	Serine X=—CH$_2$OH	Threonine $X=-CH\overset{CH_3}{\underset{OH}{\big<}}$
		Tyrosine $X=-CH_2-\!\!\left\langle\!\!\bigcirc\!\!\right\rangle\!\!-OH$
Acidic (contain —COOH in the side group)	Aspartic acid X=—CH$_2$—COOH	Glutamic acid X=—CH$_2$CH$_2$—COOH
Basic (contain —NH or —NH$_2$ in the side group)	Lysine X=—(CH$_2$)$_4$NH$_2$	Histidine $X=-CH_2-\underset{\substack{N\\ \ \ \ N\\ \ \ \ H}}{\text{(imidazole ring)}}$

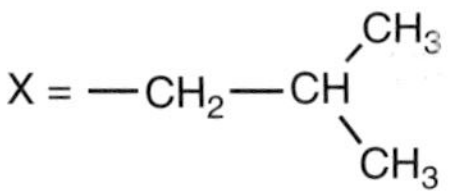

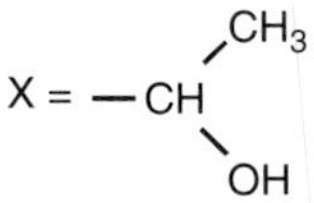

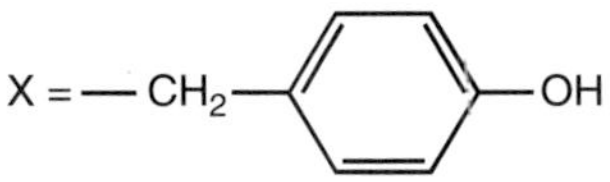

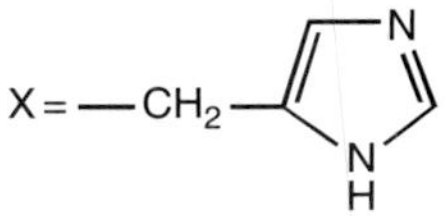

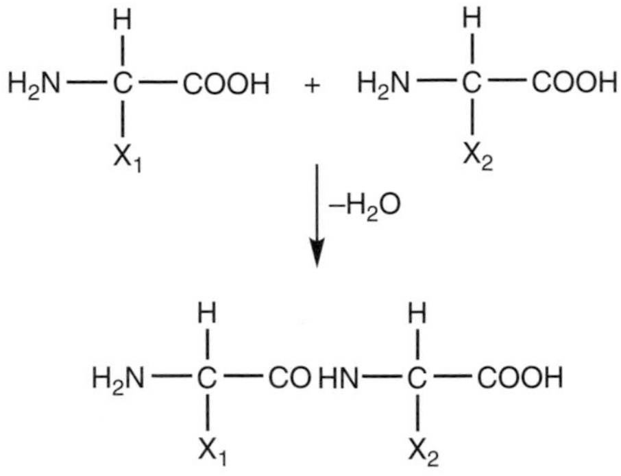

Scheme 2.3 Formation of a dipeptide from two amino acid molecules.

This process repeats at each end (Scheme 2.4).

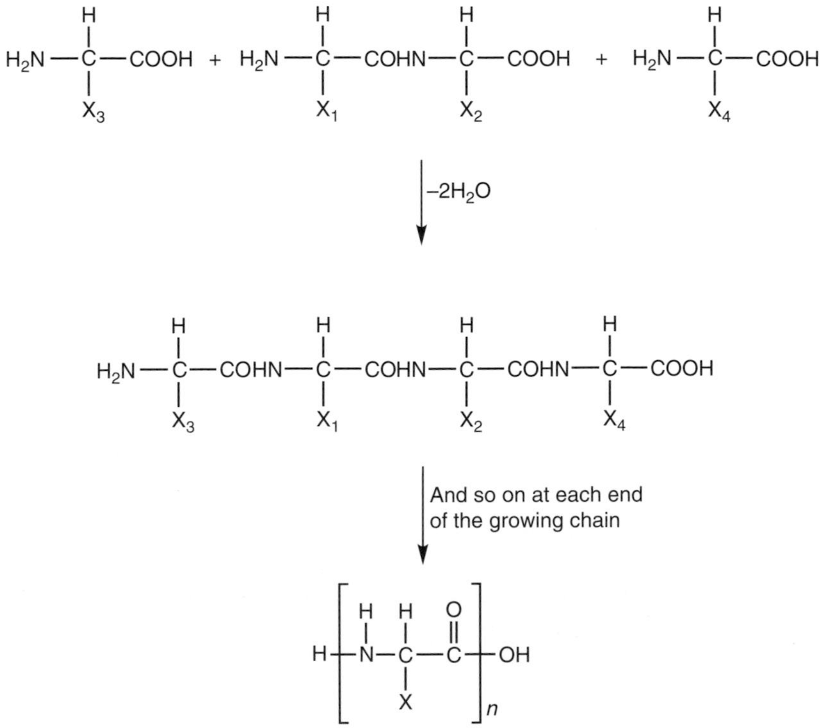

Scheme 2.4 Growth of a polypeptide chain.

The 18 different amino acids are not present in the keratin chain in equal quantities; indeed some such as methionine, tryptophan and histidine are only present in very small amounts. Of special importance in wool chemistry is the presence of the amino acid cystine (**2.7**), which contains the group $-CH_2-S-S-CH_2-$. This group is important because it links adjacent keratin chains through what are called disulphide bonds (**2.8**):

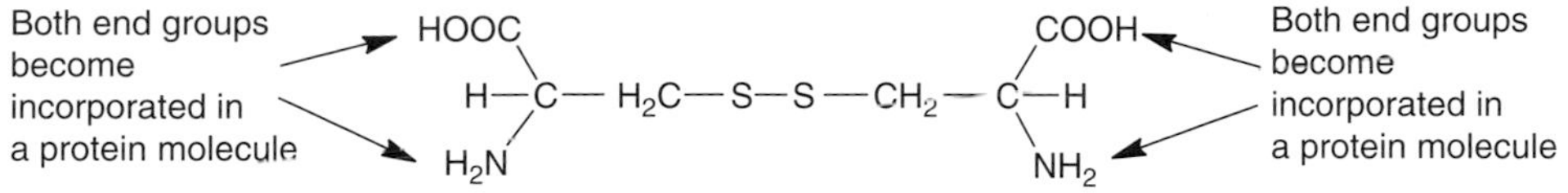

Both end groups become incorporated in a protein molecule

$$HOOC$$
$$H{-}\overset{\displaystyle |}{C}{-}\ H_2C{-}S{-}S{-}CH_2{-}\overset{\displaystyle |}{C}{-}H$$
$$H_2N \qquad\qquad\qquad\qquad NH_2$$

COOH

Both end groups become incorporated in a protein molecule

2.7

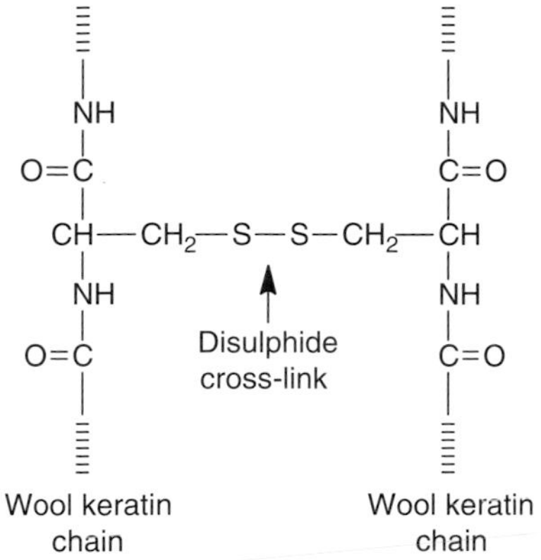

NH
O=C
CH—CH$_2$—S—S—CH$_2$—CH
NH
O=C

NH
C=O
CH
NH
C=O

↑
Disulphide
cross-link

Wool keratin chain   Wool keratin chain

2.8

The disulphide bonds can also form between different parts of the same keratin chain, and the overall effect of the cross-links is to provide stability to the fibre structure. The disulphide bonds are susceptible to hot wet conditions and alkaline solutions. Other reagents can have the same effect. If the reactions are not controlled, the fibres are damaged. For example, boiling wool fibres in alkaline solution will cause them to degrade to such an extent that they eventually dissolve. However, with care it is possible to manipulate conditions to advantage by allowing the cross-links to be broken. Then, after rearrangement of the wool fabric structure, they can be reformed in the process known as setting. In this way creases and pleats can be formed in a wool fabric.

The polypeptide chain formed is highly complex. The wool structure is stabilised by the presence of three types of cross-links. Reference has already been made to the disulphide cross-links that occur wherever the amino acid cystine is present in the chain. The other types of cross-links that can occur between the polymer chains of keratin are hydrogen bonds (**2.9**), such as what occur between serine and aspartic acid residues:

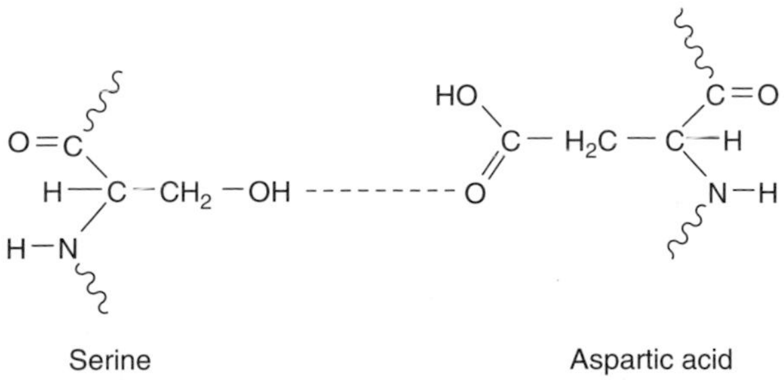

O=C
H—C—CH$_2$—OH - - - - - - - - O
H—N

HO
C—H$_2$C—C—H
C=O
N—H

Serine   Aspartic acid

2.9

and ionic bonds (**2.10**), which occur between ionised acidic and basic groups:

$$O=C \quad \begin{matrix} H \overset{|}{\underset{|}{C}} -(CH_2)_4 -\overset{+}{N}H_3 \quad ---- \quad \overline{}OOC -H_2C -\overset{|}{\underset{|}{C}} -H \\ H-NH \end{matrix} \quad C=O$$

Lysine   Aspartic acid

2.10

Additionally, the basic and acidic side groups that occur along the length of the keratin chain (e.g. where lysine and aspartic acid exist) enable wool fibres to combine with both acids and bases. Molecules with this ability are termed *amphoteric*.

At high pH values (alkaline solutions), the acidic groups ionise to form negative ions, leaving the wool with a net negative charge:

$$\text{wool} - COOH + NaOH \rightarrow \text{wool} - COO^- + Na^+ + H_2O$$

At low pH values (acidic solutions), the situation is reversed, the basic groups reacting with hydrogen ions in the solution to form positive ions and leaving the wool with a net positive charge:

$$\text{wool} - NH_2 + HCl \rightarrow \text{wool} - NH_3{}^+ + Cl^-$$

It follows that there must be a pH at which the numbers of negative and positive charges are equal. This pH is called the *isoelectric point* of the fibre and occurs at about pH of 4.6.

The isoelectric point is the pH at which a molecule carries no net electrical charge. It is at this pH that the fibre is most resistant to damage by acids or alkalis. For this reason dyeing processes that take place in the pH range 4–6 are preferred.

2.6.1.2 *Morphology of Wool*

Wool fibres have a very complex morphological structure, as illustrated in Figure 2.8a. The distinctive outer scales of wool fibres overlap in the direction of the tip of the fibre (Figure 2.8b). The scale structure gives the fibre surface its characteristic roughness, which creates friction between fibres in a yarn, thus contributing to its strength. Because of the orientation of the scales, the friction is greater in the direction of tip to root than from root to tip. This is called the *directional friction effect* and has important processing consequences.

If the fibres are mechanically agitated in the hot wet conditions of processing, they migrate preferentially in the direction of least resistance, that is, towards the tip of adjacent fibres. If the movement is allowed to go too far, individual fibres begin to form loops and become entangled, and the yarn or fabric structure becomes thicker and denser.

The effect can be desirable if controlled, as in the *milling process* to produce felted fabrics utilised in the manufacture of overcoats or baize plating surfaces for snooker tables. However, if uncontrolled, for example, during laundering, excessive shrinkage can occur.

In order to avoid shrinkage that may occur during washing, there are two methods of treating the fibres. The first is an oxidation reaction by chlorination, the most commonly

(a)

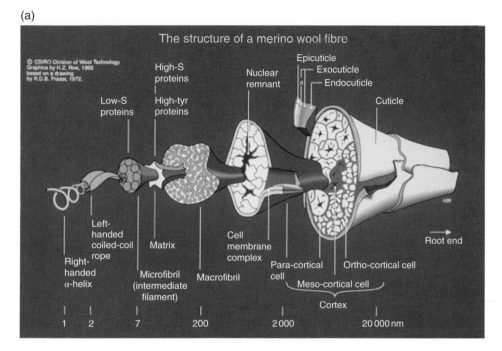

The structure of a merino wool fibre

© CSIRO Division of Wool Technology
Graphics by H.Z. Roe, 1992
based on a drawing
by R.D.B. Fraser, 1972.

(b)

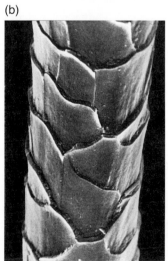

Figure 2.8 (a) Diagrammatic representation of the structure of a wool fibre. (b) Electron microscope photograph of a merino wool fibre (Images from CSIRO, Australia).

used agent being dichloroisocyanuric acid. The reaction converts the disulphide bonds in the outer cuticle layer to cysteic acid, which makes the fibre more hydrophilic. The surface proteins swell and the scales are softened.

The second method involves depositing a thin polymer coating over the scales, thereby smoothing their structure so the fibres slide over each other easily. The most widely used process for making shrink-resistant wool that is machine washable is the chlorine–Hercosett

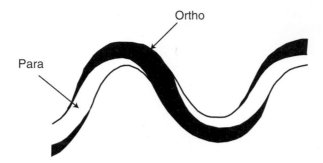

Figure 2.9   Distribution of the orthocortex and paracortex in a wool fibre.

process in which wool tops are chlorinated first and then the water-soluble resin Hercosett 215 is applied. However, chlorine-based processes are not allowed under GOTS (see Sections 2.5.1.5 and 2.6.1.4), and as an alternative, polymers are marketed that impart shrink resistance to unchlorinated wool.

The outer scales are thin structures that form the main component of the wool fibre cuticle. This surrounds the bulk of the fibre material, which is contained in the cellular cortex beneath. The cortex cells are tapering spindle-shaped structures, constructed from component macrofibrils in which the successively smaller microfibrils, protofibrils and finally the helical structure of the protein molecule may be detected. The cortex is heterogeneous and is divided into two regions along the length of the fibre, known as the orthocortex and the paracortex. The two halves spiral along the fibre axis, with the more extensible orthocortex always lying on the outer curvature of the fibre (Figure 2.9). They differ slightly in their chemical composition, and the difference is responsible for their difference in elasticity: this endows the fibres with a characteristic crimp.

The whole fibre is held together by a series of membranes surrounding the cell structure, the outermost of which is the epicuticle. The epicuticle is important from the point of view of dyeing, since it is hydrophobic and is a barrier to the entry of dyes. Because it is so thin, it is easily destroyed during preparatory processes (e.g. scouring), but the extent to which it is removed influences the ease of dye uptake.

2.6.1.3   *Properties of Wool Fibres*

Wool fibres stretch easily and recover well from the stretch, which is due to the degree of alignment of the polymer chains along the fibre axis. The keratin chains in wool fibres have a coiled configuration due to the tetrahedral bond arrangement in the constituent amino acids. This configuration is stabilised by the various types of cross-links, which can form bridges between coils within the same molecule, so that the fibre becomes an extensive molecular grid structure. When the fibres are stretched, the coils straighten out, but on release of the tensile force, the cross-links cause the chains to spring back to their original helical arrangement. With this molecular configuration, there is very little crystallisation, and any strain is distributed unevenly between molecules, which gives rise to a low breaking strength. Wool is therefore weak (tenacity only 9–15 cN/tex), but highly extensible; the balance between tensile strength and elasticity makes it a very desirable fibre.

Because the keratin chains contain so many amino acids that are polar in character (the —OH containing the acidic and basic types), which can form hydrogen bonds with water molecules, wool fibres are hydrophilic. Indeed the moisture regain value of wool is 14–18% and this absorbency contributes to the comfort of woollen garments. The crimp of wool fibres prevents their close packing in a yarn. Consequently a good deal of the volume of the yarn is air, which makes for a lofty (bulky) warm yarn, the warmth being due to the insulating properties of the static air inside the yarn.

In contrast to cotton fibres, wool is easily degraded in alkaline solutions. In fact wool fibres will dissolve in 5% sodium hydroxide solution at 100 °C. Alkalis attack the disulphide cross-links between the keratin chains and also hydrolyse the peptide links of a chain (Scheme 2.5). Consequently the use of alkaline conditions in wool processing is very restricted.

Wool fibres are much more resistant to acids and are usually dyed under acidic or near neutral conditions of pH. In the presence of acid, any amino groups at the ends of the polymer chains (or along the polymer chain in the side group X, e.g. if it is lysine) become protonated (see Section 2.6.1.1), but at pH values lower than 2.5 wool is damaged appreciably due to *hydrolysis* of the main chain.

The setting of wool to form permanent creases and pleats is an important process in the manufacture of woollen goods. The reactions are based on the rupture, then

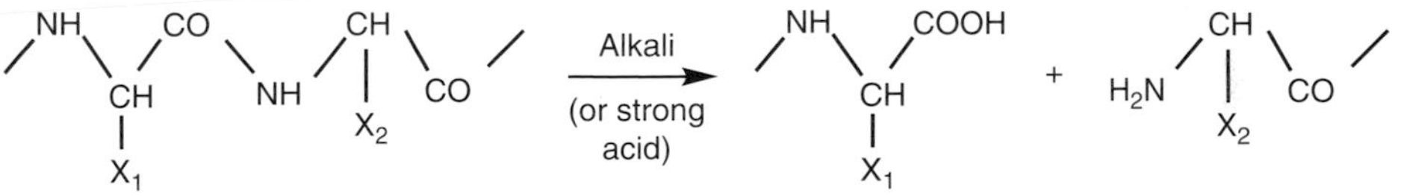

Scheme 2.5 Degradation of the polypeptide chain in wool by alkali (or strong acid).

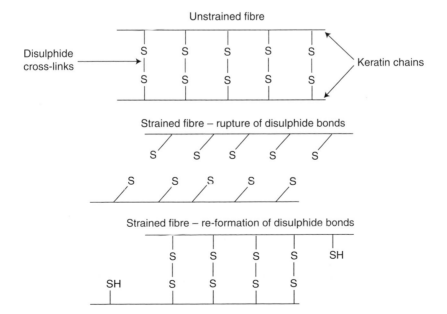

Figure 2.10 Rupture and re-formation of cross-links.

re-formation of disulphide cross-links, under hot, wet conditions. When the fibres are strained, the cross-links break and then reform in the new creased conformation (see Figure 2.10). The reactions taking place are complex, and in addition to the re-formation of disulphide cross-links, other types of cross-links can form as well. The process can be carried out by steaming woollen fabrics in their pleated shapes and the effect is permanent. These reactions can also occur during dyeing processes and can lead to a loss in fibre strength and to the formation of surface effects on woollen fabrics, such as crow's feet, in piece dyeing.

2.6.1.4 *Ecological Aspects*

As with cotton, the processing of wool, in terms of scouring, bleaching and dyeing, has ecological issues associated with it, and there is an increasing drive towards the implementation of the GOTS, mentioned in Section 2.5.1.5. The impact as far as wool is concerned is related to the use of chemical agents to kill insects and other parasites on sheep, the residues of which may still be present in the wool during processing. The types of detergents used for scouring are now mainly fatty alcohol ethoxylates, which are more readily biodegradable than the octyl- and nonylphenol ethoxylates used previously. GOTS also prohibits chlorination in textile processing, but chlorination is widely used in shrinkproofing wool, as described in Section 2.6.1.2. Consequently, alternative methods to shrinkproofing wool that are environmentally acceptable and meet the standards required of GOTS have been developed.

2.6.2 Hair Fibres

Fibres derived from animals other than sheep are referred to as hair fibres. Examples are cashmere and mohair from goats and fibres from camels, alpacas and llamas. These fibres differ from wool fibres and, from one another, in both diameter and scale structure. These fibre types are very fine (the average diameter of cashmere fibres is 19 μm) and are used for specialty (luxury) textile fabrics. However, apart from cashmere, these fibres are less durable than wool fibres.

2.6.3 Silk

Silk is also a protein fibre, and the same amino acids that make up the keratin chains in wool are also the building blocks of the polypeptide chains that constitute silk fibroin, but they are present in amounts very different from those in wool. Indeed silk fibroin is comprised mainly of the amino acids glycine, alanine and serine. Also in contrast to wool keratin, silk fibroin contains only a very small amount of cystine.

The fibre is produced as a cocoon by the larva of the silk moth *Bombyx mori*. The protein is extruded from two spinnerets, one on each side of the head of the caterpillar, as a pair of continuous filaments between 3000 and 4000 m in length, joined together by a gum called seracin. The gum is removed during processing to release the individual filaments, which have a triangular cross section and a smooth surface.

The tenacity of silk is the highest of the protein fibres, at 38 cN/tex, and the fibres are able to stretch considerably, with good recovery from stretch. They have a slightly lower moisture regain (~10%) than wool fibres.

2.7   Regenerated Fibres

2.7.1   Early Developments

Cotton, wool and silk possess a wide range of properties desirable for textile fibres. For many years, silk with its smooth handle was expensive and very much a luxury fibre, so during the second half of the nineteenth century, chemists made huge efforts to produce cellulose-based fibres that would compete with silk in terms of fineness and handle.

Regenerated fibres are so called because naturally occurring forms of fibrous polymeric materials are extracted by chemical means to dissolve them. They are then spun (see spinning methods in Section 2.9) from solution to regenerate the fibre form. By controlling the extrusion process carefully, fibres of consistent high quality can be produced. The two main naturally occurring polymer types are cellulose and protein, though initial work, and the greatest success, was achieved with cellulosic fibres.

Cellulose is the structural tissue of the cell walls of plants and is therefore a very widely available fibre resource. However it is insoluble in common solvents, so to extract it from the plant material requires aggressive chemical processing. The first successful commercial venture of this kind was made by Count Hilaire de Chardonnet in France, who converted cellulose into a viscous mass of cellulose nitrate, which was dissolved in a mixture of ethanol and ether and then forced through a spinneret to form a filament. Cellulose nitrate is highly flammable, but Chardonnet developed a process to de-nitrate the fibres which he patented in 1885. However the process was messy and inefficient and it is now obsolete.

Another solvent for cellulose that was discovered in the mid-nineteenth century is cuprammonium hydroxide. This solvent forms a soluble complex (**2.11**) with the —OH groups of the cellulose ring:

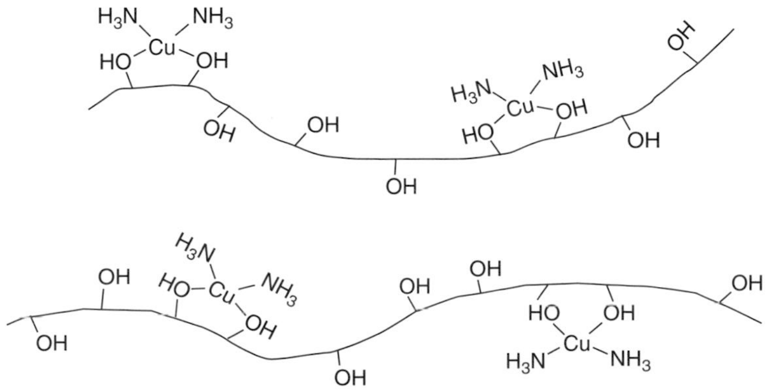

2.11

The solution of cellulose in this solvent is a clear blue colour and it is wet-spun by passing it through a spinneret, then down specially designed long thin tubes and into an acidic coagulation bath. The resulting fibres, called *cuprammonium rayon*, are regenerated cellulose and so have characteristics similar to those of viscose. However they are finer than viscose fibres and retain their strength better when wet. They are noted for their superb handle, and they became known as 'Bemberg silk'. The raw material used to make these Bemberg fibres is cotton linters, which are very short fibres of cotton form in the boll, but which are too short to be spun into yarns and so otherwise of no value. During the twentieth century cuprammonium rayon fibres were produced in Europe, America and Japan, but now it is produced only in Japan.

The most commercially important regenerated cellulosic fibres produced today are viscose and the lyocell fibre types.

2.7.2   Viscose

Cross and Bevan in England developed a different process, one based on the extraction of cellulose from the cellulose in trees. The process they developed – the viscose process – patented in 1893 was long because it required complex procedures lasting many hours. The main raw material nowadays for the manufacture of viscose is trees, mainly beech, which are obtained from managed forests.

Firstly, it is necessary to extract the cellulose from the wood and then steep the cellulose in sodium hydroxide solution, which reacts at each of the —OH groups of the cellulose ring to form 'soda cellulose':

$$Cellulose - OH + NaOH \rightarrow Cellulose - O^- Na^+ + H_2O$$

Carbon disulphide (CS_2) is then added, which forms a viscous solution of a chemical complex of cellulose called sodium cellulose xanthate, which has a bright orange colour:

$$Cellulose–O^-Na^+ \; + \; CS_2 \rightarrow Cellulose–O–\overset{\overset{\displaystyle S}{\|}}{C}–S^- \; Na^+$$

This solution has to be aged for 1–3 days to reach the correct viscosity for spinning and it is then wet-spun into an acidic bath:

$$2Cellulose–O–\overset{\overset{\displaystyle S}{\|}}{C}–S^- \; Na^+ \; + \; H_2SO_4 \; \rightarrow \; 2Cellulose–OH \; + \; 2CS_2 \; + \; Na_2SO_4$$

The fibre produced, viscose, has superb properties in terms of handle, lustre, softness and absorbency. It is because of these attractive characteristics that the fibre is still produced today, though it has meant considerable investment in factories by the fibre producers to meet the strict environmental standards required nowadays. Carbon disulphide (CS_2) is an explosive, highly toxic and very smelly liquid and is reformed in the coagulation bath, so sophisticated processes have had to be developed to recover it effectively.

Viscose fibres are almost pure cellulose and show similar chemical properties to cotton. However it differs considerably from cotton in both its molecular structure and morphology. As a result of its manufacturing method, the cellulose polymer molecules in viscose are much shorter than in cotton. Also, the cellulose molecules are arranged in a much less ordered fashion than in cotton, so the degree of crystallinity of viscose is lower and overall the fibres have a greater amorphous character. Viscose fibres do not have the complex morphological structure of cotton fibres, but the extrusion process results in them having a 'skin–core' structure, whereby the orientation of the cellulose molecules in the outer layer is greater than that in the inner core of the fibres. Having tenacities of 25–30 cN/tex, viscose fibres are much weaker than cotton fibres, and they become much weaker when wet.

During the middle part of the twentieth century, much research was carried out to improve the mechanical properties of viscose, mainly by making subtle changes to the composition of the bath into which the fibres are extruded. This research led to the development of polynosic fibres with much better strength, particularly wet strength, and of high-tenacity rayons and flame-retardant rayons for specialised uses. The high absorbency of viscose fibres has led to their use in non-textile applications as well, such as wipes, disposables and feminine hygiene products.

2.7.3   Lyocell Fibres

Lyocell fibres are also cellulosic fibres, and as with viscose, the starting material for their production is wood pulp. However the process differs from the viscose process or the cuprammonium process in that it does not involve the formation of a chemical derivative of cellulose. Instead, after extracting the cellulose from wood pulp, it is dissolved directly into a solvent, *N*-methylmorpholine-*N*-oxide (NMMO). This solvent is non-toxic and biodegradable, and in addition the process recovers over 99.5%, so very little of the NMMO is lost to the environment. The cellulose solution is dry-jet wet spun (see Section 2.9) into a water bath, and the fibres formed are washed, dried and then cut into staple lengths, most usually of around 40 mm.

The key stages in the manufacture of viscose and lyocell fibres are shown in Figure 2.11.

Lyocell fibres are produced by Lenzing AG principally in Austria but also at plants in Grimsby in the United Kingdom, the United States and China. They are sold under the trade name Tencel®. The whole process is much shorter (about 8 hours) than that of viscose manufacture (about 40 hours), and the fibres are less degraded, so they have much greater tenacities (about 36 cN/tex) than viscose with much better wet strength.

Lyocell fibres also differ from viscose in that they are much smoother and have a more circular cross-sectional shape. Another characteristic feature is their high degree of crystallinity and orientation. The high degree of orientation causes the fibres to fibrillate, that is, to form microfine surface hairs called fibrils under hot, wet processing conditions. These fibrils can be a nuisance in dyeing because they can combine to form pills on a fabric surface. Also, when heavy depths of shade are dyed, they can give a frosted effect because they are so fine that they retain a white appearance, despite the presence of dye within them. It is often necessary to remove these fibrils,

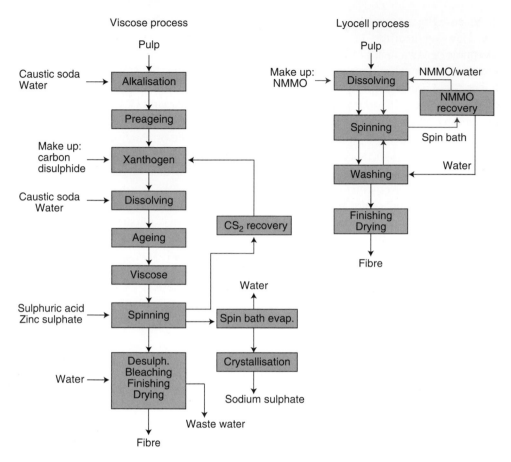

Figure 2.11 Stages of manufacture of viscose and lyocell fibres (the pink arrows show the use and recovery of the NMMO solvent).

which is achieved by treatment with enzymes. Nevertheless secondary fibrillation, producing fibrils that are much shorter and finer, can occur in subsequent processing, though they give fabric a 'peach-skin' handle, which is very acceptable for casual wear. Most lyocell fabrics are dyed in open-width machines, which prevent fibrillation occurring and leave a clean, flat fabric surface.

A cross-linked version of lyocell is produced, Tencel A100, which is resistant to fibrillation. This version is not suitable for blending with cotton however, because it has a higher dyeability than cotton and is not stable to strong alkali, so it cannot be mercerised. Another cross-linked type, Tencel LF, is more stable in alkali and has a dyeability similar to cotton, so it is more suitable for blending with cotton.

Lyocell fibres such as Tencel are used for the same types of end uses as cotton, such as clothing (e.g. shirts, blouses, leisure garments) and bedding materials (e.g. bed linen, mattresses and quilts). The fact that it is produced from a renewable resource by a process that produces very little chemical waste and is biodegradable gives it an excellent environmental credibility.

2.7.4 Cellulose Acetate Fibres

There are two types of cellulose acetate fibres, the secondary cellulose acetate and the triacetate. They are usually included in this class of regenerated fibres because they are cellulose based, but in fact they are not pure cellulose; instead they are chemical derivatives of cellulose. The triacetate fibre (**2.12**) is formed by reacting the cellulose (usually from cotton linters) with acetic anhydride with a sulphuric acid catalyst, when the three —OH groups of the cellulose rings are acetylated,

2.12

This reaction is an esterification reaction, which is explained in Section 1.8.3.4. The conversion of the hydrophilic —OH groups to the hydrophobic (water-hating) acetate (ethanoate) —OCOCH$_3$ derivative results in the cellulosic material having much lower water-liking character.

Secondary cellulose acetate fibre, often just called 'acetate', is manufactured from the triacetate by partially de-acetylating it, so that on average only 2.4 of the three hydroxyl groups on the glucose rings remain acetylated. Since some of the glucose rings still possess an —OH group, the acetate fibre is slightly less hydrophobic (water-hating) than the triacetate. These two fibres were first marketed in the early 1920s. Indeed the secondary cellulose acetate form had been produced in large quantities during World War I when it was applied as a dope in acetone solvent to the fabric wings of aeroplanes. When the acetone evaporated, the acetate residue in the fabric made the wings impervious to air. At the end of the war, the Dreyfus brothers, who had established the process for making it, developed a process for extruding the dope by the dry spinning process (see Section 2.9) to produce fine fibres.

Both the secondary and the triacetate types met with particular interest when they were first introduced. The secondary cellulose acetate had a superbly soft handle and silk-like quality, and the triacetate with its hydrophobic character made it the first fibre to be available with quick drying properties. However the fibres have only low crystallinity and are not very strong, with tenacities of about 13–15 cN/tex. The secondary acetate softens easily and has a low melting point (230 °C), so care is required when ironing it. The triacetate has a much higher melting point (290 °C), so ironing is less of a problem. Additionally it can be heat-set, so it will hold pleats.

These two fibre types have much less importance nowadays than when they were first introduced, but they had an important impact on the dyestuff industry because at the time of their introduction none of the dyes available had satisfactory substantivity for them. It was necessary for the dye manufacturing companies to develop suitable dyes for them, dyes that are less water soluble and more hydrophobic in character. The dyes developed were called disperse dyes, and this application class of dyes (see Section 3.5.3.1) became more important when nylon and polyester fibre types came on to the market in the 1940s and 1950s.

2.7.5 Polylactic Acid Fibres

Considerable interest has developed in recent years with fibres synthesised from polylactic acid (PLA), because of their high biodegradability and the fact that they are produced from a renewable resource (corn starch, maize, sugar or wheat). PLA fibres are not strictly regenerated fibres, because their formation does not involve a naturally occurring polymeric material being regenerated in textile fibre form. Instead PLA fibres are produced by bacterial fermentation, in which natural sugars and starches are firstly converted into lactic acid and the lactic acid is then polymerised (Scheme 2.6). The final PLA polymer has a different chemical structure from the original starting material. In fact PLA fibre is an aliphatic polyester, but it is included in this section on regenerated fibres because the manufacturing sequence is based on the conversion of naturally occurring material to a textile fibre.

PLA fibres were first marketed in 2003 with the trade name Ingeo®. They are similar in characteristics to standard polyester in some ways and to polypropylene in other ways. The fibres have low moisture absorption (moisture regain 0.4–0.6%) and excellent wicking properties, making them suitable for sportswear. They have a lower density than natural fibres, enabling light fabrics to be produced, with soft handle and touch. As with standard polyester, PLA fibres are dyed with disperse dyes, though to avoid too much degradation, dyeing is best carried out for 15–30 minutes at 110–115 °C.

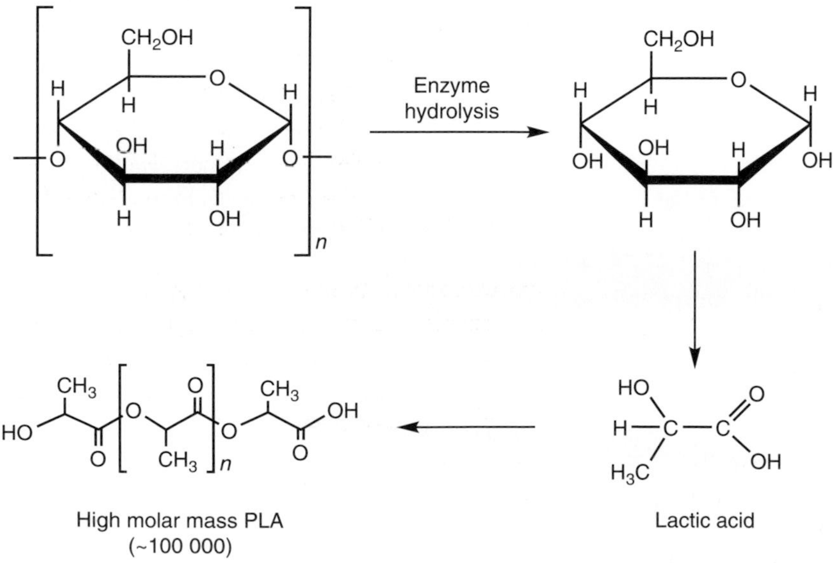

High molar mass PLA
(~100 000)

Lactic acid

Scheme 2.6 Synthetic route for the formation of polylactic acid.

2.8 Synthetic Fibres

Synthetic fibres are so called because they are synthesised from organic compounds extracted from crude oil. In contrast to natural fibres, they are therefore derived from a non-renewable resource, and further, they are generally non-biodegradable. However, that said, they are an important class of fibre because they can be produced with different properties (such as tenacity, cross-sectional shape, etc.) to serve a wide range of applications. The production methods ensure that the fibre qualities are very consistent from batch to batch.

One of the main considerations that determine the commercial viability of a synthetic fibre is the ease with which the source chemicals can be obtained and their price, bearing in mind that other chemical industries, in addition to the fibre industry, may be competing in the marketplace for the same compounds. The diagram in Figure 2.12 illustrates the complex sequences of production from crude oil to final fibre for some synthetic fibre types. For the nylons especially, it is clear that many chemical processes have to be carried out just to obtain the monomers required for the production of the polymer.

There are two main types of polymerisation reaction by which fibres are formed, these being *condensation polymerisation* and *addition polymerisation*. The first of the synthetic

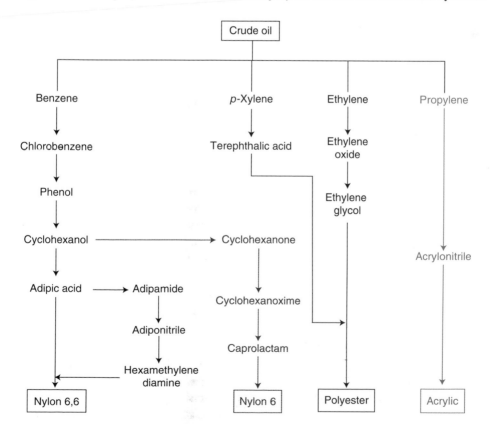

Figure 2.12   Routes to the formation of synthetic fibres.

fibres to be marketed were polyamide (nylon) in the early 1940s and then polyester in the late 1940s, both being condensation polymers. Later, in the 1950s, a fibre formed by addition polymerisation, an acrylic fibre, was marketed.

2.8.1 Condensation Polymers

Condensation polymerisation takes place between two difunctional organic compounds, examples of which are given in Section 1.8.3. The reaction is called condensation polymerisation because, in addition to the polymer being formed, a small molecule (usually water) is eliminated. A useful analogy to the formation of condensation polymers is a system of hooks and eyes (Figure 2.13).

2.8.1.1 Polyamide (Nylon) Fibres

The first synthetic fibres to be marketed were polyamide fibres and their discovery is attributed to Wallace Carothers and his research team at DuPont in the United States. Amides are made by the reaction of a carboxylic acid with an amine (see Section 1.8.3.7), but Carothers carried out the reaction with the corresponding difunctional reagents, which are compounds containing identical functional groups at both ends of the molecule. Thus, by reacting a dicarboxylic acid with a diamine, a polyamide is formed (Scheme 2.7), and Carothers experimented with a wide range of different dicarboxylic acids and diamines (i.e. different numbers x and y of $-CH_2-$ methylene groups).

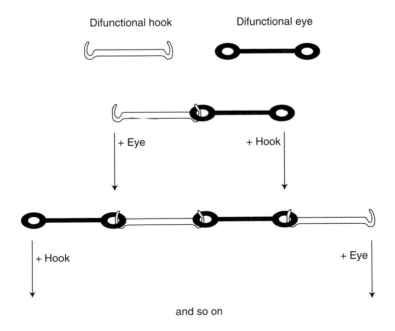

Figure 2.13 The stepwise construction of a 'hook and eye' chain.

$$HOOC–(CH_2)_x–COOH + H_2N–(CH_2)_y–NH_2$$

$$\downarrow$$

$$HOOC–(CH_2)_x–CONH–(CH_2)_y–NH_2 + H_2O$$

Scheme 2.7 Reaction between a dicarboxylic acid and a diamine.

$$H_2N–(CH_2)_y–NH_2 + HOOC–(CH_2)_x–CONH–(CH_2)_y–NH_2 + HOOC–(CH_2)_x–COOH$$

$$\downarrow$$

$$H_2N–(CH_2)_y–\mathbf{NHCO}–(CH_2)_x–\mathbf{COHN}–(CH_2)_y–\mathbf{NHOC}–(CH_2)_x–COOH + 2H_2O$$

Scheme 2.8 Formation of a nylon polymer.

$$n\,HOOC–(CH_2)_4–COOH + n\,H_2N–(CH_2)_6–NH_2$$

$$\downarrow$$

$$HO[–OC–(CH_2)_4–CONH–(CH_2)_6–NH–]_nH + n\,H_2O$$

Scheme 2.9 Formation of nylon 6.6.

The reaction continues at both ends (Scheme 2.8) and the polymer chain keeps growing until eventually all the reactants are used up. In practice it is necessary to control the length of the molecule (and therefore the molecular weight of the polymer), which is done by adding a monofunctional amine or carboxylic acid at the appropriate stage. If the polymer chains are too short, the fibres will be too weak to be useful in textiles. On the other hand, if they are too long, the polymer will become unmanageable because it cannot be melted at a temperature low enough to avoid decomposition during melt spinning (see Section 2.9).

The chemical joining group shown in bold in Scheme 2.8 is an amide group, and because it occurs regularly in the polymer chain, the polymer is called a polyamide. Its chemical formula is represented by

$$HO\left[-OC-\left(CH_2\right)_x-COHN-\left(CH_2\right)_y-NH-\right]_n H$$

where the section in the square brackets is called the *repeating unit*.

The polyamide that was found by Carothers to be most suitable for forming fibre filaments was that synthesised by reacting 1,6-hexanedioic acid (also called adipic acid) with 1,6-diaminohexane (also called hexamethylenediamine), shown in Scheme 2.9.

This polyamide was named nylon 6.6 because the dicarboxylic acid used contains six carbon atoms, as does the diamine. It was found to be the best polyamide for fibre manufacture because the ten methylene groups in the repeating unit give a good balance between their hydrophobic character and the more hydrophilic character of the amide groups. Also, if there are fewer methylene groups in the repeating unit, the polyamide formed is not as stable under alkaline conditions. Another consideration is the availability of the monomer compounds 1,6-hexanedioic acid and 1,6-diaminohexane. It is necessary to synthesise

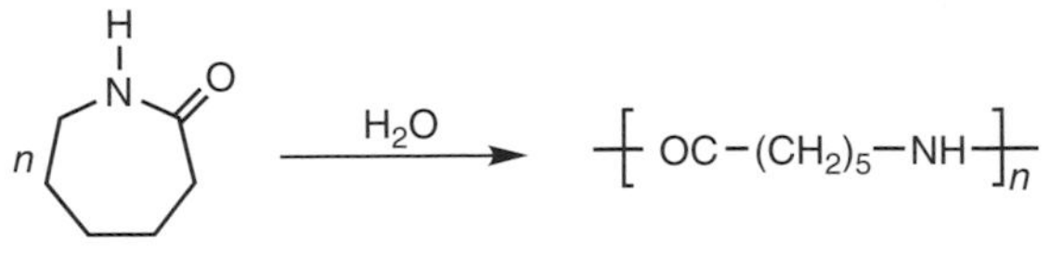

Scheme 2.10   Formation of nylon 6.

them from benzene, and a complex sequence of reactions is required, but synthesising the monomers for other polyamides is considerably more difficult.

   Another type of polyamide that is marketed is nylon 6. This nylon is manufactured by a completely different method, involving the polymerisation of a chemical called caprolactam (Scheme 2.10). Carothers and his colleagues had not been successful in polymerising this compound, but a chemist at I.G. Farbenindustrie in Germany, Paul Schlack, not only found a way of doing so but also found that the resulting polymer formed strong fibres.

   Production of nylon 6 also began in the early 1940s. Its structure is very similar to that of nylon 6.6. Both have $-CH_2-$ methylene groups in between the amide groups; whereas there are five methylene groups in nylon 6, there are alternately four and then six in nylon 6.6. In both types of nylon, one end of the polymer chain ends with a $-COOH$ group, whilst the other end ends with a $-NH_2$ group. Under acidic conditions, the terminal amino group is protonated, forming $-NH_3^+$, and this positively charged group acts as a site at which the negative (coloured) anion of an acid dye can be adsorbed. The $-CH_2-$ methylene groups of the fibres also provide hydrophobic character and enable attraction with the hydrophobic groups in dyes. Consequently polyamide fibres can be dyed with disperse dyes, as well as acid dyes. They can also be dyed with reactive dyes.

   Nylon fibres are noted generally for their strength, with tenacities in the range 40–60 cN/tex (up to 90 cN/tex for specialised high-tenacity yarns). They have very good abrasion resistance, low density and moisture regain of about 4.5%. Nylon 6.6 fibres were an instant success in the hosiery trade but are now widely used for clothing, especially workwear and active sportswear, as well as for ropes and other industrial textiles. Its excellent abrasion resistance makes it a useful component in blends with wool for carpet manufacture. Nylon 6 is used extensively in tyre cord, where its good adhesion to rubber and thermal resistance is an advantage. Although nylon 6 is easier and cheaper to make, nylon 6.6 is preferred for clothing because it has a higher melting point (263 °C for nylon 6.6 but only 215 °C for nylon 6).

2.8.1.2   *Aramid Fibres*

Aramid fibres are polyamides that are made from aromatic monomers. There are two types, *m*-aramids and *p*-aramids, the two most well known being Nomex® and Kevlar®, respectively. Nomex is synthesised (Scheme 2.11a) by the condensation polymerisation of 1,3-diaminobenzene (*m*-phenylenediamine) and isophthaloyl chloride, whilst Kevlar is synthesised (Scheme 2.11b) using 1,4-diaminobenzene (*p*-phenylenediamine) and terephthaloyl chloride.

   Although the carbon atoms of the $-CH_2-$ methylene groups in the formulae for nylon 6 and 6.6 are shown as straight chains, the actual structure of a row of four or six carbon atoms is a linear zigzag arrangement, in which the four bonds of each carbon atom point to the corners of a tetrahedron (see Figure 1.5). This allows the component carbon atoms to rotate

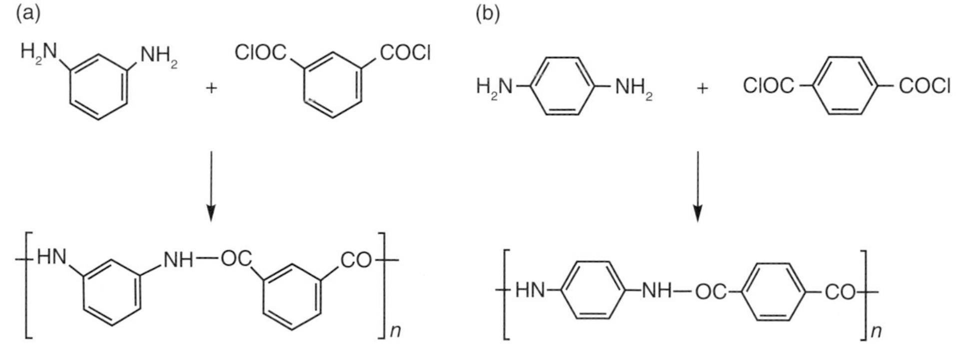

(a)   (b)

Scheme 2.11 Synthesis of (a) a *m*-aramid, for example, Nomex, and (b) a *p*-aramid, for example, Kevlar.

about the carbon–carbon bonds, and it is this freedom of movement that gives polymer molecules the flexibility usually required of a fibre, as discussed in Section 2.2. In aramid fibres the flat planar benzene rings enable adjacent polymer chains to align very closely, resulting in a high level of inter-chain bonding, which produces a very rigid structure.

Due to this highly ordered structure, aramid fibres possess very high thermal stability. For example, the thermal degradation of Nomex starts at 375 °C, whilst Kevlar, which has an even higher degree of order of the polymer chains, only starts to degrade at 550 °C. Nomex only has the tenacity similar to regular nylon and polyester fibres (about 40–50 cN/tex), but Kevlar is superbly strong, with tenacities in the range 190–240 cN/tex. Other textile properties of Kevlar® are less good however, such as abrasion resistance. Nevertheless its high tenacity makes it invaluable for reinforcing fibres for highly stressed articles (e.g. heavy-duty tyres for aircraft and lorries) and for protective clothing such as bulletproof vests, helmets and vehicle armour. The excellent resistance to chemicals of Nomex makes it suitable for protective clothing as well, and because of its high flame resistance, it is widely used in the manufacture of fire-resistant fabrics, especially for use in public buildings, aircraft and trains.

The two types of fibres can be combined to make a fabric (called Nomex III) that is a blend of 95% Nomex and 5% Kevlar. It has exceptional fire and heat resistance and better strength than 100% Nomex during exposure to flames. The very low shrinkage of Kevlar prevents garments of Nomex III from bursting open in flames and ensures insulating protection. This is a good example of a *synergistic* effect, that is, one in which the performance of the components in combination substantially exceeds that which is to be expected from an assessment of their individual properties. Nomex III is used for firemen's garments, racing car drivers' garments and fireblocker liners in aircraft seats.

The highly ordered structure of these fibres, especially the *p*-aramid type, makes them very difficult to dye, because dye molecules are unable to diffuse easily into the polymer matrix. Dyeing of the *m*-aramid type is possible with cationic dyes however.

2.8.1.3 Polyester Fibres

The reaction of an organic acid with an alcohol gives an ester (see Section 1.8.3.4). If a dicarboxylic acid and diol are used, a long polymer molecule is formed, called a polyester. Carothers and his research team tried to make polyesters using aliphatic organic monomers,

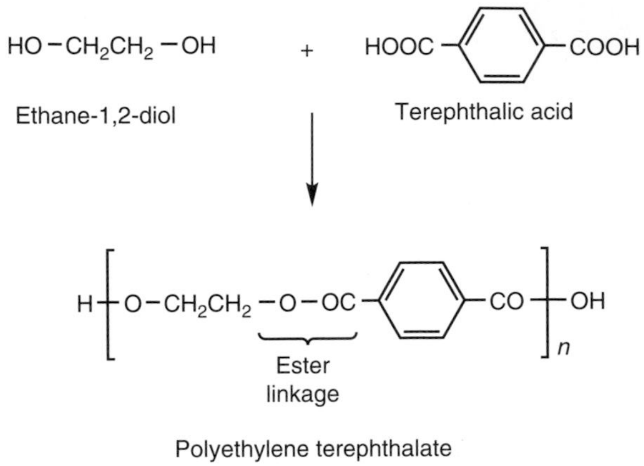

$$HO-CH_2CH_2-OH \quad + \quad HOOC-\bigcirc-COOH$$

Ethane-1,2-diol   Terephthalic acid

$$H-\left[O-CH_2CH_2-\underbrace{O-OC}_{\substack{\text{Ester}\\\text{linkage}}}-\bigcirc-CO\right]_n-OH$$

Ester
linkage

Polyethylene terephthalate

Scheme 2.12   Synthesis of polyester.

but they were unsuccessful because the polymers formed had melting points that were too low for practical use as textiles and also hydrolysed very readily in alkaline solutions.

The first commercially successful polyesters were developed by J R Whinfield and J T Dickson in the 1940s, and like polyamides, they rapidly became established as useful textile fibres. The starting materials in this case were ethane-1,2-diol, also called ethylene glycol (an aliphatic diol), and benzene-1,4-dicarboxylic acid, also called terephthalic acid (an aromatic dicarboxylic acid). A condensation polymerisation reaction between these two compounds is carried out (Scheme 2.12).

As in the case of polyamides, polyester is melt-spun into fibre form. Again, it is necessary to control the polymerisation reaction so that the polymer chains have the optimum length to give the required textile properties.

The presence of the benzene ring in the repeating unit means that adjacent polymer chains can approach closely, giving fibres with a high degree of crystallinity. Whilst giving fibres of high strength (tenacity 35–55 cN/tex), they have a high glass transition temperature (about 105 °C), making them difficult to dye. Because of its hydrophobic character (the moisture regain is only 0.4%), polyester fibres must be dyed with non-ionic disperse dyes, but even these dyes have to be applied at 130 °C because at the boil they are unable to diffuse into the polymer matrix. Polyester has inferior elastic recovery to polyamides, and its abrasion resistance, whilst high, is not quite as good as that of polyamides either. However it has good chemical resistance and polyethylene terephthalate (PET) garments can withstand vigorous washing cycles.

Other types of polyester fibres are marketed, such as those synthesised using propane-1,3-diol or butane-1,4-diol instead of ethane-1,2-diol, but they are of much lesser commercial importance than PET. PET fibre has become far and away the most important fibre (see Table 2.2), having applications in a whole host of areas, including clothing, home textiles, medical and industrial. It is also commonly found in blends with other fibre types, such as cotton and wool. The fibre is produced in a wide range of variants, including specialised high-tenacity types (up to 100 cN/tex) and different cross-sectional variants such as circular and hollow, and at different fineness, including microfiber versions.

2.8.1.4 *Elastomeric Fibres*

Elastomeric fibres have the ability to recover rapidly and completely from very high levels of stretch and over repeated stretch/relaxation cycles. There are four main chemical types: elastane, elastodiene, elastomultiester and elastolefin, but of these elastane is commercially the most important type. Elastanes are marketed as Spandex® in the United States and Lycra® in Europe.

Elastanes are based on a class of polymers called *segmented polyurethanes*, which are block copolymers of two alternating segments. One of these segments is a 'hard' component, typically an aromatic–aliphatic polyurea, and the other segment is a 'soft' component, typically either an aliphatic polyether or a polyester. An example of a polyether type is shown in **2.13**. This soft segment is a highly amorphous, randomly coiled structure that provides flexibility and stretch. The hard structure has strong inter-chain bonding with the hard structure of neighbouring chains, a feature that prevents molecular slippage during stretching and therefore holds the fibre structure together.

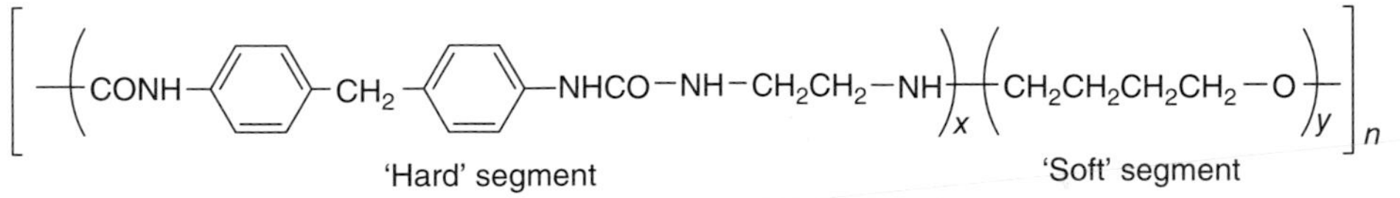

'Hard' segment   'Soft' segment

2.13

Elastanes are blended with other fibres, usually at a level of around 2–4%, so that the resulting fabrics have a certain amount of stretch. Such fabrics are used for activity wear, as well as hosiery, intimate apparel and casual outerwear such as denim jeans and shirts.

2.8.2 Addition Polymers

Other well-known synthetic fibres are synthesised using a type of monomer that is very different in behaviour from those involved in condensation polymerisation. Acrylic and polyolefin fibres are synthesised through the reactions of *vinyl* groups, in which two carbon atoms are joined by a double bond. The polymerisation reaction is called addition polymerisation, and it occurs because of the different character of the two bonds. The two shared electrons forming one of the covalent bonds are firmly bound in a σ covalent bond. As explained in Chapter 1, the second bond (a π bond) is much more strained and therefore more easily broken, making the bond more reactive.

2.8.2.1 *Polyolefin Fibres*

Olefin is the name originally given to the class of hydrocarbons called alkenes, so strictly the fibres formed from them should be named polyalkenes. However in the textile industry the name polyolefin has remained in use. To further complicate matters, compounds derived from them are often referred to as *vinyl* compounds. The simplest monomer is the alkene ethene, **2.14** (also called ethylene):

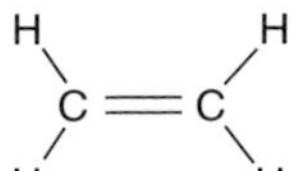

2.14

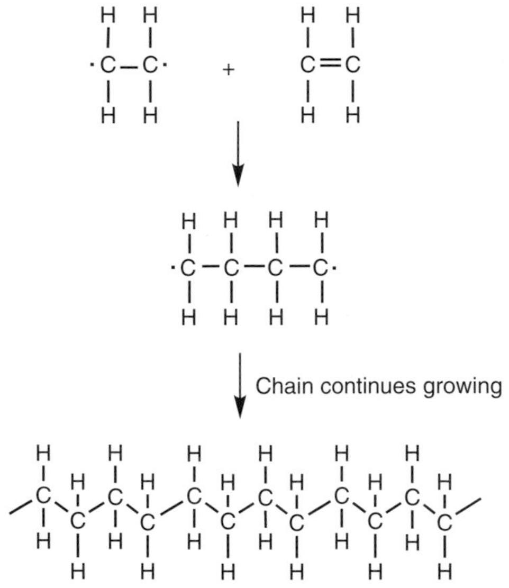

Scheme 2.13   Polymerisation of ethene to give polyethylene.

Molecules of ethene can be made to react together on the surface of a catalyst where the two electrons of the less stable second bond (the π electrons) become unpaired and react rapidly with corresponding unpaired electrons in adjacent ethene molecules (Scheme 2.13). This results in the formation of a stable single bond joining the two molecules together. The process continues with other neighbouring molecules, resulting in the formation of a long flexible chain, some tens of thousands of carbon atoms in length.

The polymer formed is called polyethylene, even though the repeat unit is simply the $-CH_2-$ methylene group. Polyethylene polymers have very high molecular weight and the fibres produced from them have uses in high-performance applications rather than commodity textiles. Because they are entirely hydrocarbon, they lack moisture absorbency, but are very chemically resistant, hence their use in protective clothing for chemical workers.

A closely related polyolefin fibre is polypropylene (**2.15**), made by the addition polymerisation of propene (propylene). In this case the polymer has $-CH_3$ methyl groups on every alternate carbon atom along the polymer chain.

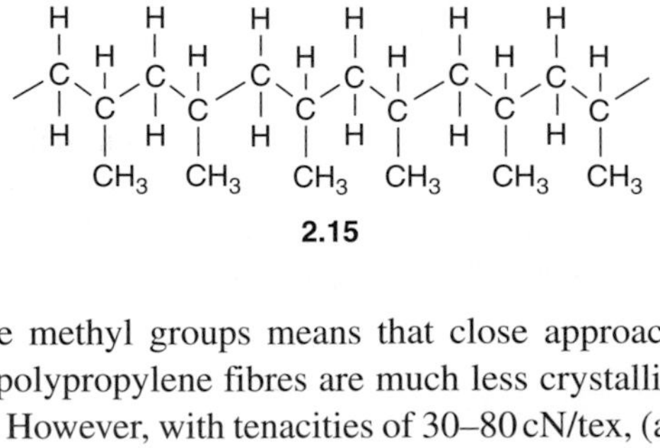

2.15

The presence of the methyl groups means that close approach of adjacent polymer chains is inhibited, so polypropylene fibres are much less crystalline and therefore not as strong as polyethylene. However, with tenacities of 30–80 cN/tex, (and up to 100 cN/tex for the high-tenacity variants), polypropylene fibres are still very strong indeed. The fibres

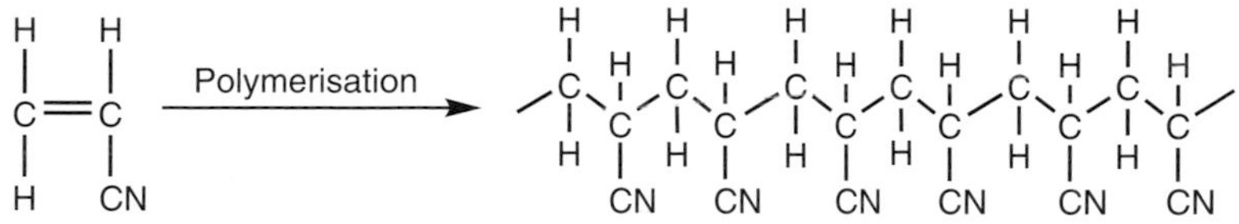

Scheme 2.14   Addition polymerisation of cyanoethene.

have low melting points and virtually zero absorbency and are difficult to dye, so they have only limited use in apparel manufacture. They are being increasingly used in activity clothing such as cycle shorts, swimwear and lightweight outerwear for climbers, for which not only water repellency is needed but also good wicking properties. Their technical performance is very high given their relative low cost of production, so the main uses of polypropylene are for carpets, ropes, sacks, bags and netting.

2.8.2.2   *Acrylic Fibres*

An important type of vinyl fibre that was developed during the 1950s is that based on the addition polymerisation (Scheme 2.14) of cyanoethene (acrylonitrile or vinyl cyanide).

Pure polyacrylonitrile proved difficult to convert in fibre form because it could not be melted (it decomposes before melting), so it was necessary to spin fibres from solution in a solvent. However polyacrylonitrile is insoluble in most of the common solvents that were available at that time, and also the fibre properties were poor. These problems were largely overcome by incorporating another monomer (a comonomer), of which various compounds are used. The acrylic fibres manufactured today are typically copolymers of acrylonitrile and another monomer such as vinyl acetate (**2.15**), methyl acrylate (**2.16**) or methyl methacrylate (**2.17**).

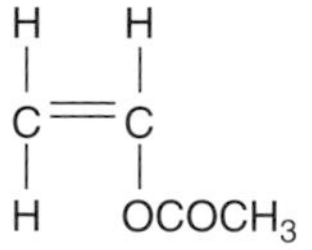

2.15 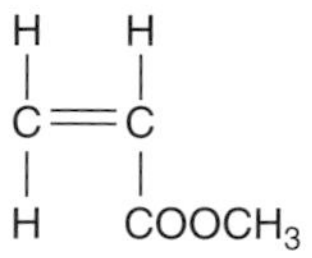　　　　**2.16** 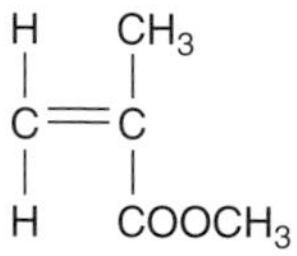　　　　**2.17**

The incorporation of these types of comonomer molecules along the polymer chain opens up the polymer structure and enables dye molecules to penetrate. For example, if vinyl acetate is used in the manufacture of the polymer, **2.18**, acetate ($-COCH_3$) groups, which are bulky groups, will occur at intervals along the polymer chain:

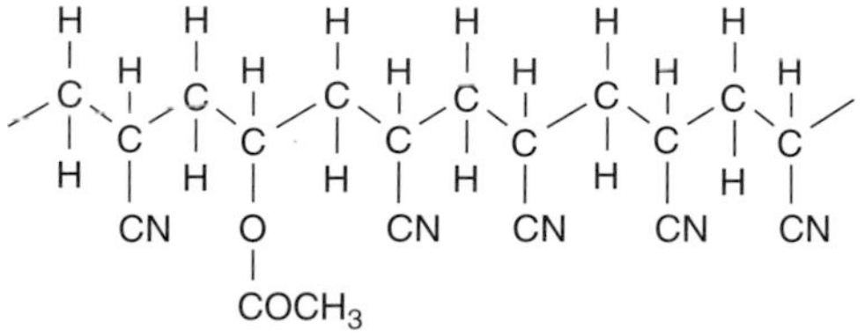

2.18

Also the copolymer formed is soluble in a solvent, such as dimethylacetamide, so it can be wet-spun to form fibres. The amount of the second comonomer in the final polymer is much less than acrylonitrile, but for a fibre to be called an acrylic fibre, it must contain at least 85% of acrylonitrile and no more than 15% of the second comonomer. Fibres that contain between 35 and 85% acrylonitrile are termed *modacrylic*. Such fibres are produced because they have modified properties such as enhanced flame resistance and can be used in protective clothing and furnishings.

A structural feature of acrylic polymers is that at the ends of the polymer chains there are anionic groups such as sulphate ($-SO_4^{2-}$) or sulphonate ($-SO_3^-$) that are residues from the initiator used in the polymerisation reaction. These groups provide substantivity for cationic (basic) dyes, and it is this application class of dyes that are most used to dye acrylic (and modacrylic) fibres.

Acrylic fibres possess an inherent warmth and soft handle and come closer to wool than any other man-made fibre. Indeed they are superior to wool in terms of resistance to chemical and microbiological attack and in physical properties such as tenacity. They have good wicking properties, so they are used for making active sportswear clothing. They are also used in carpet yarns and knitting yarns.

2.9 Conversion of Synthetic Polymers into Fibre Filaments

Once the polymer reaction is completed, the polymer is formed as granules, and it has to be converted into fibre form, either as a continuous filament or as staple fibres (fibres of short length, typically about 40 mm). Firstly the polymer granules are either melted or dissolved in a solvent and then extruded under pressure through a spinneret, a process in this part of the textile industry called 'spinning'. In this context spinning is quite different from the use of the term in yarn manufacture, where individual fibres or filaments are twisted together. The spinneret contains many hundreds of tiny holes so that the molten polymer (or polymer solution) emerges from the spinneret as many very fine strands running in parallel. In the case of molten polymer, as the strands meet the cold air surrounding them, the polymer solidifies into fibre filaments. If the polymer is in solution, the solvent has to be removed, and in both cases the filaments formed are collected as a multifilament yarn which is wound on to bobbins.

There are four different methods of producing fibre filaments – melt spinning, dry spinning, wet spinning and dry-jet wet spinning – all of which are represented in Figure 2.14.

Melt spinning (Figure 2.14a) is appropriate for polymers, such as polyester or polyamide, that do not decompose at their melting temperature. Molten polymer is pumped through a spinneret, and as the tiny jets of molten polymer exit through the holes, they encounter the cold surrounding air and solidify into continuous fibre filaments. The filaments are wound up on to bobbins.

Dry spinning (Figure 2.14b) is used for polymers, such as acetate and some acrylic types, that are more difficult to melt but can be dissolved in a volatile solvent to form a viscous solution. This solution is extruded through the spinneret, and once the fine stream

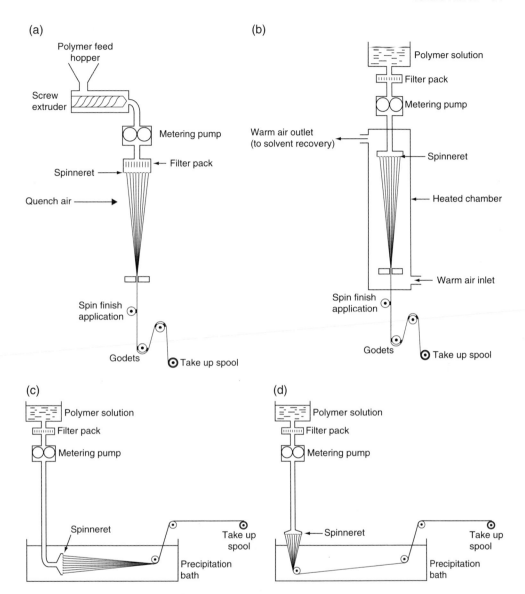

Figure 2.14 Different methods of fibre spinning: (a) melt spinning, (b) dry spinning, (c) wet spinning, (d) dry-jet wet spinning.

of 'dope' reaches the surrounding current of warm air, the solvent evaporates, leaving fine fibre filaments that are wound up on to take-up bobbins.

Wet spinning (Figure 2.14c) is used for those polymers that need to be regenerated by some form of chemical action after extrusion. Typical of this method is the spinning of viscose fibres in which the solution of sodium cellulose xanthate is extruded directly into a bath containing acid, the acid reacting with the xanthate derivative to reform the cellulose. The extruded cellulose fibres (viscose) are then dried and wound on to take-up bobbins. Acrylic fibres are also produced by wet spinning, usually from solution in sodium thiocyanate or dimethylacetamide, though in these cases the polymer solution is extruded into water when dilution of the solvent occurs and the fibre filaments form.

Dry-jet wet spinning (Figure 2.14d) is a variation of the wet spinning method. In dry-jet wet spinning, the polymer solution is extruded through the spinneret that is suspended approximately 1–2 cm above the precipitation bath. The polymer solution therefore has to pass through an air gap, which allows the polymer molecules to align more parallel with each other before entering the bath. The fibre filaments formed then have a high degree of crystallinity and orientation and therefore high tenacity. This method is used for extruding high-performance fibres such as Kevlar (see Section 2.8.1.2), but it is also used for the extrusion of lyocell fibres, thereby conferring superior tenacity to these cellulosic regenerated fibres.

Synthetic fibres can be produced in versions that are either 'bright' or 'delustred'. The *bright* fibre types are 100% polymer without additives. Fabrics made from these fibres transmit a certain amount of light, which is not always desirable in a garment. The difficulties are eliminated by making opaque or delustred versions (usually called *dull fibres*) by incorporating a white pigment into the fibre at the extrusion stage. Titanium dioxide is used for this purpose, usually at a level of about 2%. The particles of pigment are easily visible in such fibres under the optical microscope, when they appear as dark specks.

2.10   Fibre Cross-Sectional Shapes

Man-made fibres are produced in a variety of cross-sectional shapes, finenesses and tenacities and elasticity to serve the needs of different applications. Extrusion of the fibre filament is the stage at which the choice of cross-sectional shape is made by choosing an appropriate shape for the holes in the spinneret (Figure 2.15). The cross-sectional shape of a man-made fibre is a function of both the method of production and the dimensions of the spinneret holes. The holes can be circular, crenellated, trilobal, convex, triangular and hollow, and each shape imparts particular characteristics. For example, ultrafine polyester filaments can

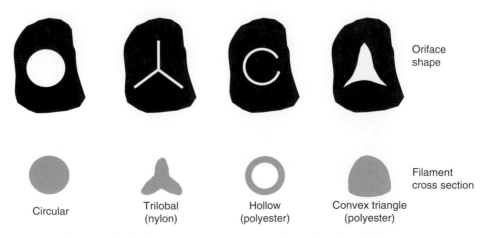

Figure 2.15   Orifice shape and the cross sections of the resulting filaments.

be woven into breathable fabrics; since the cross-sectional area of the filaments is exceptionally low, the space between the filaments is so small that surface tension prevents water droplets from penetrating whilst allowing water vapour to pass through.

Multilobal nylon fibres are useful in carpet piles because they have a reduced lustre; moreover, some shapes increase their bulk and recovery from compression, in addition to being better able to hide soiling within the pile. The disadvantage is that dirt particles may become stuck in the crevices, thus making cleaning more difficult. The latter problem has been tackled by producing fibres with a smooth surface but four air-filled channels running down the length of the fibre. This reduces the visibility of the soiling through the multiple reflections inside the fibre, and at the same time it produces a lighter fibre. Fibres with grooves along their length, especially the more hydrophobic fibre types, have excellent wicking properties and are used in active sportswear clothing for transmitting perspiration away from the body.

Fibres with a Y-shaped cross section are more resistant to bending and have better recovery than those with a circular or bean-shaped cross section, which makes them particularly suited to use as carpet pile. Hollow polyester fibres have good insulation properties because of the air trapped within them, making the staple fibres (cut filaments) particularly suited for use as fillings in pillows, duvets and sleeping bags. Hollow fibres are also used in garments where both warmth and breathability are required, such as for mountaineering activities.

2.11 Microfibres

As stated in Section 2.4.2, microfibers have very small diameters, typically having counts of 0.3–1.0 dtex, roughly equivalent to diameters of 5–10 µm. They are usually made from polyamide or polyester and have a number of desirable characteristics such as soft handle, lustre and excellent drapability, and with their large surface areas, they can be woven tightly into fabrics that are breathable, yet water repellent, for outdoor wear.

There are three main methods for making microfibers. Firstly, they can be melt-spun directly by extruding through spinnerets with very fine holes, though the filaments produced have limited fineness. The second method involves extruding bi-component fibres, that is, fibres of two chemically distinct polymer types (e.g. polyester/polyamide), through a specially designed spinneret so that the two constituent polymers each extend along the length of the fibre. At this stage the fibres have the same fineness as ordinary melt-spun fibres, and they are woven or knitted into fabric form. After weaving or knitting, and before dyeing, the two fibres are split. This is achieved by treatment in sodium hydroxide solution, in which the polyamide and polyester components swell to different extents and split apart. The third method, often called the 'sea-island' method, involves extruding two different polymer types through a specially designed spinneret, so that several 'island' component fibres are embedded in a 'sea' of the other component. The 'sea' component is then removed by dissolving it out. Typically the 'sea' component is an alkali-soluble polyester and the 'island' component is normal polyester. Again, after weaving a fabric of the blend, the 'sea' component is dissolved away using sodium hydroxide solution, leaving only the normal polyester fibres as microfiber. Microfibres produced by this method are finer than those produced by direct melt spinning.

2.12   Absorbent Fibres

Modification of both the external and internal surface areas of fibres can also powerfully affect the absorption of water. The formation of regenerated cellulosic fibres with a hollow structure provides a method of meeting demand for high-absorbency fibres for medical uses such as swabs, babies' nappies and feminine hygiene products. Viscose with its inherent highly hydrophilic character is a popular fibre type for these applications, and it is usually constructed in the form of non-woven assemblies of the fibres. Viscose fibres have a skin and core structure, with small pores in the skin but larger pores in the core. The jagged cross-sectional shape of viscose fibres also aids water holding capacity because they help build cavities within a non-woven fibre assembly. By extruding viscose fibres through special spinnerets, trilobal or star-shaped cross sections can be obtained, which influences the size of the voids between the fibres and thereby increases absorptive capacity.

2.13   Drawing of Synthetic Fibre Filaments

The fibre filaments formed after the extrusion stage are weak and dimensionally unstable. If they are pulled, they stretch irreversibly. In this 'raw' state the polymer chains are disoriented, and the structure lacks any coherent molecular arrangement. The degree of molecular alignment of the polymer chains along the length of the fibre axis influences the strength of the fibre. If a synthetic fibre is stretched after formation, the disordered polymer chains straighten and become extended along the fibre axis. During stretching the molecules slip past each other, but with the increasing order that the stretching imposes, they eventually become close enough for intermolecular attractions to develop and to hinder further slippage without fracture. At this stage the filament has developed its maximum strength and the stretching is stopped. The individual polymer chains are now closely aligned and have become bonded to each other.

A process corresponding to this stretching is carried out by the fibre manufacturer, called the *drawing* operation. The filaments are passed around consecutive rollers, each of which rotates faster than the one before; the relative speeds of the rollers can be adjusted to impart the required degree of stretch to the filament (Figure 2.16). In this way the length of a filament can be increased some fivefold, whilst its diameter is reduced and its breaking strength increased.

At this stage it is impossible to change one physical property without affecting others, so the degree of stretch needs to be controlled. The close alignment of the polymer chains may allow some crystallinity to develop, and this increases the brittleness of the fibre. Although crystallinity increases the strength of the fibre, if the fibre becomes too crystalline, it will fracture when bent and will have poor abrasion resistance. The fibre will also be more difficult to dye or to be penetrated by chemical reagents or water (permeability to water is an important factor in the comfort of clothing).

Thus by adjusting the *draw ratio* (the ratio of the final length to the original length), these various properties can be controlled to give fibres with the characteristics required for

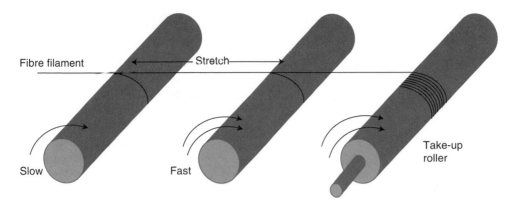

Fibre filament

Stretch

Slow

Fast

Take-up roller

Figure 2.16   Drawing fibre filaments.

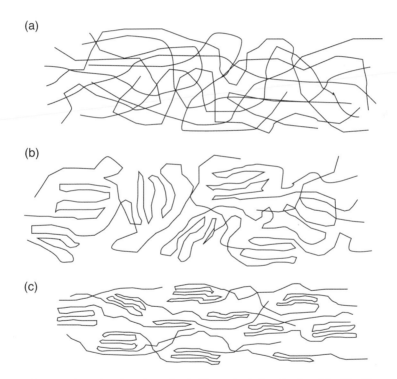

(a)

(b)

(c)

Figure 2.17   Changes in molecular orientation in filaments during drawing: (a) at the filament emerges from the spinneret, orientation is mainly random (amorphous); (b) on cooling and stretching some crystalline order appears, with a degree of general orientation along the fibre axis; (c) further stretching aligns amorphous and crystalline regions along the fibre axis.

their intended end use. In the final state therefore, the fibres are composed of polymer chains aligned roughly parallel to the fibre axis with varying degrees of molecular order along the length of the fibre, ranging from crystalline to highly oriented non-crystalline to areas of low orientation (Figure 2.17).

2.14 Conversion of Man-Made Fibre Filaments to Staple

Once the filaments have been obtained, they may be twisted together to form a yarn for use in knitting or weaving. Often however, the properties required for the yarn are better obtained using a staple fibre of a predetermined length, for example, for the purpose of blending with natural fibres. The man-made fibres will be cut to correspond closely with the staple length of the natural fibre so that the blended fibres can be processed easily on spinning machines to make yarns. For other purposes, such as the preparation of fibres for *flocking* (a process in which very small fibres are deposited on to a surface), the chosen length may be as short as a few millimetres. For staple fibres intended for general textile applications, the cutters are helical-bladed (similar in shape to those of a lawn mower), and the staple length is governed by the spacing of the cutter blades. The cut is made at an angle to the direction of travel of the parallel filaments so the cut ends are not coincident with each other. The web of staple fibres is loosened and rolled diagonally to form a continuous sliver for the subsequent conversion into yarn.

2.15 Imparting Texture to Synthetic Fibres

Since the introduction of the first synthetic fibres, efforts have been made to meet consumer demands for synthetic fabrics with greater comfort in wear and pleasanter handle, more akin to those associated with fabrics made from natural fibres.

A major influence on the texture of a fabric is the *crimp* of the fibre, and the 'loftiness' or 'bulkiness' of wool has been imitated with some success by introducing a crimp into the synthetic fibre. This improves both the appearance and insulation properties of the fabrics subsequently produced, whilst the easy drying properties are preserved. Furthermore, the greater airspace in the fibre allows moisture to be transported more easily from the body to the surrounding air, thus helping to overcome the discomfort often associated with the hydrophobic nature of synthetic fibres worn close to the skin.

Many of the methods for texturising synthetic fibres depend on their *thermoplastic* nature. When synthetic fibres are heated, they do not melt sharply at a certain temperature, as simpler substances do, since the extreme length of the polymer chains prevents them from separating easily. Instead the onset of melting is marked by a change in physical properties. In particular the fibres become more readily deformed by mechanical stress. This is due to a general increase in the molecular movements of the polymer chains, which at lower temperatures (i.e. temperatures below their glass transition temperature, T_g; see Section 2.2) are 'frozen' in their conformations. At temperatures above their T_g, the polymer chains are free to take up new positions in response to an external force. Consequently methods have been devised in which fibres held above their T_g are crimped mechanically and then allowed to cool. Such changes in a thermoplastic material are reversible but will remain unaffected as long as the temperature of subsequent processing or aftercare treatments remains below T_g.

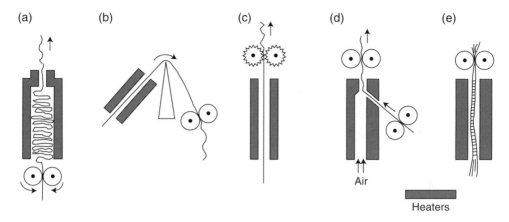

Figure 2.18 Schematic diagrams of fibre crimping methods: (a) stuffer box, (b) knife edge, (c) Gear crimp, (d) air jet, (e) false twist.

Some of the methods of crimping fibres are illustrated schematically in Figure 2.18. In the stuffer box method (Figure 2.18a), yarn is carried into a heating chamber at a faster rate than that at which it is removed, and the compressive action during heating leads to a bulky yarn. Another way of imparting texture is to drag the yarn over a heated knife edge (Figure 2.18b), an action that creates an unsymmetrical cross section to the filaments or yarns.

When the fibres are subsequently relaxed in hot water, curling results, providing them with a stretch. In gear crimping (Figure 2.18c), the heated yarn is passed between heated intermeshing cog wheels before cooling. Another method relies on the disorganisation produced in the yarn structure by a jet of air being forced through the yarn (Figure 2.18d). In a widely used method, a twisted yarn is passed through heaters to set the crimp, and this is followed by an untwisting operation that separates the filaments as distorted coils (Figure 2.18e).

Crimped fibres have also been produced by spinning dual-component fibres, made up of two different polymers lying alongside each other down the length of the fibre. Subsequent heat treatment shrinks both polymers, but to different extents, and the differential effect leads to the development of a crimp. Such fibres are known as bi-component, conjugate, composite or hetero fibres.

2.16   Fibre Blends

In the manufacture of clothing, single fibre types are often used, such as cotton or viscose for underwear and T-shirts or polyester for sportswear where wicking properties are required. However it is also common for blends of different fibre types to be used. By blending fibres of different types, the deficiencies of one fibre can be compensated for by strengths of another, so that fabrics with the optimal characteristics for their intended end uses can be obtained.

There are four main benefits of producing fibre blends:

(1)   The physical properties of the resulting yarns or fabrics can be enhanced. Properties such as tenacity, moisture regain (related to absorbency), elasticity and elastic recovery can be tailored to achieve the functionality of the fabric that is needed.

(2)   The resilience and durability can be improved when the presence of a strong durable fibre may compensate for a fibre that is weaker, though perhaps softer and warmer.

(3)   Different optical effects can be obtained, whether it be in terms of lustre or in colour such as can be obtained in blending differently coloured fibres of the same type rather than different fibre types.

(4)   There is the issue of price, and for increased market penetration, a luxury fibre such as cashmere can be blended with a less expensive fibre such as lambswool or silk to produce fabrics with almost the same desirable aesthetic characteristics. That said, price is not usually the main driving force for blending; it is usually the enhancement of performance that matters most.

Most blends employed for optimising the physical characteristics are usually a mixture of a natural fibre with a synthetic fibre. A very common fibre blend is that of cotton/polyester. In this blend the cotton component provides absorbency, whilst the polyester improves the crease resistance. However the blend proportion is important because at around 75% cotton/25% polyester, the blend is a lot weaker than either of the two fibre types alone. It is necessary that the blend contains at least 40% polyester so that this problem is avoided. Other widely used blends of this type include wool/nylon and wool/polyester. A good example of this is the blending of nylon (at a level of about 20%) with wool to improve the durability of carpets. A wool/nylon blend is also good for socks, where the wool provides warmth and the nylon adds strength, abrasion resistance and durability. Another very widely used blend is that of cotton and elastane. Fabrics (such as denim) produced from elastane/cotton blends possess a degree of stretch that considerably enhances comfort of wear, even though only about 2–4% of elastane fibre is required in the blend. Not all blends are two-component types; it is sometimes advantageous to construct three- or even four-component blends, especially for high-performance functional garments for active sportswear. For example, some trekking socks are made of blends comprising wool (for warmth), nylon for durability, polyester and polypropylene for wicking.

The majority of blends are intimate blends produced by mixing staple yarns of the components at the carding stage of yarn manufacture. On spinning a yarn the two fibre types are then intimately mixed and evenly distributed. Another type of blend is the core-spun type, where one fibre type forms the core of the yarn, with the other fibre type spun around it, giving a sheath. A variation of this is the plied yarn type where two yarns of different fibre types are wound round each other. Whatever the method of producing fibre blends, their coloration provides challenges to dyers because the application classes of dyes suitable for the fibre types in the blend have to be used. This can mean that two separate dyeing procedures appropriate to the classes of dyes are necessary.

2.17 Textile Manufacturing

A detailed explanation of textile manufacturing processes is outside the scope of this book. However, it is useful for dyers to be aware of the forms of textile they may encounter, so this section gives a simplified description of the formation of textiles. Additionally, some commonly used terms in textile manufacturing are given in the Appendix.

2.17.1 Yarns

In the yarn spinning operation, the staple fibre filaments (previously carded to introduce a large degree of parallelisation) become locked together by twisting them round each other to form a continuous yarn. The longer the staple length, the fewer fibre filaments required to be twisted together to produce a yarn of the required strength. Thus finer yarns are produced from fibre filaments that have long staple lengths and small diameters. Conversely, to produce a yarn of the required strength with fibre filaments of lower staple length, more fibres will be required, and they will need to be twisted to a greater extent. The result will be a thicker, coarser yarn.

Wool yarns are produced by the worsted or woollen methods. Worsted yarns are spun from fibres of long staple length (short fibres are removed), and the spinning process removes natural crimp from these fibres. The resulting yarns are leaner and used for suitings.

In the formation of woollen yarns, short fibres are not removed; the fibres are not aligned as parallel, so the resulting yarns are bulkier. Woollen yarns are warmer, hairier and softer than worsted yarns and used for mainly for knitwear.

Yarns of the various fibre types, whether natural or man-made, can vary widely in their diameter. They also vary in their density, and since yarn is sold by mass, mass alone is insufficient to indicate just what length of yarn is being purchased. Over the years different methods of expressing the fineness of a yarn (the *count* or *yarn number*) have been developed, these traditional systems giving rise to a confusing situation. The system recommended by the ISO is now the universally adopted *tex* system, which is explained in Section 2.4.2.

Another important constructional detail of yarns is the amount of *twist* inserted when spinning the yarn. The degree of twist influences several properties, including strength, compactness, compressibility and lustre. The twist is expressed as turns per centimetre and the direction of twist is indicated by the term 'S-twist' when the fibres slope from left to right across the main axis of the fibre and 'Z-twist' when the slope is from right to left (Figure 2.19).

2.17.2 Fabrics

Textile fabrics are formed by either weaving or knitting processes. Both are a type of *fabric construction* where horizontal and vertical yarns are interlaced. However, this process of interlacing the yarns is carried out in very different ways:

- Woven fabrics are an interlacing of yarns that remain straight in either the *warp* or the *weft*.
- In knitted fabrics the yarns are looped over each other to form *courses* or *wales*.

Figure 2.19   S-twist and Z-twist.

Figure 2.20   Winding from bobbins on to a beam.

Within both the woven and knitted fabric types, there are a large number of different fabric constructions produced by varying the techniques used to interlace the yarns.

2.17.2.1   *Woven Fabrics*

Woven fabrics are formed by interlacing *warp* and *weft* yarns at right angles to each other. Warp yarns run the length of the fabric and are placed on the loom before weaving in the process called *warping*. Prior to warping the yarns have to be wound onto a beam—the *winding* operation (Figure 2.20). Figure 2.21 shows the beam mounted at the back of a loom.

Figure 2.21   Rear view of a loom showing the beam and warped threads.

Warp yarns are generally thinner and more tightly twisted than weft yarns. Weft yarns run the width of the fabric in the loom and traditionally run horizontally across garments. Generally there are more warp yarns (called 'ends') than weft yarns (called 'picks') in a given area of woven fabric. The interlacing can be made in different ways, leading to a variety of weave structures, the main types being plain weave, twill weave and satin weave, but within each of these types, there are variations.

Plain weave is the most common type (Figure 2.22), giving a smooth surface that is ideal for printing and wears well. Variations in the number of ends and picks in a given area mean that fabrics of different density and porosity can be obtained. Examples are calico, gingham and muslin.

Twill weaves (Figure 2.23) are described by the length of the warp float $\times$ the length of the weft float. For example, the 2×1 weave means two warp floats and one weft float on the surface, whilst 2×2 weave means two warp floats and two weft floats on the surface. A characteristic of twill weaves is diagonal lines across the fabric. The twill weave structure has fewer yarn interlacings, which are woven tightly. They are soft yet strong, have good abrasion resistance and are pliable with good drape qualities. Examples of twill weaves are chino, drill, denim, gabardine and serge.

In satin woven fabrics (Figure 2.24), the warp yarns 'float' over four or more weft yarns and vice versa. They are woven tightly, with generally high thread counts, giving good wind resistance, but are susceptible to snagging and wear from abrasion. They have a smooth surface and are frequently used for interlinings. They shed soil easily. Examples of satin fabrics are antique satin, crepe-back satin and double-faced satin.

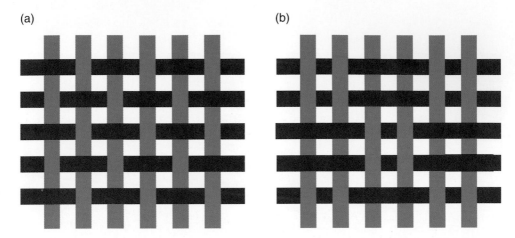

Figure 2.22 Plain weave types (a) plain weave – one over one then one under one, (b) hopsack two over two then two under two.

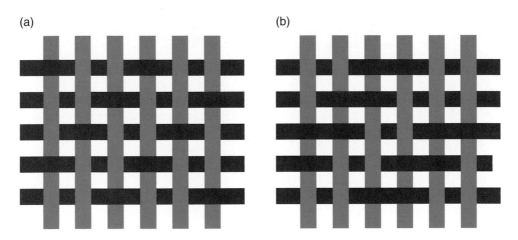

Figure 2.23 (a) 2×1 twill weave, (b) 2×2 twill weave.

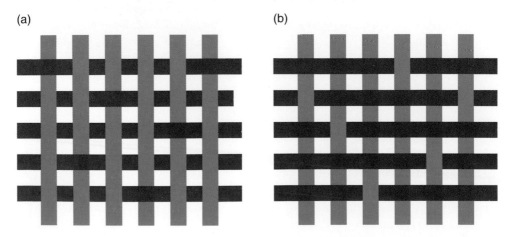

Figure 2.24 Satin weaves (a) warp-faced satin, (b) weft-faced satin.

2.17.2.2 Knitted Fabrics

There are two main categories of knitted fabrics:

- Weft knits, in which the loops run the width (horizontally) across the fabric (Figure 2.25)
- Warp knits, in which the loops run the length (vertically) of the fabric, similar to the warp threads in a woven fabric (Figure 2.27)

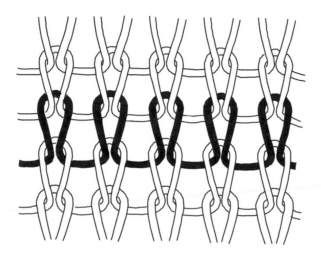

Figure 2.25   Structure of a weft knitted fabric.

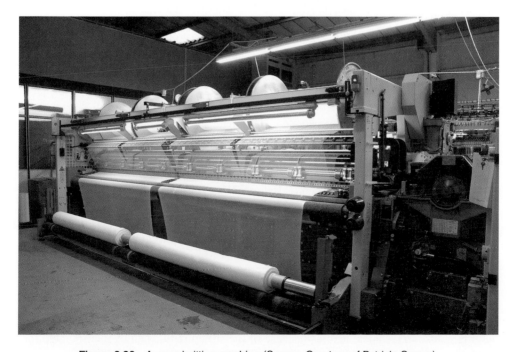

Figure 2.26   A warp knitting machine (Source: Courtesy of Patricia Sawas).

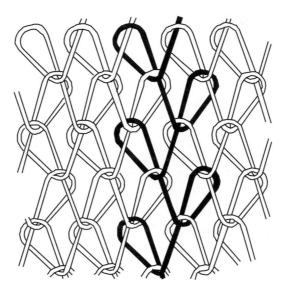

Figure 2.27   Structure of a warp knitted fabric.

In knitting the vertical line of the fabric is made up of *courses* and the horizontal line is made up of *wales*.

Weft knits are produced by circular or flatbed machines. The yarn is fed to the needles from banks of yarn packages close to the machine. When produced on circular machines, the fabric can be processed as a tube or split before processing. A large proportion will be split before finishing as open-width fabric though some can be finished as a tube.

Interlock, single and double jersey are common weft knits used for sweaters, men's underwear and hosiery; *rib knits*, used for cuffs and the bottom edges of sweaters; and *purl knits*, used for sweaters and scarves. Weft knits have a stretch in the horizontal in line with the direction of the loops of yarn.

Fully fashioned knitting machines have the capability of creating a full 3-D garment with no need for further sewing to create the garment. These garments are therefore seamless and colour must be incorporated at the yarn stage and used as coloured yarn on the machine. By using coloured yarn on any type of weft knitting machine, patterns can be introduced to the fabric, which can be incorporated into the finished garment.

Warp knitted fabrics are produced on flat machines (Figure 2.26) and the yarn is fed to the machine from a warp beam similar to that used in weaving. However, the process of forming the fabric is very different from weft knitting. The yarn from the beam is guided through a needle that forms a loop and passes it through the loop previously created. To give stability from left to right, this single yarn that forms a loop must move from left to right whilst forming vertical loops (Figure 2.27).

Warp knitted fabrics can be produced from many types of yarn and tend to be lightweight fabrics. Many of the fabrics produced by this technique are made from synthetic continuous filament yarns. Warp knitting allows the production of technical fabrics and has a large use in technical textiles because of the structure that can be incorporated into the design.

Warp knits have a stretch in the vertical in line with the direction of the loops of the yarn. Common warp knits are Tricot knits, used for lingerie, dresses and lightweight women's blouses, and Raschel knits, used for nets, fine laces, thermal underwear and foundation garments. Raschel knits have more texture and open spaces and are made from heavier yarns. Pattern and form can be incorporated into the fabric by using coloured yarns. By introducing different yarns on the beam, the physical properties of the fabric can be altered to create different end uses.

Suggested Further Reading

K Bredereck and F Hermanutz, Man-made cellulosics, *Review of Progress in Coloration*, **35** (2005) 59.

R R Mather and R H Wardman, *The Chemistry of Textile Fibres*, 2nd Edn., The Royal Society of Chemistry, Cambridge, 2015.

R W Moncrieff, *Man-made Fibres*, Newnes Butterworth, Oxford, 1975.

R H Peters, *Textile Chemistry, Volume 1*, Elsevier, London, 1963.

S B Warner, *Fiber Science*, Prentice Hall, New York, 1995.

3

Chemistry of Dyes and Pigments

3.1 Introduction

First, it is important to distinguish between dyes and pigments, both of which are a type of *colorant*. In simple terms, dyes are soluble in water (or can be made soluble for their application), whilst pigments are insoluble, and this distinction leads to significant differences in the way they are applied to textiles.

Both dyes and pigments can be applied to textile materials but using very different techniques, giving different effects. When dyes are applied to fabrics, either by dyeing or printing, they penetrate the fibres and are attracted to them by primary forces (i.e. ionic or covalent bonds) and secondary forces, such as hydrogen bonds (see Section 1.4.3). When pigments are applied to textiles, mostly through printing, they are mechanically bonded to the fibre surface by resins called *binders*. Pigments have no affinity for textile fibres.

For a colorant to be commercially viable, it must meet the following requirements:

(1) It must have a suitable colour and a high tinctorial strength. If a colorant molecule does not absorb light strongly, then it will be difficult to produce heavy depths of shade unless large quantities are applied to a fabric.
(2) It must be lightfast, that is, it must resist chemical degradation on exposure to sunlight, with its high ultraviolet (UV) component.

In the case of a dye, additional requirements are:
(3) Good substantivity for the fibre, that is, it must transfer readily from the dyebath into the fibre, in a reasonably short period of time.
(4) Good fastness properties when adsorbed by a fibre. This means it should not wash out by the action of water, detergents or other chemicals the material may encounter in use, such as dilute acids, alkalis or perspiration.

3.2 Classification of Colorants

Colorants can be classified in two ways – by their method of application or by their chemical constitution. The classification of colorants by their method of application has evolved gradually over the years, as new fibre types have been introduced. A colorant of a given chemical type may be applied by different application methods, or vice versa. For example, the azo chemical types are commonly found in the acid, direct and basic dye application

An Introduction to Textile Coloration: Principles and Practice, First Edition. Roger H. Wardman.
© 2018 John Wiley & Sons Ltd. Published 2018 by John Wiley & Sons Ltd.

classes. In this chapter, the focus is on the chemical types of colorants, and examples will be given of the most important of them.

The most comprehensive classification of colorants is the *Colour Index*™, produced by the Society of Dyers and Colourists. The Colour Index is now available online, having previously been produced as a hard copy publication [1]. The value of the Colour Index lies in the fact that it is used internationally by colorant manufacturers and users, as well as researchers in fields such as the development of colorants and investigation into historical colorants. It is used by legislators and regulators as an identification system for chemicals.

The Colour Index lists the *essential colorant* within a commercially available product. Many commercial products that are placed on the market are not just 100% of the substance responsible for the colour. They usually contain, in addition, quantities of other substances (generally referred to as additives) that improve the product's application properties, for example, its dispersibility, flow or flocculation resistance. In some cases, significant amounts of these additives may be present.

The agreed definition for the essential colorant is *a dye or pigment responsible for the colour of the product in the absence of additives*.

The Colour Index uses a dual classification system:

* Colour Index Generic Name (often abbreviated to CIGN) – This is the prime descriptor, the one most commonly used in discussions by colorant users and the easier to remember. This is related to the application process.
* Colour Index Constitution Number (often abbreviated to CICN) – This is related to the chemical structure.

The CIGN describes a commercial product by its recognised application class, its hue and a serial number (which simply reflects the chronological order in which related colorant types have been registered with the Colour Index), for example, C.I. Acid Blue 52, C.I. Direct Red 122, C.I. Pigment Yellow 176 and C.I. Solvent Black 34. The full list of application classes used in Colour Index is given in Table 3.1.

Dyes in each of the usage subdivisions are grouped according to their colour and assigned a CIGN. For example, yellow dyes of the acid dye type are classified as C.I. Acid Yellow 1, C.I. Acid Yellow 2 … through acid oranges, reds and so on, ending with blacks. Similarly, direct dyes are labelled C.I. Direct Yellow 1, C.I. Direct Yellow 2 and so on.

In addition, each dye is given a five-digit CICN, which serves as an unambiguous reference. For example, C.I. Acid Red 1 (**3.10**) has the CICN 18050, C.I. Basic Red 12 (**3.17**) has the CICN 48070 and C.I. Direct Blue 1 (**3.36**) has the CICN 24410.

The CICN is directly linked to the chemical structure of the essential colorant. The numbering system used places colorants with the same essential structure within the same group. Therefore the CICN can identify many application classes of dyes under similar numbers as all being azo. There are 31 major classes that are then subdivided even further. Full details are available publicly on the Colour Index website [1].

The Colour Index gives only the specification of the essential colorant and not the commercial product, so it does not claim to compare the commercial products listed; instead it simply lists all those that contain essentially the same chemical structure. The strength, final shade, fastness and so forth must be tested by the end user. Due to the complex synthetic

Table 3.1   Colour Index application classes.

classification of colorants according to usage
Acid dyes
Azoic dyes
Basic dyes
Direct dyes
Disperse dyes
Fluorescent dyes
Food dyes
Leather dyes
Mordant dyes
Natural dyes
Pigments
Reactive dyes
Reduced vat dyes
Solubilised sulphur dyes
Solubilised vat dyes
Solvent dyes
Sulphur dyes
Vat dyes

Table 3.2   Example of the information given in a CIGN entry in the Colour Index.

Reactive Red 83

Constitution Number:	18230
Chemical Class:	Monoazo
Shade:	Red
CAS Numbers:	61969-26-4
EC Number:	
Discoverer:	Ciba Geigy
First Manufacturer:	Ciba Geigy
Classical Name:	Lanasol Red G
First Product:	Lanasol Red G

routes by which many dyes are produced, there are inevitably minor impurities, such as isomers, in the final commercial products. For this reason, there can be minor variations in the shade of a dye with a given CICN that is produced by different manufacturers.

Under the listings of the CIGN, commercial products that have been registered by the manufacturers are listed, as mentioned in Table 3.2. There is also the capability to see historical data listing colorants no longer available commercially but which have been listed previously.

In the remaining sections of this chapter, examples are given of the structures of dyes and pigments typical of the various chemical and application classes. For each of the examples that have been chosen, their CIGN are given.

3.3 Colour in Organic Molecules

The absorption of light energy by an organic dye or pigment molecule causes an electron to 'jump' to a higher energy level and the molecule moves into an 'excited' state. It is easier for a pi (π) electron from a double bond to jump into a higher energy level, and it is easier still for the transition to occur if alternate double and single bonds (i.e. a conjugated double bond system) exist in the same molecule. Consequently, as the excitation of an electron becomes easier, the required spectral energy moves from the invisible UV region into the longer wavelengths of the visible spectrum. For example, in **3.1** there are eight alternating single and double bonds. This compound absorbs violet light at around 410 nm and as a result is perceived as yellow in colour. Conjugated double bonds are present in the molecules of all dyes and pigments.

3.1

Extended conjugated systems occur in several natural colouring matters, including carotene in carrots and fruit juices and crocin found in saffron. Such compounds are not very substantive to fibres however and are unsuitable as starting points for dye synthesis.

The absorption of visible light energy by the compound promotes electrons in the molecule from a low energy state (the *ground state*) to a higher energy state (the *excited state*). The molecule is said to have undergone an electronic transition during this excitation process. Particular excitation energies correspond to particular wavelengths of visible light.

It is a pi (π) electron (an electron in a double or triple bond) that is promoted to the excited state. Even less energy is required for this transition if alternate single and double bonds (i.e. conjugated double bonds) exist in the same molecule. The excitation of the electron is made even easier by the presence of aromatic rings because of the enhanced delocalisation of the pi electrons (see Section 1.8.1.2). By altering the structure of the compound, colour chemists can alter the wavelength of visible light absorbed and therefore the colour of the compound.

The molecules of most coloured organic compounds contain two parts:

(1) A single aryl (aromatic) ring such as benzene or a benzene ring with a substituent. Alternatively there may be a *fused ring system* such as naphthalene or anthracene (see Section 1.8.1.2).
(2) An extensive conjugated double bond system containing unsaturated groups, known as *chromophores*, such as

The intensity of colour can be increased in a dye molecule by addition of substituents containing lone pairs of electrons to the aryl ring such as

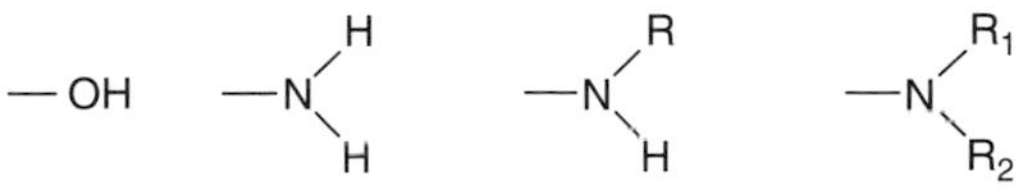

$$— OH \qquad —N\overset{H}{\underset{H}{\big\langle}} \qquad —N\overset{R}{\underset{H}{\big\langle}} \qquad —N\overset{R_1}{\underset{R_2}{\big\langle}}$$

These groups are called *auxochromes*.

Whilst auxochromes influence the colour of the dye molecule, they also play an important part in giving dyes substantivity for fibres. They are polar in character and are able to form hydrogen bonds and other forces of interaction (such as dipolar forces; see Section 1.4.3) with polar groups in fibres. In the case of disperse dyes, the presence of these groups provides a small but important solubility in water at the temperatures used for their application.

Another important requirement of dye molecules for many application classes (such as acid, direct and reactive dyes) is solubility in water. This property is conferred by the introduction of the sulphonate group, usually in the form of its sodium salt $-SO_3^-Na^+$. Another group sometimes used for this purpose is the carboxylic acid group, $-COOH$ (or its sodium salt $-COO^-Na^+$). These groups have no significant influence on the colour of dye molecules.

Chromophores give the potential for molecules to be coloured, that is, they absorb light in the visible region of the spectrum, because they confer the possibility of *resonance* through the creation of charged structures. When a molecule absorbs light, it is raised from its ground state to an electronically excited state, and the energy required to cause this transition determines the wavelength of the light absorbed. The relationship between the energy required for the transition (ΔE) and the wavelength of light (λ) is

$$\Delta E = \frac{hc}{\lambda} \tag{3.1}$$

where h is Planck's constant (6.626×10^{-34} J/s^{-1}) and c is the speed of light (3×10^8 m/s).

Since ΔE is inversely proportional to λ, then if the energy difference between the ground and excited states is small, the light absorbed will be at longer wavelengths. Shorter wavelengths will be reflected, and the colour a human sees will correspond to a light of low wavelength, that is, blue. If the energy difference is high, then the light absorbed will be at lower wavelengths and the longer wavelengths will be reflected.

Resonance amongst charged structures lowers the energies of both ground and excited states, but lowers that of the excited state the most. This means that the greater the resonance between charged structures, the longer the wavelength of light absorption. No one resonance structure describes the actual structure of a molecule; instead the actual form is a hybrid of the various charged structures.

The azo group ($-N{=}N-$) is a very commonly used chromophore in dyes and **3.2** shows some of the resonance forms that contribute to the hybrid of azobenzene:

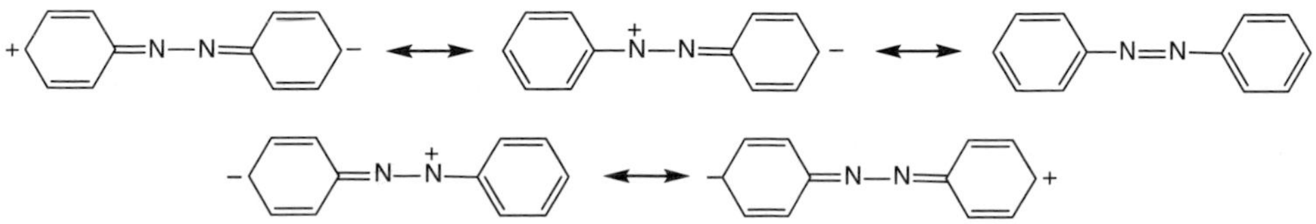

3.2

A large amount of energy is required for the formation of these resonance forms and azobenzene absorbs in the UV region, at around 320 nm. Consequently it has only a very weak yellow colour.

A good example of the contribution by an auxochrome to resonance creating charged structures is provided by *p*-aminoazobenzene (**3.3**). This compound has one charged structure involved in the resonance. The energy required is less than that for **3.2**, so it absorbs at a slightly longer wavelength (see Equation 3.1) though again mainly in the UV region (at 332 nm) and still has a weak yellow colour:

3.3

Auxochromes can be classified as either electron donor or electron acceptor groups, and the amino group in *p*-aminoazobenzene acts as an electron donor.

Other electron donor groups are $-NR_2$, $-NHR$, $-OR$ and $-OH$.

Examples of electron accepting groups are $-NO_2$, $-CN$, $-SO_2R$ and $-COOH$.

If a molecule contains an electron-attracting group at one end and an electron-donating group at the other end, even less energy is required to form the various resonance structures. The absorption bands shift from the UV into the visible region. The lower the energy required, the more the light absorbed moves from the violet/blue part of the spectrum (higher energy) to the red end (low energy), as dictated by Equation 3.1. Because the visible colour is the complementary of the absorbed light, as the energy of light absorbed decreases, the colour seen follows the series:

$$\text{Yellow} \rightarrow \text{orange} \rightarrow \text{red} \rightarrow \text{purple} \rightarrow \text{violet} \rightarrow \text{blue} \rightarrow \text{green}$$

The effect is greatest when the two types of groups (electron attracting and electron donating) are as far apart in the molecule as possible. Thus, for example, the molecule **3.4**, which is C. I. Disperse Orange 3, is more deeply coloured than **3.3**, due to the 'push–pull' action of the two auxochromes. The energy absorbed by **3.4** is now much lower and it absorbs light at around 480 nm, giving it an orange colour:

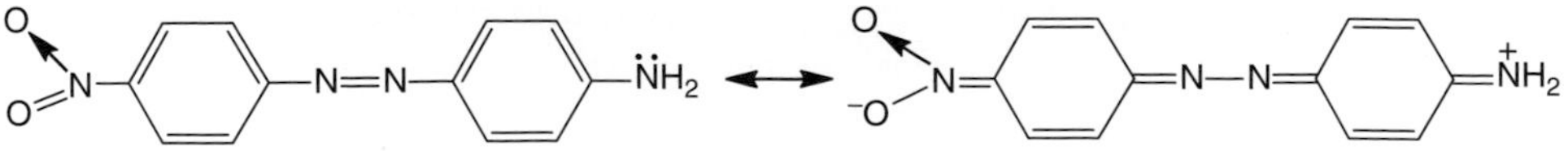

3.4

Another chromophore group is the carbonyl group (C=O), present in a class of compounds called *quinones*. Quinones contain two carbonyl groups, the most commonly used being dyes based on anthraquinone (**3.5**), which can form a variety of charged structures, such as **3.6**:

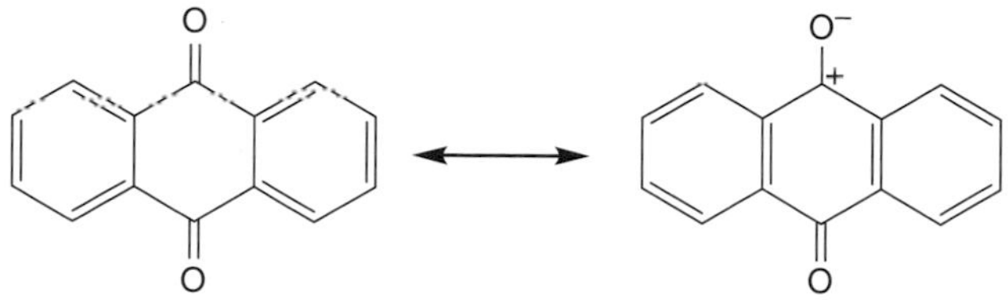

3.5   **3.6**

3.4 Classification of Dyes According to Chemical Structure

There is a wide range of chemical classes of dyes, though some of them are of minor importance in commercial dyeing operations. Some of the more important classes are described in the following sections. Fuller accounts of all of the classes can be found in References [2, 3].

3.4.1 Azo Dyes

Azo dyes are the most common chemical class. The chromophore present in these dyes is the azo group (–N=N–), and there may be more than one azo group in a dye molecule depending on its complexity. Thus azo dyes are referred to as *monoazo*, *disazo*, *trisazo* or *tetrakisazo*, depending on whether they contain one, two, three or four azo groups respectively in the molecule.

The formation of an azo dye involves a chemical reaction between a *diazonium salt* and a *coupling component*. The diazonium salt is formed by the diazotisation of a primary amine that is then reacted with the coupling component, usually a phenol or an aromatic amine.

The diazotisation reaction involves treating the primary amine with nitrous acid at low temperature and under conditions of acidity carefully controlled by the presence of hydrochloric acid. The overall reaction can be represented by the equation (Scheme 3.1)

$$Ar–NH_2 + NaNO_2 + 2HCl \rightarrow Ar–N=N^+Cl^- + 2H_2O + NaCl$$

Scheme 3.1   Diazotisation of primary amines to form a diazonium salt.

where Ar represents an aromatic ring, such as benzene or naphthalene. The important reacting species, nitrous acid, is formed from reaction between sodium nitrite and hydrochloric acid. The diazonium species formed, $Ar–N=N^+$, is highly reactive and unstable at room temperatures, so the solution has to be kept cold to avoid decomposition.

The diazonium species is an electrophile and reacts preferentially at electron-rich sites in amines and phenols. These sites are generally at the ortho- and para- positions on their aromatic rings, though the preferred coupling position depends on the pH conditions used in the coupling reaction. For example, coupling to phenol and amines usually occurs preferentially at the para- position, as shown in Scheme 3.2.

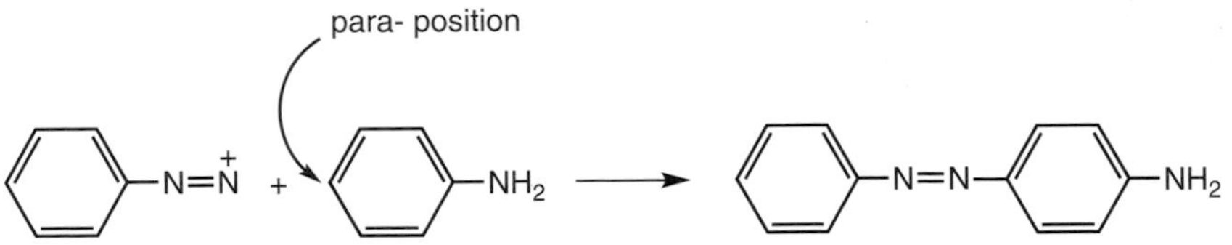

para- position

Scheme 3.2   Diazo coupling to form an azo dye.

Some coupling components possess both amino and hydroxyl groups, and the position of coupling of the diazonium compound depends on whether acidic or alkaline conditions are used. For example, coupling to the coupling component known as H-acid can yield two products (Scheme 3.3).

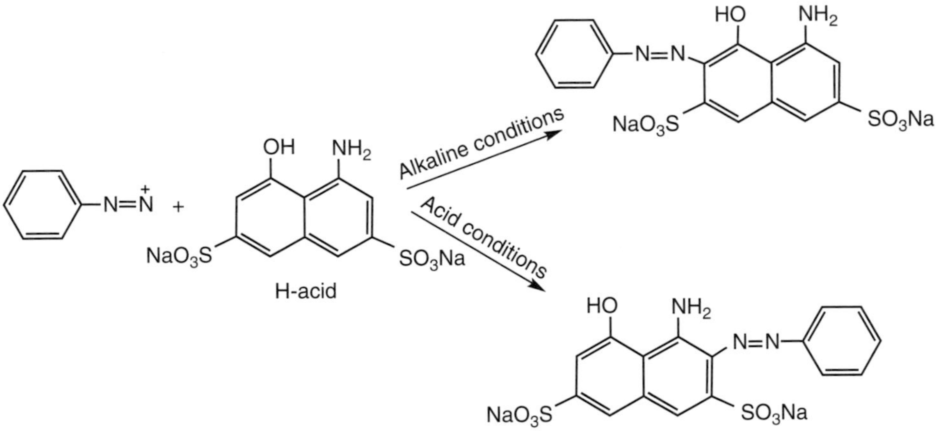

H-acid

Alkaline conditions

Acid conditions

Scheme 3.3   Coupling reaction under acid and alkaline conditions.

A disazo dye can be synthesised by firstly coupling under one set of conditions (say, alkaline conditions) and then coupling under acid conditions. Using the example in Scheme 3.3, the disazo dye formed will have the structure **3.7**:

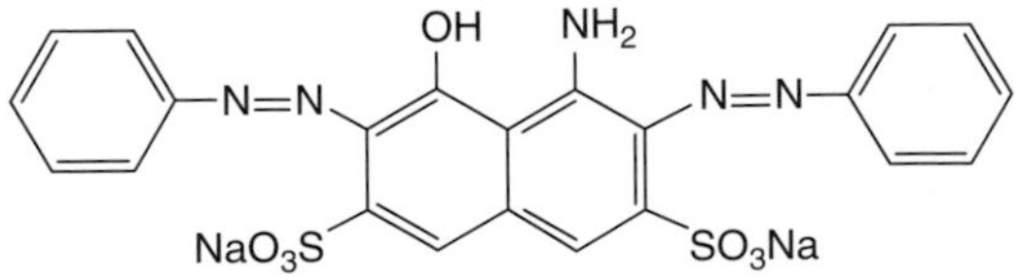

3.7

The simplest azo molecule, phenylazobenzene (**3.2**, sometimes simply called azobenzene), has a weak yellow colour. However, as explained in Section 3.3, the colour of the molecule may be modified and increased in intensity of colour by introducing auxochromes, as illustrated by structures **3.8** and **3.9**:

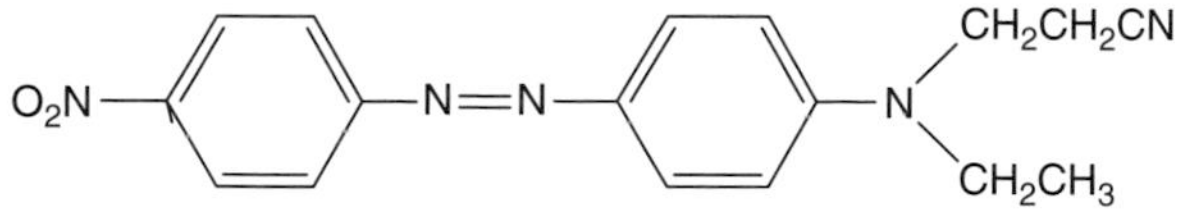

C.I. Disperse Orange 25

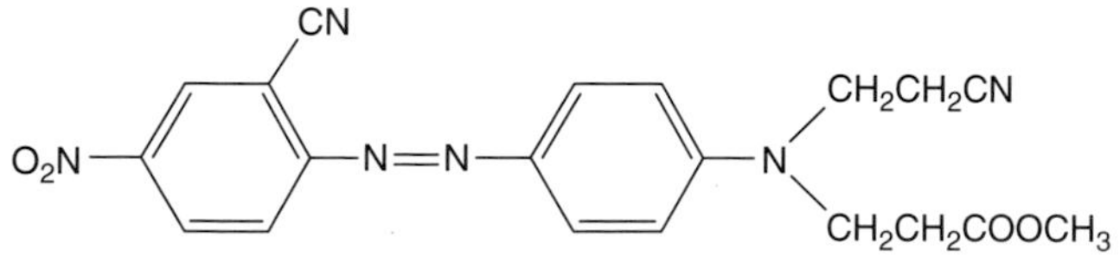

3.8

C.I. Disperse Red 90

3.9

Many dyes are based on naphthalene, such as C.I. Acid Red 1 (**3.10**) and the disazo dye C.I. Acid Black 1 (**3.11**):

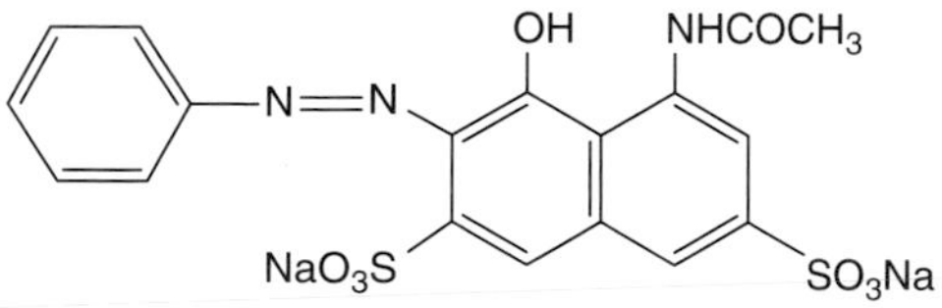

3.10

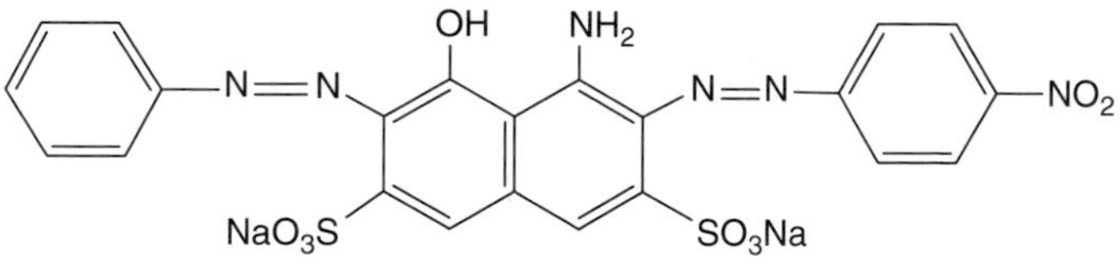

3.11

An example of a trisazo dye is C.I. Direct Black 166 (**3.12**):

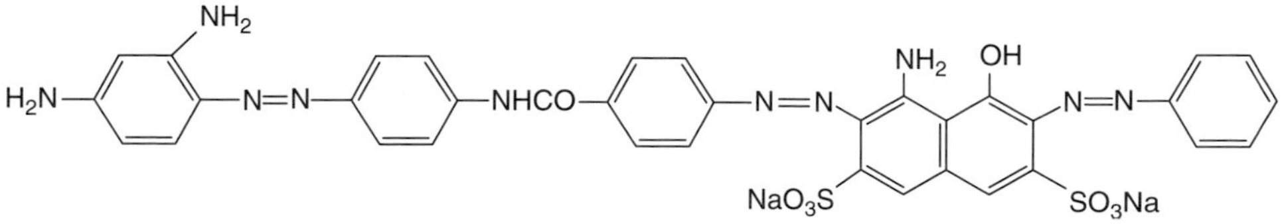

3.12

Tetrakisazo dyes (dyes with four azo groups) do exist, but are much less common.

3.4.2 Anthraquinone Dyes

Anthraquinone (**3.14**) can be synthesised by the oxidation of anthracene (**3.13**) (Scheme 3.4).

Anthraquinone is used as the basis for the manufacture of a wide range of acid, vat and disperse dyes, mainly in the shade range violet, blue and green. Typical of an acid dye is

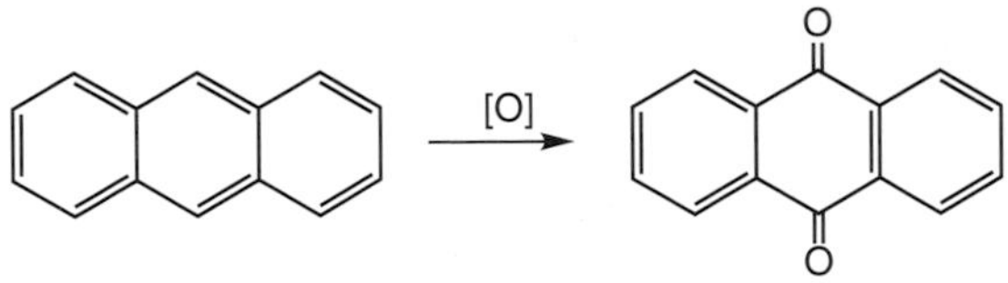

3.13   **3.14**

Scheme 3.4   Synthesis of anthraquinone by oxidation of anthracene.

C.I. Acid Blue 45 (**3.15**), a classic acid dye for wool, and typical of a disperse dye is C.I. Disperse Red 60 (**3.16**), an important dye for polyester:

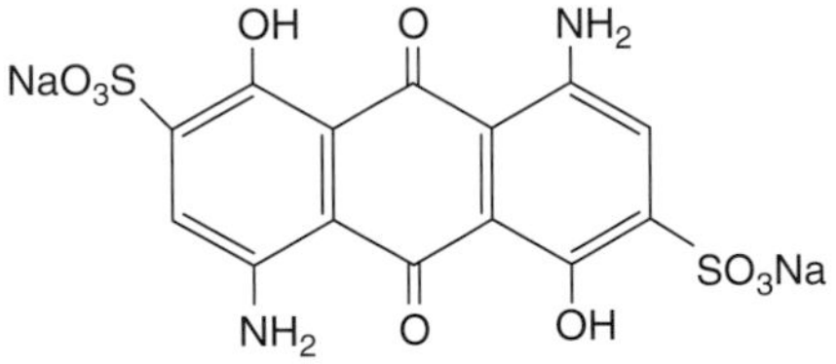

3.15 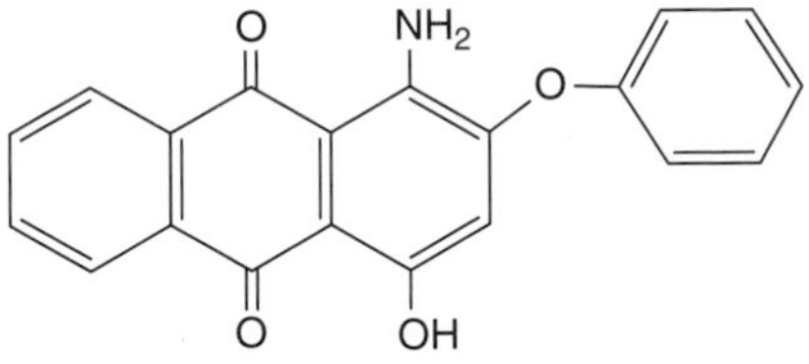   **3.16**

3.4.3   Methine and Polymethine Dyes

The characteristic structural feature of these dyes is a conjugated system of alternate double and single bonds created through methane (–CH=) groups. It is this conjugated double bond structure that is responsible for the colour. Cationic polymethine dyes are used for the dyeing of acrylic fibres and give bright shades with good fastness properties, such as C.I. Basic Red 12 (**3.17**), but neutral polymethines are also used, such as C.I. Disperse Blue 354 (**3.18**), a widely used dye for polyester:

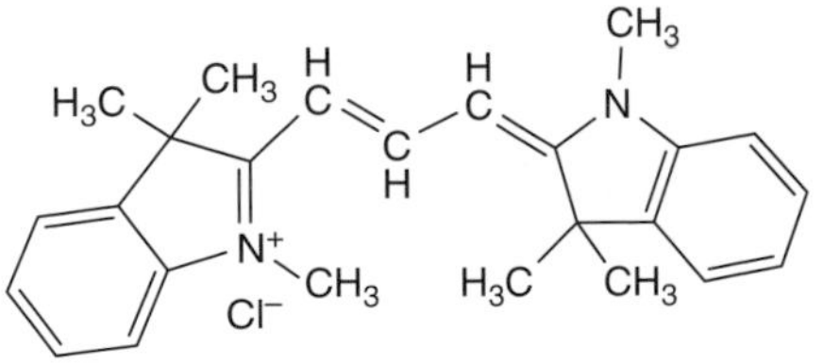

3.17 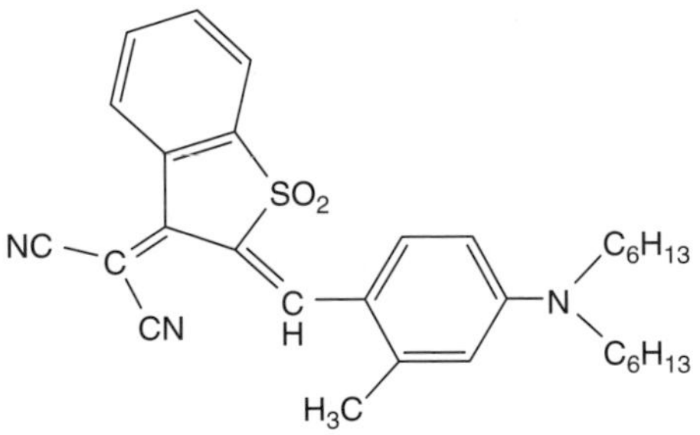   **3.18**

3.4.4   Nitro Dyes

This is a small group of dyes in which the nitro (–NO$_2$) group, normally incorporated in dye molecules as an auxochrome, acts as the chromophore. An example of such a dye is C.I. Disperse Yellow 1 (**3.19**):

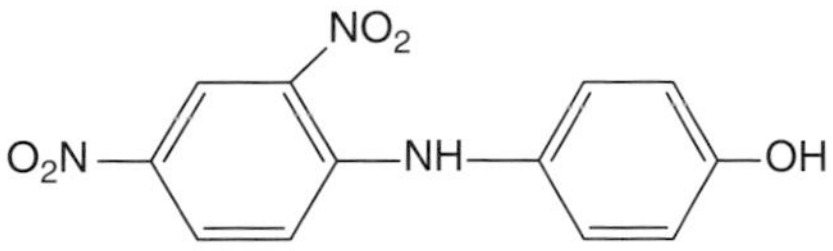

3.19

3.4.5   Triarylmethane Dyes

There are relatively few triarylmethane dyes used commercially. Those that are available are acid and basic dyes, mainly of violet, blue and green hues. Examples are C.I. Acid Blue 1 (**3.20**) and C.I. Basic Violet 3 (**3.21**):

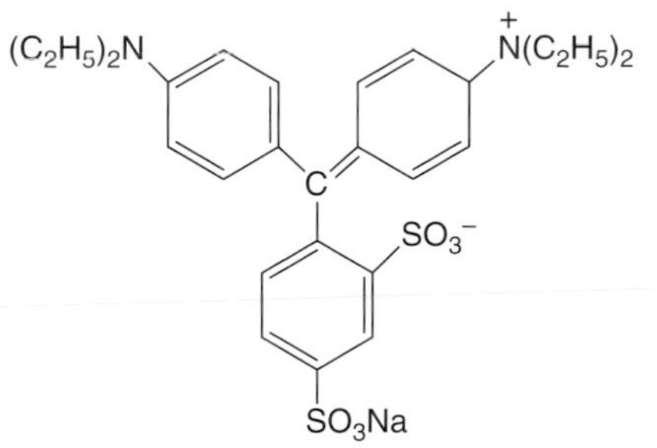

3.20 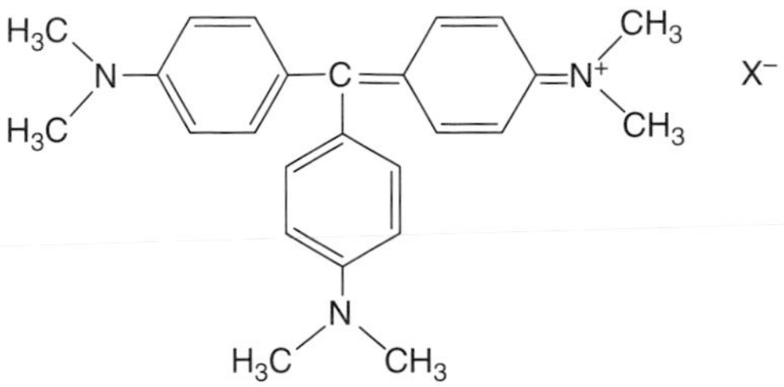   **3.21**

3.5   Classification of Dyes According to Application Class

In illustrating the important chemical types of dyes, examples have already been given of dyes belonging to some of the application classes. In this section, the characteristic features of dyes belonging to the various application classes are explained.

3.5.1   Dyes for Protein Fibres

3.5.1.1   Acid Dyes

The term 'acid dye' refers to the need for an acid dyebath for its application rather than the chemical character of the dye.

Acid dyes are anionic dyes with substantivity for protein fibres (wool, silk and nylon). They are soluble in water, and like simpler electrolytes, their molecules dissociate into positively and negatively charged ions (see Section 1.4.1). Acid dyes are sufficiently soluble to allow direct application from an aqueous solution. The solubility is due to the presence of sulphonate groups (as their sodium salts $-SO_3^-Na^+$) in the negatively charged ion (the anion), which is the coloured ion.

Most acid dyes are azo dyes, such as **3.10** and **3.11**, but there are some based on anthraquinone, and some acid dyes are triarylmethane dyes, though triarylmethane dyes are of declining importance. Well-established anthraquinone acid dyes are C.I. Acid Blue 25 (**3.22**) and C.I. Acid Blue 45 (**3.15**), and an example of a triarylmethane acid dye is C.I. Acid Blue 1 (**3.20**). Some acid dyes have complex structures and high molecular weights, such as C.I. Acid Violet 34 (**3.23**) and C.I. Acid Red 111 (**3.24**):

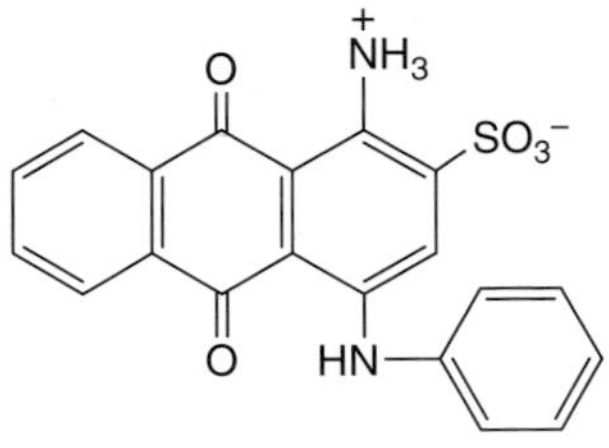

3.22

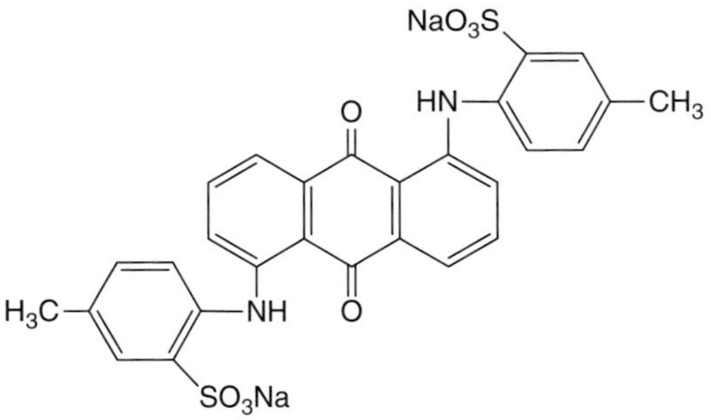

3.23

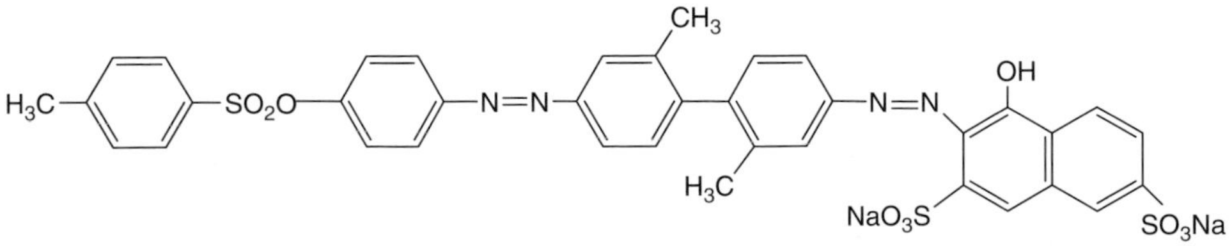

3.24

The substantivity of acid dyes for protein fibres is due to the electrostatic attraction between the negatively charged dye anion (due to the sulphonate group, $-SO_3^-$) and the positively charged groups (mainly $-NH_3^+$) in the fibre. This is especially the case when simple, low molecular weight dyes are applied. Such dyes are referred to as *equalising acid dyes*, because they are the easiest to dye level. C.I. Acid Blue 45 (**3.15**) is an example of an equalising acid dye. This type of acid dye requires the presence of a strong acid such as sulphuric acid in the dyebath to give a pH value of about 2; otherwise the important salt linkages between the fibre and the dye cannot form.

Equalising acid dyes usually remain unaggregated in water. They have low anion substantivity and so they migrate readily. Unfortunately their good migration, which makes equalising acid dyes suitable for the production of level-dyed materials, also means they may be removed easily from wet fabric. They therefore do not possess a particularly high degree of fastness to wet treatments.

Acid dyes with more complex structures, such as **3.24**, are able to form secondary forces of attraction, such as hydrogen bonds and van der Waals forces, with polar groups in the fibre in addition to electrostatic forces. These dyes generally have better wet fastness properties and are called *milling acid dyes*. Accordingly they can be applied at higher pH values of around 4.5–5.5 by using methanoic (formic) or ethanoic (acetic) acid instead of sulphuric acid.

Milling acid dyes tend to aggregate in solution, though the aggregates break down as the temperature of the dyebath is raised. Milling is a process in which woollen goods are treated in a weakly alkaline solution with considerable mechanical action to promote felting. Dyes of good fastness to milling are essential to avoid bleeding of colour during this process. Milling acid dyes give bright shades, as do equalising acid dyes, but their wet fastness properties are better. Their levelling properties are adequate for woollen fabrics intended for dresswear, pale- to medium-depth knitting yarns and some types of upholstery.

A type of acid dye that possesses wet fastness in between that of equalising acid and milling acid dyes is the *fast acid* dye. Dyes of this type usually only contain one sulphonate group and have slightly higher molecular weights than equalising acid dyes, so their migration properties are not as good. They are applied in dyebaths of slightly higher pH, obtained using ethanoic (acetic) acid at 1–3% owf rather than sulphuric acid, and contain 5–10% owf Glauber's salt.

Yet another type of acid dye is the *neutral dyeing acid dye*, sometimes referred to as *supermilling* dyes. These dyes have very high anion affinity and require a minimum of acid for their application. Typically they are applied at pH 5.5–6.5. Dyes in this group tend to be more individual in character in terms of the precise conditions required for their application. Two examples are C.I. Acid Orange 63 (**3.25**) and C.I. Acid Red 138 (**3.26**):

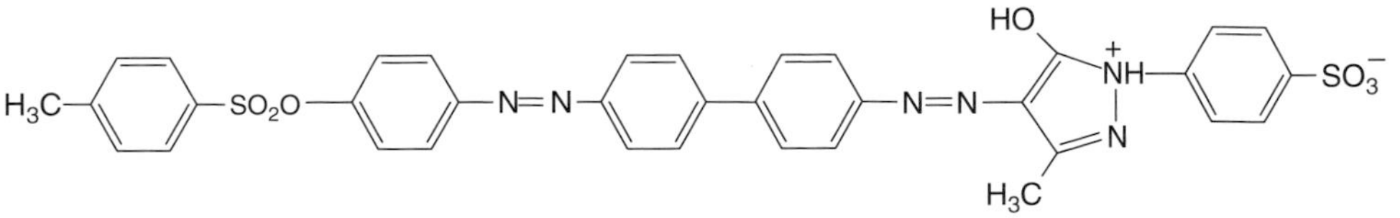

3.25

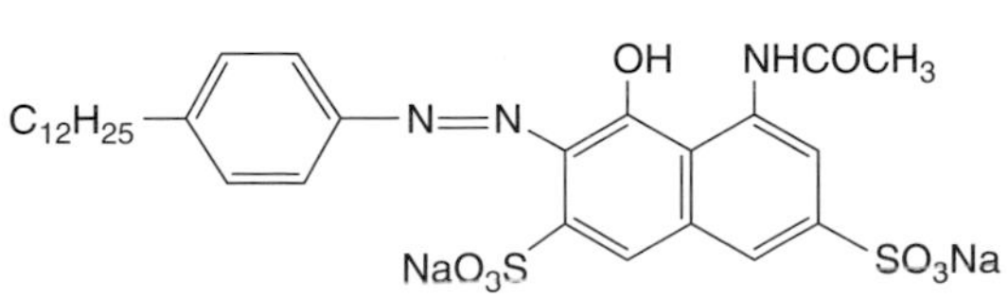

3.26

If the structure of C.I. Acid Red 138 (**3.26**) is compared with that of the equalising acid dye C.I. Acid Red 1 (**3.10**), it will be noticed that the only difference is that the former contains a dodecyl hydrocarbon chain ($-C_{12}H_{25}$) at one end of the molecule. This dodecyl group provides the opportunity for van der Waals forces to form with the hydrophobic

Table 3.3 Characteristics of acid dyes.

	equalising dyes	milling dyes	supermilling dyes
Acid used	Sulphuric	Methanoic or ethanoic	Ammonium sulphate
Dyebath pH	2.5–3.5	4–5.5	5.5–6.5
Migration ability	High	Low	Very low
Wet fastness	Poor–fair	Very good	Excellent
Molecular weight	Low	High	Very high
Dye solubility	High	Low	Low
State in solution	Monomolecular	Aggregated	Highly aggregated
Substantivity (pH 6)	Very low	High	High

regions of wool, silk and nylon fibres, so consequently the substantivity is considerably higher. C.I. Acid Red 138 is therefore much less mobile, so it is difficult to apply level, but it has far superior fastness to wet treatments than C.I. Acid Red 1.

In general, large molecules containing few sulphonate groups have lower water solubility, migrate least during dyeing and give dyeings of better wash fastness. Conversely, small dye molecules with many sulphonate groups migrate very well but have poor wash fastness. Dyes of low molecular weight have poor exhaustion when the pH of the dyebath is above 4. The characteristics of the types of acid dyes are summarised in Table 3.3.

3.5.1.2 *Mordant Dyes*

Whilst metals such as cobalt, copper, tin and aluminium can be used as mordants, the metal most commonly used is chromium. A requirement of the dyes used is that they must contain suitably disposed chemical groups with which the chromium can combine. Most chrome dyes resemble acid dyes but are capable of reacting with the chromium ion Cr^{3+} in a reaction called *chelation*. The chromium complex of the dye so formed has a substantially increased molecular size, which prevents the outward migration of dye molecules even during severe wet treatments.

It is therefore possible to have the advantages of equalising acid dyes during the exhaustion stage of the dyeing process, together with the improvement in fastness properties through after-treatment with a solution of a chromium salt such as potassium dichromate. For many years this after-chrome process has been popular because it yields a product with excellent fastness properties more cheaply than most of the alternatives. It was favoured for the production of very deep shades on loose stock and of blacks and navies on yarn and fabric.

In practice the advantages are now outweighed by the pollution of dyehouse effluent by chromium residues, in particular concerns about the toxicity of the hexavalent form of chromium contained in the dichromate ion $Cr_2O_7^{2-}$. Whilst the chromic cation Cr^{3+} is non-carcinogenic, the hexavalent form of chromium is carcinogenic. A number of strategies have been developed to reduce the discharge of chromium in effluent, such as using only the minimum amount of dichromate required to react stoichiometrically with the dye, control of pH to the region 3.5–3.8 and a temperature to around 90 °C. The addition of various chemical auxiliaries to the bath has also been suggested.

In addition, as noted in Section 2.6.1.4, the use of heavy metals in the processing of wool is forbidden within the regulations of GOTS, so chrome-dyed wool does not qualify for GOTS labelling. Also the levels of chromium in chrome-dyed wool exceed the maximum permitted levels of extractable chromium set by the Oeko-Tex® 100 human ecology standard (see Section 8.6). There are strict controls on the handling of dichromate, which means its availability to the dyeing industry is limited. For these reasons chrome dyeing processes are in significant decline, though still used to a certain extent because the other dye application classes are unable to provide heavy shade depths of some colours with the required fastness standards.

3.5.1.3 Pre-metallised (or Metal-Complex) Dyes

These dyes, also referred to as metal-complex dyes, contain the metal already combined with the dye molecules by the dye manufacturer. The metal used is usually chromium though some contain cobalt. In understanding the chemistry of the formation of a metal-complex dye, it is useful to consider first the electronic structure of the chromium atom and its chromic ion, Cr^{3+}. As illustrated in Figure 3.1, the chromium atom can lose three valence electrons to give Cr^{3+}, which in turn can take up six pairs of electrons in its outer shells by the formation of coordinate bonds with ligands. A ligand is a molecule or ion containing a lone pair of electrons that can be donated to another molecule or ion to form a coordinate bond.

The requirement of the dye molecule is the presence of o,o'-disubstituted azo compounds of the type shown in structure **3.27**, when the dye acts as a tridentate ligand, with the oxygen atoms of the –OH group and one of the nitrogen atoms of the azo group forming bonds with the metal ion. The substituents in the o,o' positions do not necessarily need to be –OH groups: groups such as –COOH or $–NH_2$ are also suitable:

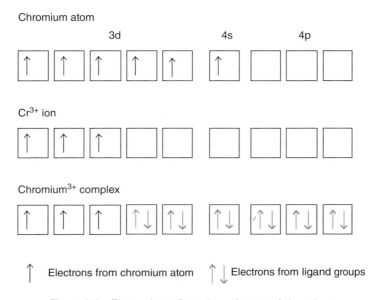

↑ Electrons from chromium atom   ↑↓ Electrons from ligand groups

Figure 3.1 Electronic configurations of states of chromium.

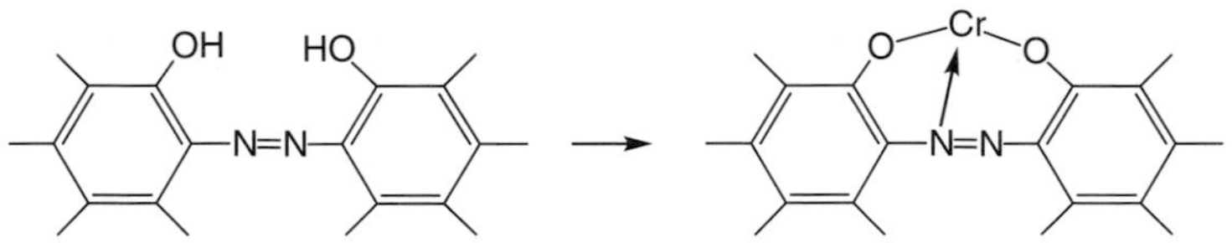

3.27

The remaining three valencies may be accounted for by the presence of ligands such as water or by groups such as –OH, –NH$_2$ or –SH in the protein fibre.

The earliest complexes incorporated one chromium atom into each dye molecule to give what is called a 1:1 complex, of which C.I. Acid Red 180 (**3.28**) is an example. These dyes are soluble in water and may be applied in the same way as equalising acid dyes because they are anionic in nature. However, they require a strongly acidic dyebath (pH ~ 2), which can cause fibre damage and consequent difficulties in spinning yarns. For this reason 1:1 complexes are less common than the 2:1 complexes:

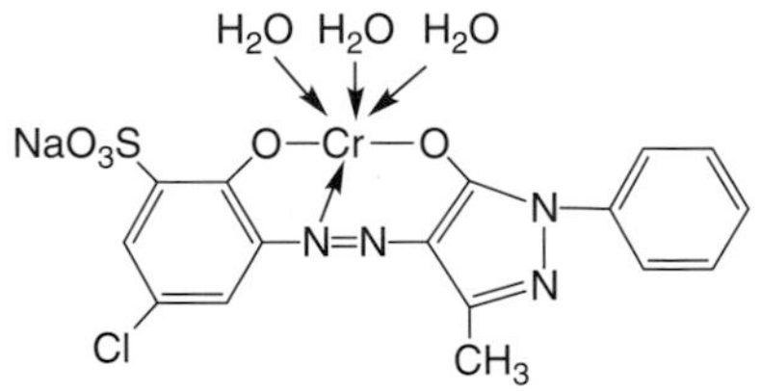

3.28

Instead of the three remaining coordination sites being occupied by colourless ligands such as water (as in **3.28**), they can be occupied by a second tridentate dye molecule to yield a 2:1 dye–metal complex. Since the metal atom is fully coordinated with the two dye ligands, it cannot form bonds with groups in the wool fibre. However the molecule is very large, and therefore van der Waals forces of attraction exist, which enable a high degree of wet fastness.

There are two types of 2:1 pre-metallised dyes – weakly polar and strongly polar. An example of a weakly polar type is C.I. Acid Violet 78 (**3.29**):

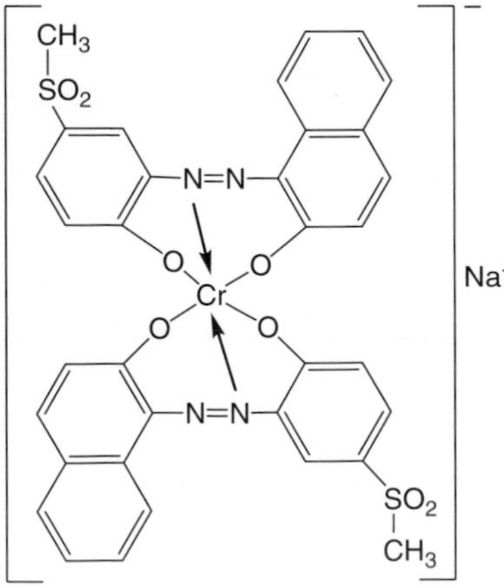

3.29

The weakly polar 2:1 dye–metal complex dyes do not contain water-solubilising groups such as the sulphonate (–SO$_3$Na) or carboxyl (–COONa) commonly associated with acid dyes for wool. Instead these dyes contain non-ionic hydrophilic groups such as methylsulphonyl (–SO$_2$CH$_3$) or methylsulphonamide (–SO$_2$NHCH$_3$) to confer a limited solubility and decrease the rate of dyeing. The strongly polar types usually contain a sulphonate group (some contain two), though in each case few structures have been disclosed. Although pre-metallised dyes are used widely for wool dyeing, the stringent demands for wet fastness with machine-washable wool are best achieved using reactive dyes.

3.5.1.4   *Reactive Dyes*

One of the biggest challenges is the coloration of shrink-resistant-treated wool that may be washed in a domestic washing machine. Much of the wool so designated is produced using after-treatments that deposit polymeric compounds on the fibres. These have the effect of reducing the wet fastness of most conventional wool dyes, just where greater resistance to wet treatments is needed. This deficiency can be effectively covered by the use of dyes chemically identical with, or similar to, the reactive dyes intended for cellulose (see Section 3.5.2.4). Reactive dyes for wool cover a wide shade gamut and give attractive bright colours.

Reactive dyes are necessarily applied to cellulosic fibres under alkaline conditions, but alkaline conditions are usually inappropriate for wool because of the sensitivity of the disulphide cross-links (see Section 2.6.1.1). Fortunately, the chemical nature of the functional groups in wool enables the reactive dyes to react with the fibres under mildly acidic conditions. Particular care is needed to achieve adequate levelness because, once the dye has reacted with the fibre, migration is clearly impossible. Furthermore, any unfixed dye remaining on the fibre still remains sufficiently substantive to cause staining problems during use, so careful washing off after dyeing is needed. One other disadvantage is that their lightfastness in pale shades is less good than that of 2:1 pre-metallised dyes.

As with reactive dyes for cellulose, reactive dyes for protein fibres react by either nucleophilic addition or nucleophilic substitution. The general structure of reactive dyes for wool is the same as that for reactive dyes for cotton, as illustrated in Figure 3.3. There are three types of reactive groups that are mainly used in reactive dyes for protein fibres: α-bromoacrylamido (**3.30**), chlorodifluoropyrimidine (**3.31**) and blocked vinyl sulphone (**3.32**):

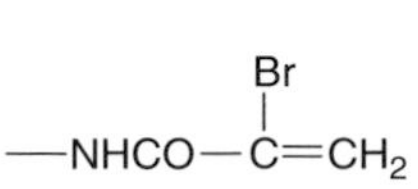

3.30 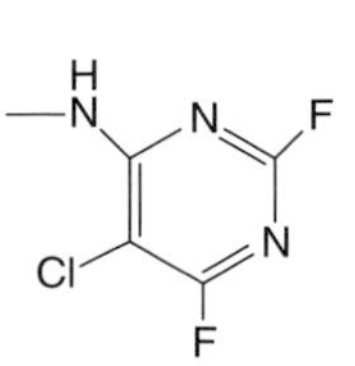   **3.31**   **3.32**

The α-bromoacrylamido dyes have a bromine leaving group on the dye molecule. It was thought that these dyes formed a covalent bond with the fibre by a nucleophilic substitution reaction, but it has been shown that they can also form covalent bonds by a nucleophilic addition reaction. Covalent bonds can be formed with both thiol (–SH) and amino (–NH$_2$) groups in protein fibres.

Some α-bromoacrylamido dyes have two α-bromoacrylamido groups in the molecule, for example, C.I. Reactive Red 83 (**3.33**):

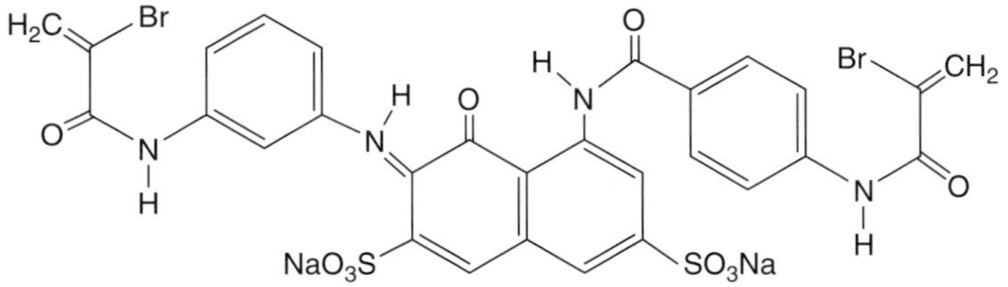

3.33

The level of fixation of these dyes is very high, giving excellent wet fastness. Their mechanism of reaction with protein fibres is shown in Scheme 3.5.

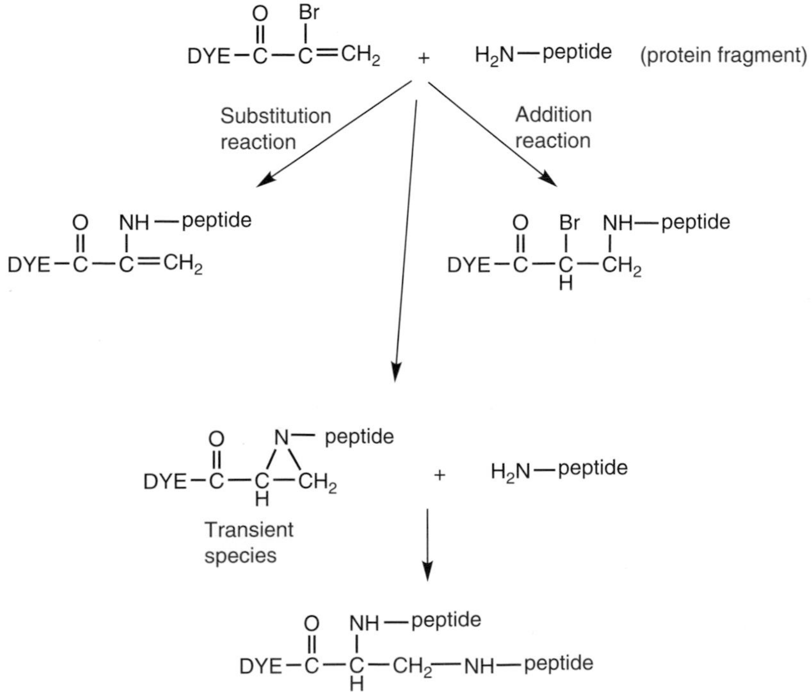

Scheme 3.5 Nucleophilic substitution and nucleophilic addition reactions of α-bromoacrylamido reactive dyes with protein fibres.

Chlorodifluoropyrimidine dyes are bifunctional dyes since both fluorine atoms on the pyrimidine ring can form covalent bonds with groups in the wool fibre. An example of this type of reactive dye is C.I. Reactive Red 147 (**3.34**):

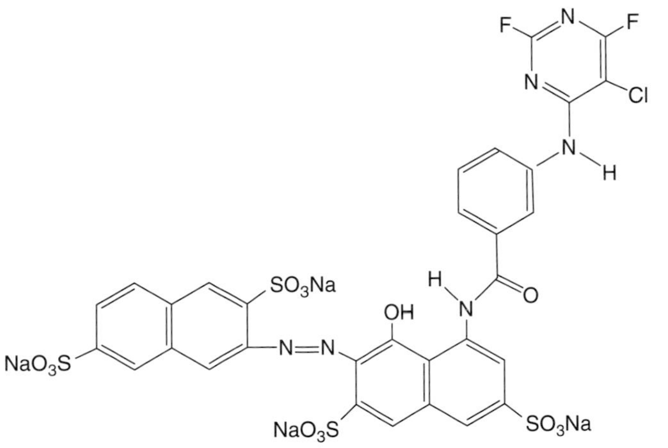

3.34

Dyes of this type form covalent bonds with protein fibres by the nucleophilic substitution reaction shown in Scheme 3.6.

The blocked vinyl sulphone derivatives gradually activate to the reactive vinyl sulphone at elevated temperatures under acidic conditions. This has the advantage of improving dye levelness due to suppression of the dye–fibre covalent bonding reaction at temperatures

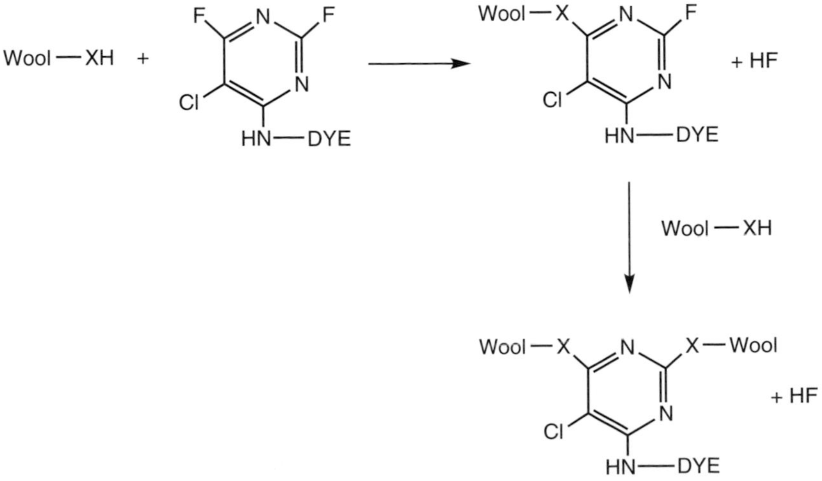

Scheme 3.6 Nucleophilic substitution reaction of chlorodifluoropyrimidine reactive dyes with protein fibres.

below the boil. An example of this type of dye is C.I. Reactive Red 180 (**3.35**) in its sulpha-toethylsulphone form:

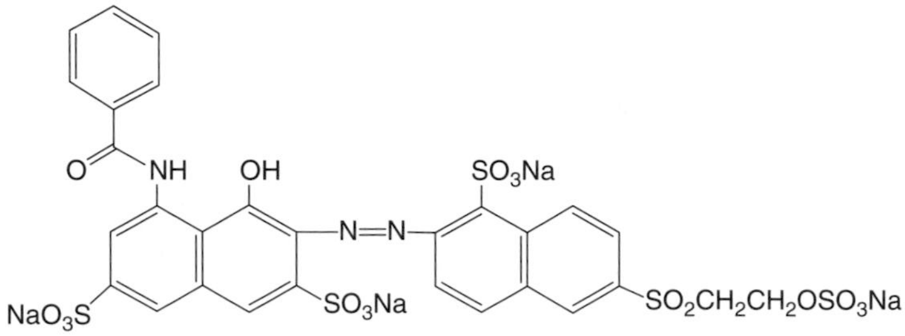

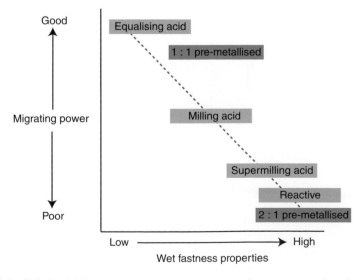

3.35

During the dyeing process the sulphatoethylsulphone group forms the reactive vinyl sulphone group that then reacts with groups such as $-NH_2$ in the fibre (Scheme 3.7).

$$Dye-SO_2-CH=CH_2 \ + \ H_2N-peptide \ \rightarrow \ Dye-SO_2-CH_2-CH_2-HN-Dye$$

Scheme 3.7 Nucleophilic addition reaction of vinyl sulphone reactive dyes with protein fibres.

3.5.1.5 *Summary*

The different application classes of dyes available for the dyeing of protein fibres have different characteristics in terms of their ease of application, levelling and fastness properties, in particular fastness to wet treatments. The relationship between the migration and fastness of the classes is shown in Figure 3.2. Whilst acid and reactive dyes cover a wide range of shades and are bright, the pre-metallised dyes are rather duller and cover a more restricted range of shades.

Figure 3.2 Relationship between migrating power and wet fastness properties of acid dyes.

3.5.2   Dyes for Cellulosic Fibres

3.5.2.1   Direct Dyes

Historically, in order to dye cotton with the dyes available, it was necessary to treat the fabric with a mordant that would 'fix' the dye within the fibres. When direct dyes were introduced, they could be applied to cellulosic fibres by a simple one-bath process (i.e. directly) in the absence of a mordant. For this reason they were called 'direct' dyes.

The molecules of direct dyes are similar in structure to those of acid dyes, but they are larger, that is, they generally have higher molecular weights. Another important distinction between direct and acid dyes is in molecular shape. Whilst acid dyes with bulky structural shapes can diffuse successfully into protein fibres, it is necessary for direct dyes to have long, narrow planar structural shapes, so they can penetrate the more highly crystalline cellulosic fibres.

Direct dyes generally conform to the formula $Y_1-N=N-X-N=N-Y_2$, and typical examples are C.I. Direct Blue 1 (**3.36**) and C.I. Direct Orange 25 (**3.37**):

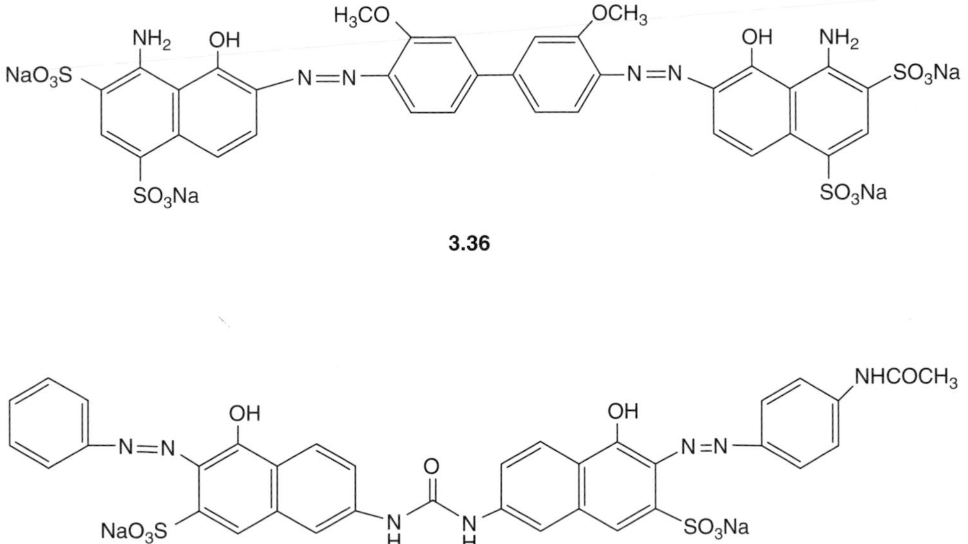

3.36

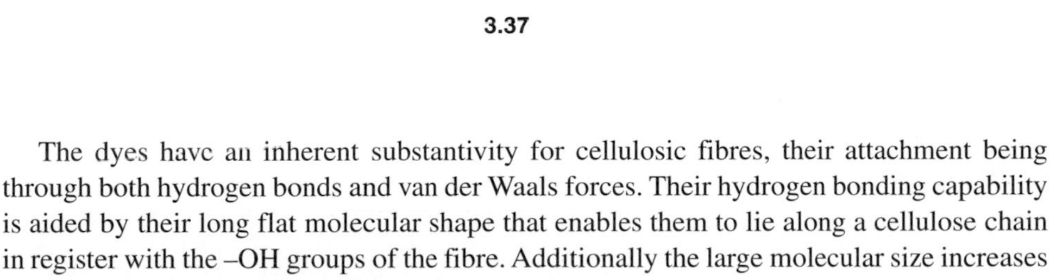

3.37

The dyes have an inherent substantivity for cellulosic fibres, their attachment being through both hydrogen bonds and van der Waals forces. Their hydrogen bonding capability is aided by their long flat molecular shape that enables them to lie along a cellulose chain in register with the –OH groups of the fibre. Additionally the large molecular size increases the opportunity for van der Waals forces to form.

The function of the sulphonate groups is limited to conferring water solubility on the molecule because, unlike protein and polyamide fibres, cellulose contains no basic groups with which the coloured dye anions can form electrostatic linkages. The molecules contain just enough sulphonate groups to give the dyes adequate solubility without aggregation being too much of a problem, and they are distributed evenly across them.

Compared with other application classes of dyes for cellulose, the wet fastness properties of direct dyes are usually poor, and where repeated washing of the dyed fabric is likely, as with towelling or lingerie, they are generally of use in pale shades only. A variety of after-treatments is available to make the dye on the fibre less soluble in water to give improved fastness performance (see Section 4.2.2.2), but unfortunately such improvements are accompanied by a change of shade or reduced fastness to light. The merits of direct dyes are their simplicity of application, the wide range of shades they can provide and a cost that usually compares very favourably with other alternatives.

3.5.2.2 *Vat Dyes*

Vat dyes are used in the dyeing and printing of all types of cellulosic fibres, as well as their blends with polyester. In their coloured form vat dyes are insoluble pigments, so their application requires them to be rendered soluble. In the dyebath the pigment is converted into a water-soluble form using a strong alkaline solution of a reducing agent, usually sodium dithionite. This forms the sodium 'leuco' compound of the dye, which is soluble in water but often different in colour from the original pigment. It is then allowed to dye the cellulose in this water-soluble form. Once exhaustion is completed, the leuco compound is oxidised back to the original coloured, insoluble form.

The vast majority of modern vat dyes are based on the anthraquinonoid or the indigo (or thioindigo) chromophores. Indigo, one of the oldest dyes still in use, remains popular through the wide use of indigo-dyed denim. Several of the anthraquinone dyes are complex polycyclic quinones (see Section 3.3), and they all possess two carbonyl (C=O) groups linked by alternate single and double bonds in a conjugated chain. This molecular arrangement is responsible for the easily reversible redox reactions (see Section 1.7) on which the application of vat dyes depends, the reduction reaction creating the soluble C–O⁻Na⁺ leuco form. In this reduced form the dye anion has some substantivity for cellulose through van der Waals forces, but generally vat dyes contain few other functional groups that would increase substantivity because they can be sensitive to the redox reactions. However, once the dyeing process is complete and the 'parent' insoluble dye has been reformed, the dye is held fast in the fibre and wet fastness properties are excellent.

In earlier centuries, when all textile colorants were obtained from natural sources, indigo plants were steeped in a large vat. It is from this ancient vatting process that the term 'fermentation vat dyes' is derived. Fermentation converts one of the plant constituents into the soluble leuco dye, which diffuses out of the plant.

The replacement of natural indigo by synthetic indigo (**3.38**) at the end of the nineteenth century gave the impetus to research on other syntheses, with the result that many synthetic vat dyes have since followed. Typical of these dyes are Indanthrone (C.I. Vat Blue 4, **3.39**), Flavanthrone (C.I. Vat Yellow 1, **3.40**), benzanthrone acridone (C.I. Vat Green 3, **3.41**) and anthraquinone carbazole (C.I. Vat Orange 15, **3.42**):

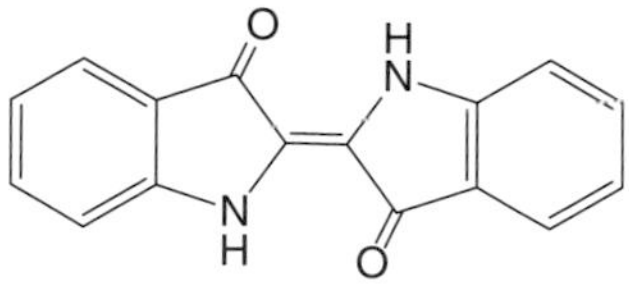

3.38

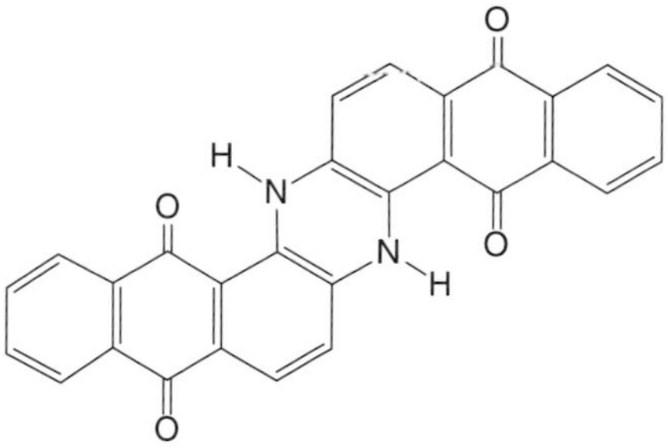

3.39

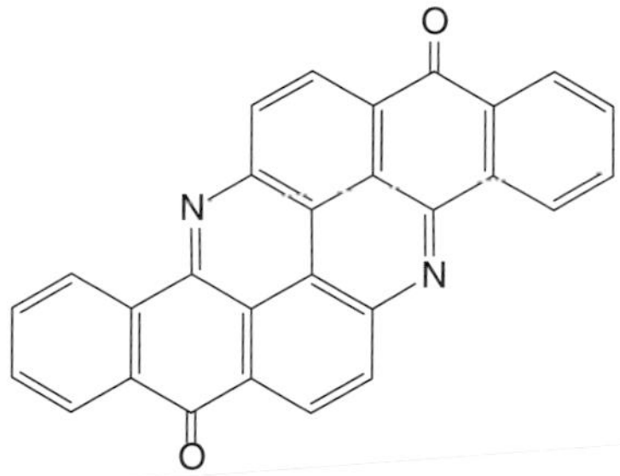

3.40

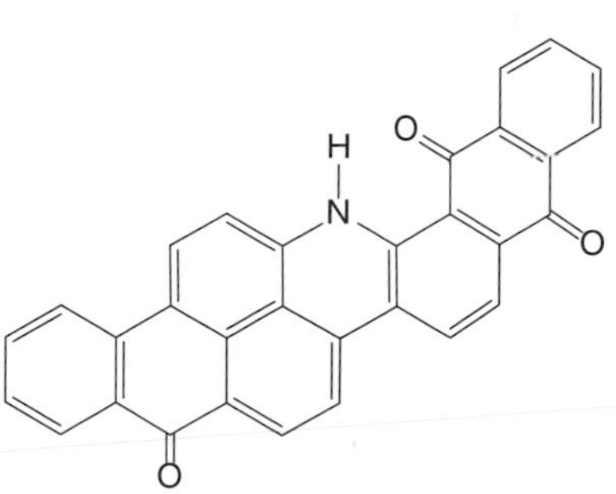

3.41

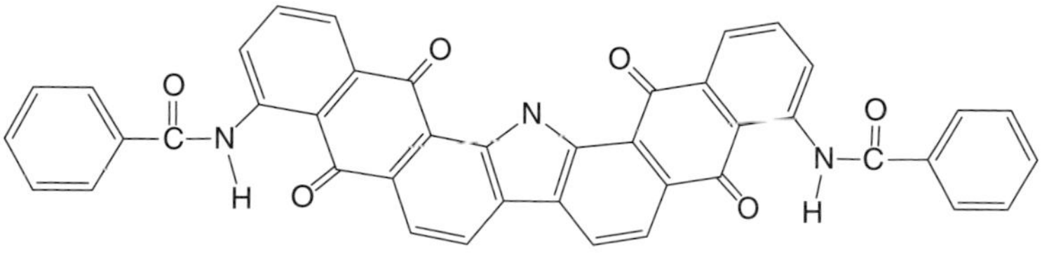

3.42

A structural feature that is common to many vat dyes is the close proximity of a –N–H group to a –C=O group. Through hydrogen bonding, a ring may form, which adds to the stability (especially the light fastness) of the molecule, as exemplified by Indanthrone (**3.39**) in Scheme 3.8.

3.5.2.3   *Solubilised Vat Dyes*

The need to reduce vat dyes before use makes their application a cumbersome process. Although it is possible to isolate the reduced form of the dye, it is too readily oxidised in air for the manufacturer to provide the dyer with leuco compounds. It is possible, however, to convert the leuco acid into the leuco ester, a derivative that has greater resistance to

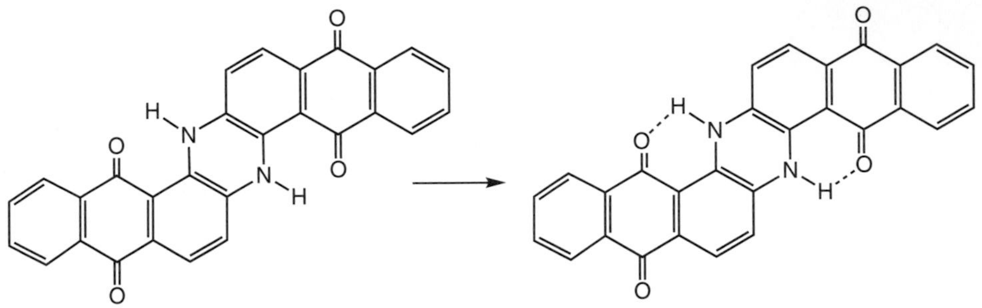

Scheme 3.8   Intramolecular hydrogen bonding in a vat dye molecule.

oxidation and greater solubility in water. Such esters can be formed by the reaction of a hydroxyl group of a leuco acid with sulphuric acid, forming a sulphuric ester (Scheme 3.9):

$$\text{Dye–OH} + \text{H}_2\text{SO}_4 \rightarrow \text{Dye–OSO}_3\text{H} + \text{H}_2\text{O}$$

leuco vat   sulphuric ester

Scheme 3.9   Formation of a sulphuric ester of a leuco vat dye.

The sodium salts of such esters are stable and can be stored until required for use. Since the ester group is only weakly attached to the rest of the dye molecule, it is easily removed by the action of sodium nitrite in dilute sulphuric acid. The regenerated leuco compound may then be oxidised back to the pigment form after dyeing.

Solubilised vat dyes are less rapidly taken up than are the more conventional vat dyes and are mainly used for the production of pale shades. As with ordinary vat dyes, application under alkaline conditions is essential, thus eliminating wool from the list of possible substrates because alkaline conditions damage wool. The low uptake and higher cost of solubilised vat dyes make them uneconomical for deep shades, however, and for these normal vat dyes have to be used.

3.5.2.4   *Reactive Dyes*

Until 1956 the only known way of obtaining very high wet fastness on cellulosic fibres was through the deposition of water-insoluble pigments within the fibre. But at this time Rattee and Stephen discovered that dye molecules containing certain chemical groups (*reactive groups*) could react chemically with cellulose under alkaline conditions. Thus for the first time it became possible to make a dye react with the fibre and become part of it, rather than remaining as an independent chemical entity within the fibre.

Rattee and Stephen's discovery was followed by the commercial introduction of reactive dyes that illustrated the technical possibilities of producing bright shades of high fastness through a variety of application methods. Since then many similar dyes have become available. The current ranges of reactive dyes include many that have a broad spread in their level of reactivity and substantivity. Since their introduction, these dyes have played a dominant role in the dyeing of cellulosic fibres.

Many reactive dyes possess a resistance to daylight, which was previously only associated with vat dyes, and this is reflected in their use for top-quality curtains, furnishings and awnings. They also contribute to the colour quality of many domestic goods that require frequent washing, such as towelling, and are used extensively for shirting, tapes, ribbons, dress goods and knitted sportswear.

The general structure of a reactive dye molecule is shown in Figure 3.3. The chromogen is attached, via a bridging group (usually a –NH– group), to the fibre-reactive group.

The structures of the fibre-reactive groups (**3.43–3.50**) typically used by the various dye manufacturers have different characteristics insofar as their reactivity and the temperature required for their application are concerned. These characteristics are summarised in Table 3.4:

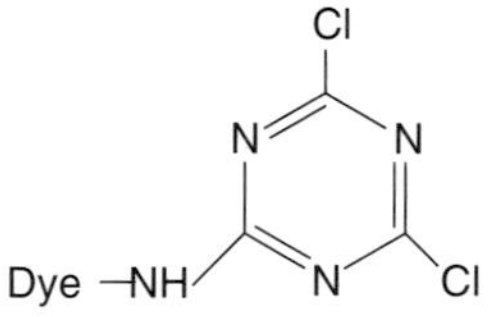

Dichlorotriazinyl (DCT)

3.43

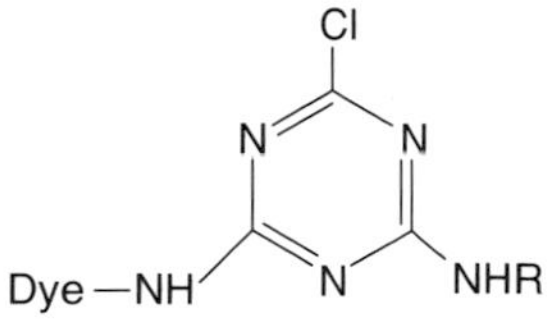

Monochlorotriazinyl (MCT)

3.44

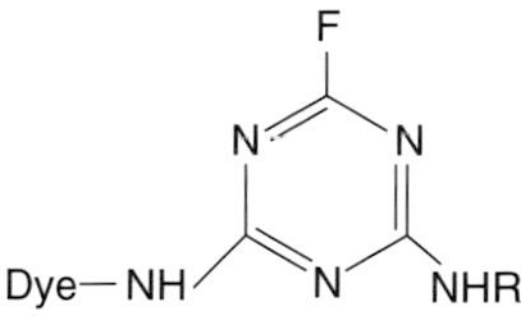

Monofluorotriazinyl (MFT)

3.45

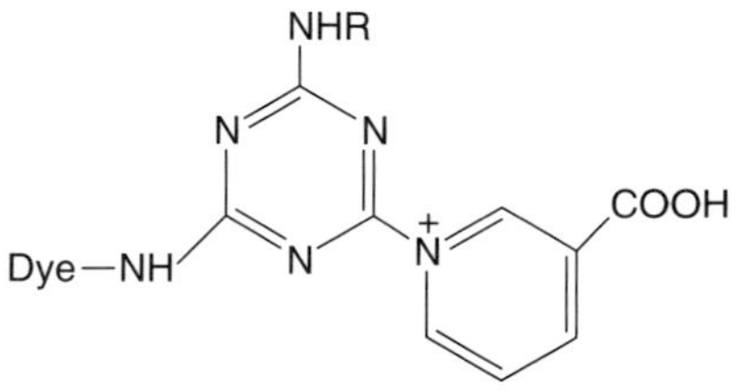

Nicotinyltriazine (NT)

3.46

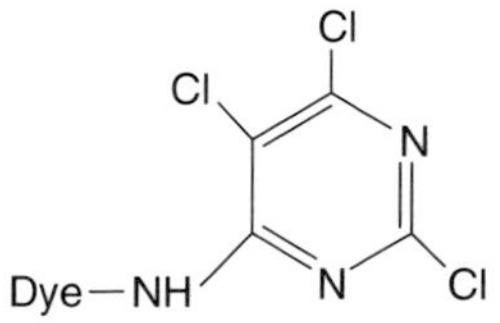

Trichloropyrimidine (TCP)

3.47

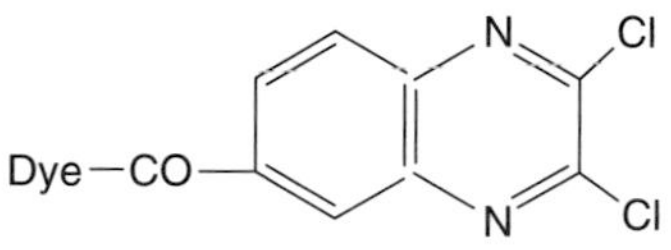

Dichloroquinoxaline (DCQ)

3.48

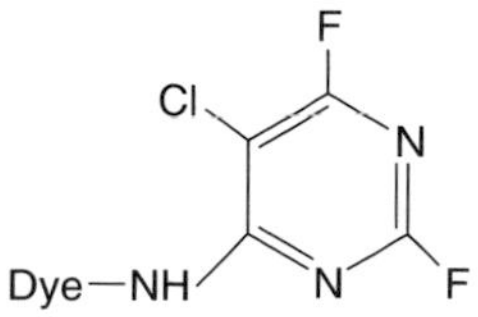

Difluorochloropyrimidine (DFCP)

3.49

Vinyl sulphone (VS)

3.50

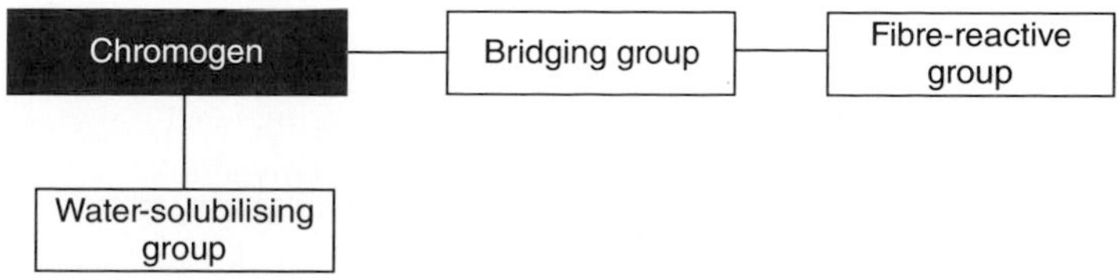

Figure 3.3 General structure of a fibre-reactive dye.

Table 3.4 Characteristics of fibre-reactive groups.

reactive group	reactivity	exhaust dyeing temperature (°C)
Dichlorotriazinyl	High	25–40
Difluorodichloropyrimidine	Moderate to high	30–50
Monofluorotriazinyl	Moderate	40–60
Vinyl sulphone	Moderate	40–60
Dichloroquinoxaline	Low	50–70
Monochlorotriazinyl	Low	80–85
Trichloropyrimidine	Low	80–95
Nicotinyltriazine	Moderate to high (react under neutral conditions)	100–130

Dyes whose molecules contain one reactive group are called *monofunctional*, an example of which is the vinyl sulphone dye C.I. Reactive Blue 19 (**3.51**):

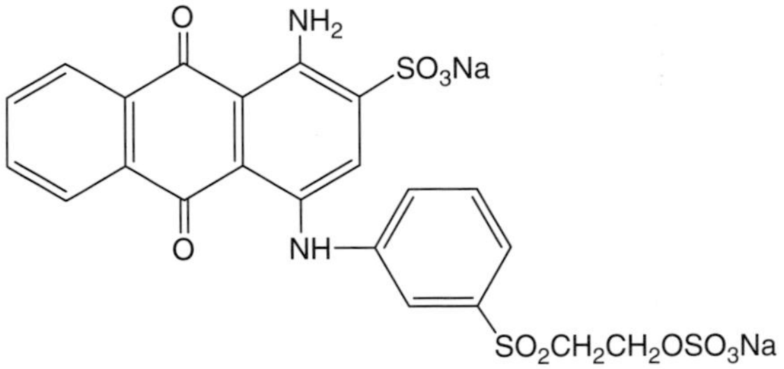

3.51

Dyes with two reactive groups of the same type are called *homo-bifunctional*, and an example of this type is the widely used C.I. Reactive Black 5 (**3.52**):

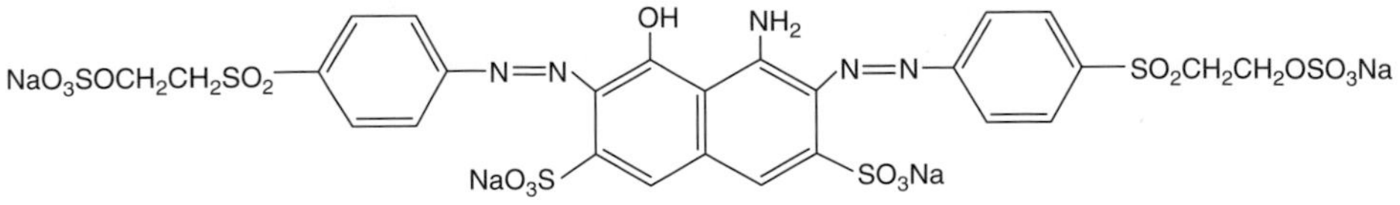

3.52

Dyes whose molecules contain two chemically different reactive groups are called *hetero-bifunctional* reactive dyes, and an example of this type is C.I. Reactive Red 194 (**3.53**), which is a bifunctional chlorotriazine sulphatoethylsulphone dye:

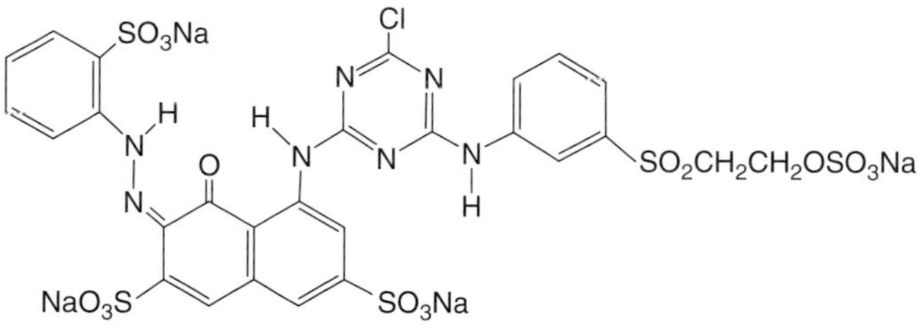

3.53

As with reactive dyes for protein fibres, reactive dyes for cellulosic fibres can react by nucleophilic substitution or by nucleophilic addition. In the former, the leaving group, usually a chlorine or fluorine atom, is only weakly bound to the ring due to the electron-attracting influence of the nitrogen atoms in the ring. The chlorine (or fluorine) atom is easily substituted by the strong nucleophilic Cellulose–O⁻ ion (Scheme 3.10).

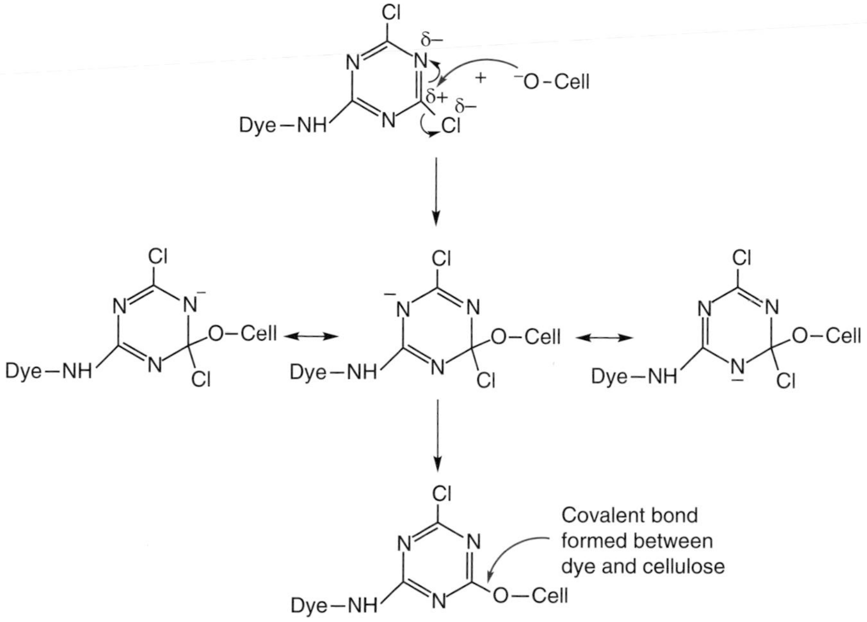

Scheme 3.10 Nucleophilic substitution reaction of chlorotriazinyl reactive dyes with cellulosic fibres.

Unfortunately, the alkaline conditions required to generate the Cellulose–O⁻ ion involve the presence of a high concentration of OH⁻ ions. These are also strong nucleophiles and are able to compete with the Cellulose–O⁻ ions for reaction with the dye. If the OH⁻ ions react, what is termed hydrolysed dye is formed (**3.54**). Once this has occurred, the dye molecule cannot form a covalent bond with the cellulose, and it behaves as a simple direct dye molecule with little substantivity.

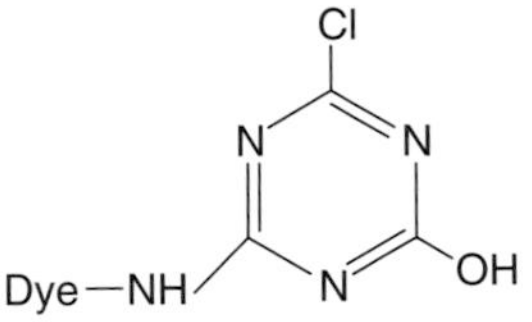

3.54

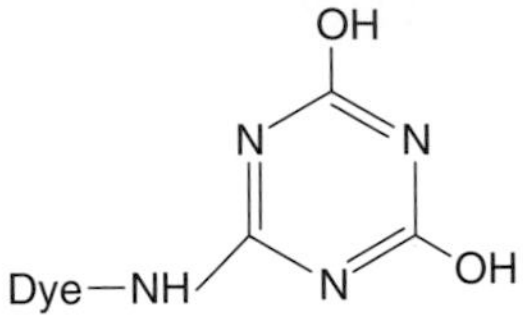

3.55

   Potentially, in the case of dichlorotriazine reactive dyes, reaction with cellulose can also occur by substitution of the second chlorine atom. This is desirable and can serve to cross-link the polymer chains of the cellulose. However, by the same token, hydrolysis can also occur at both sites, giving the structure **3.55**.

   Vinyl sulphone reactive dyes react with cellulose by a nucleophilic addition reaction. The steps involved in this reaction are shown in Scheme 3.11. All being well, the dye forms a covalent bond with the fibre (**3.56**), but again, a competing reaction is that with OH⁻ ions giving hydrolysed dye (**3.57**).

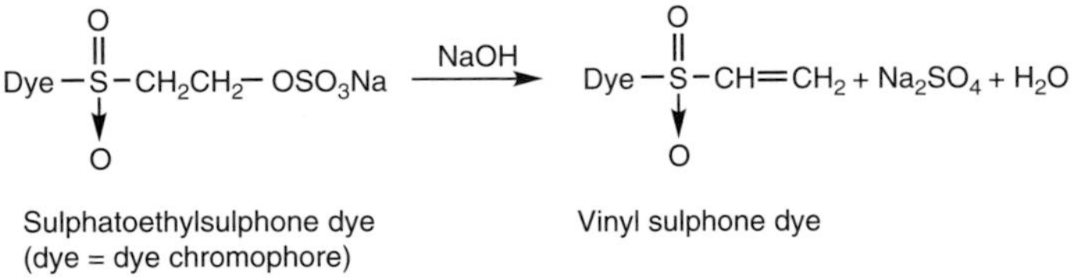

Sulphatoethylsulphone dye
(dye = dye chromophore)

Vinyl sulphone dye

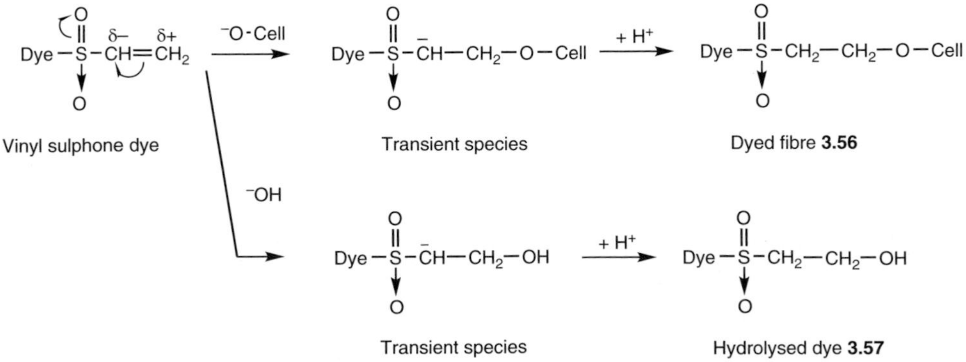

Scheme 3.11   Nucleophilic addition reaction of vinyl sulphone reactive dyes with cellulosic fibres.

3.5.2.5   *Sulphur Dyes*

Sulphur dyes were first made from organic compounds of known constitution in the 1890s. Various blacks could be obtained by melting certain nitro-substituted phenols and ary-lamines with sulphur and sodium sulphite. In 1891 a particularly economical black was discovered by refluxing an aqueous mixture of 2,4-dinitrophenol with sodium polysulphide (Scheme 3.12). The product C.I. Sulphur Black 1 was one of the earliest sulphur dyes – a high-quality dye of excellent fastness and low cost.

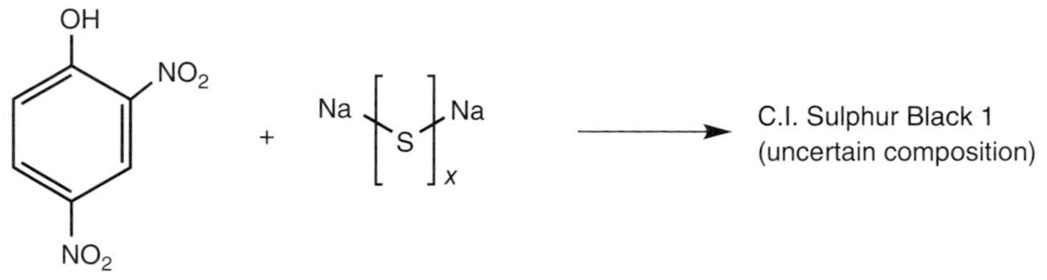

OH
NO$_2$

+

Na$\left[\text{S}\right]_x$Na

⟶

C.I. Sulphur Black 1
(uncertain composition)

NO$_2$

Scheme 3.12   Synthesis of C.I. Sulphur Black 1.

In spite of their long-established use, little is known of the precise constitution of sulphur dyes. They are likely to contain atoms of sulphur not only in the sulphide (–S–), disulphide (–S–S–) and polysulphide (–S$_x$–) forms but also in heterocyclic ring systems. They are normally distinguished in terms of the organic intermediates from which they are formed and the process of sulphurisation used in their manufacture.

The classification of sulphur dyes in the colour index has become somewhat confusing owing to deletions from and additions to ranges and newly developed product types, but the main subdivisions are still clear. There are four subdivisions:

(1) C.I. sulphur dyes: The definition of a sulphur dye (group 1) is a water-insoluble dye, containing sulphur both as an integral part of the chromophore and in attached poly-sulphide chains, normally applied in an alkaline reduced (leuco) form from a sodium sulphide solution and subsequently oxidised to the insoluble form within the fibre.

(2) C.I. leuco sulphur dyes: A leuco sulphur dye (group 2) has the same chemical consti-tutional number as the parent dye (group 1) but is a powder or liquid brand containing the soluble leuco form of the parent dye and reducing agent, in sufficient quantity to make the dye suitable for application either directly or with the addition of only a small amount of extra reducing agent.

(3) C.I. solubilised sulphur dyes: A solubilised sulphur dye (group 3) has a different chemical constitution number because it is the thiosulphuric acid derivative of the parent dye, non-substantive to cellulose but converted to the substantive alkali-solu-ble thiol form during dyeing. These dyes are sometimes referred to as *Bunte salts*.

(4) C.I. condensed sulphur dyes: A condensed sulphur dye (group 4), although contain-ing sulphur, their constitution and method of manufacture bear little resemblance to those of traditional sulphur dyes. They still require sodium sulphide or polysulphide for dyeing, but conventional sulphur dyeing methods are unsuitable.

The range of sulphur dyes available is fairly small and they cover a limited shade range, but they are important for black, navy, mauve, olive, green, bordeaux and reddish-brown colours in medium to heavy depths. They are widely used on cellulosic fibres and their blends, for example, with polyester. Sulphur black and navy blue dyes are also used to dye cotton denim.

Sulphur dyes can be difficult to oxidise back to the parent form, different oxidants producing variations in hue and fastness properties. Since reducing agents are required, the effluent contains large amounts of sulphides. The discharge of sulphides to drain is not permissible because of the ecological damage they can cause. If oxidation is carried out using potassium dichromate, chromium is formed in the dyebath, and there are severe legislative restrictions on chromium in effluent. Much research has been conducted to develop a more eco-friendly process, such as the use of organic reducing agents, such as reducing sugars. One such development is in the application of sulphur dyes to warp yarns for denim [4].

3.5.2.6   Azoic Dyes

The process of azoic dyeing involves creating an insoluble azo dye in situ within the fibre (usually cotton). As explained in Section 3.4.1, the formation of an azo dye involves a chemical reaction between a diazonium salt and a coupling component. The azoic dyeing process involves applying the coupling component to the fabric, drying it and then passing the fabric through an ice-cold solution of a diazonium salt. As a result, an azo dye is synthesised within the fabric.

Both coupling component and diazonium salt are relatively small molecules and can easily penetrate the pores of cellulosic fibres, even in the cold. The particular diazonium salts and coupling components used in the process are selected so that the resulting azo dye is insoluble in water. Therefore, once the dye is formed, it is trapped in the fibre.

An important coupling component used is C.I. Azoic Coupling Component 2 (**3.58**), which is sold under the trade name Naphthol AS. This compound has some substantivity towards cellulosic fibres that contributes to the effectiveness of the coupling reaction and subsequent affinity of the final dye:

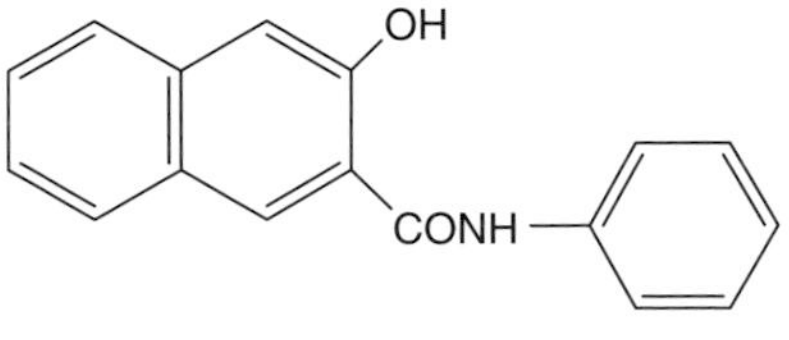

3.58

Other similar coupling components are produced as well. Ranges of suitable diazotisable amines are marketed as well, often sold as the hydrochloride, for example, C.I. Diazo Component 4 (**3.59**):

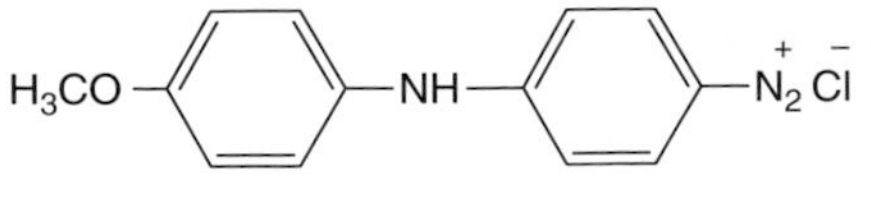

3.59

The dye formed by reaction between **3.58** and **3.59** will have the structure **3.60**. The dye molecule contains no water-solubilising groups so is insoluble, giving high wet fastness:

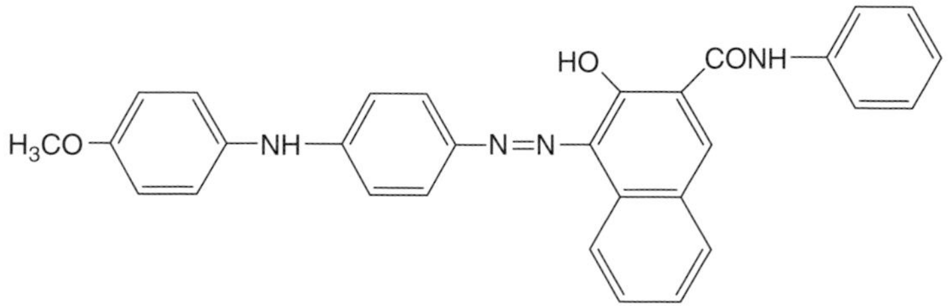

3.60

In subsequent soaping the dye particles aggregate into larger particles, giving dyed materials with very good fastness to wet treatments, light and bleach. However the shade range is limited to reds, oranges and yellows, with very few examples of blues or greens. Also, the process is complex, and it is possible to obtain dyed materials of equal quality using reactive dyes. Consequently azoic dyeing has declined considerably in recent years.

3.5.3   Dyes for Synthetic Fibres

The most commonly used synthetic fibres are polyester, nylon and acrylic, of which polyester is by far the most important (see Table 2.2). All can be dyed with disperse dyes. Nylon can be dyed with acid and reactive dyes as well. The same theoretical principles apply to the dyeing of nylon with acid and reactive dyes as for protein fibres, though there are slight differences in the application processes. Acrylic fibres are usually dyed with basic dyes.

3.5.3.1   Disperse Dyes

When cellulose secondary acetate and cellulose triacetate fibres (Section 2.7.4) were first produced in the late 1920s, they presented a serious problem to the dyer. Unlike all other known fibres of the time, they were hydrophobic and could not be dyed by water-soluble dyes. This threatened to limit the uses of the fibres, but the problem was eventually overcome by developing a new class of dye. The dyes concerned were non-ionic though they did possess a very small but important solubility in water. By incorporating a surface-active agent, it was possible to prepare the dye as a dispersion in water and produce a uniform distribution of dye in the dyebath, allowing uniform dyeing to be obtained.

Because the dyes are applied from a dispersion, they are called *disperse dyes*. The particles of disperse dyes are usually in the range 0.5–1.0 μm. They are now the main class of dyes for polyester.

The main difference between water-soluble dyes discussed previously and disperse dyes is that the latter do not contain the chemical groups (such as –SO$_3$Na or –COONa) commonly incorporated to confer water solubility. They do, however, contain polar groups such as –NH$_2$, –OH, –CN and –NO$_2$. Consequently disperse dyes are non-ionic and as such are only sparingly soluble in water, even at high temperatures. These hydrophobic dyes are capable of 'dissolving' in the hydrophobic fibres. It is as though the fibre acts like an organic solvent, extracting the dye from the water. This is analogous to the simple solvent extraction of organic compounds from water by shaking up the aqueous phase with an appropriate water-immiscible solvent.

Azo disperse dyes, of which **3.8** and **3.9** are two examples, are the most common type. Some azo types are synthesised using heterocyclic diazo components, which provide bright, intense colours, of which C.I. Disperse Blue 339 (**3.61**) is an example:

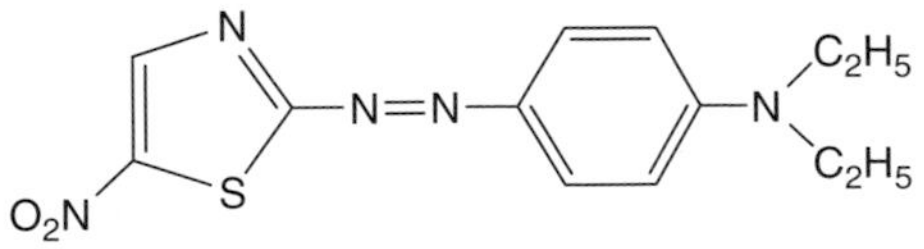

3.61

In addition, disperse dyes based upon anthraquinone, such as **3.16**, are common, and other types such as methines, exemplified by **3.18**, are also used.

The inclusion of surface-active agents (see Section 1.8.5.2) in the dyebath is a crucial factor in the application of disperse dyes. The surface-active agents used with these dyes are usually anionic in nature. Once such a compound is added to water, its dual character results in the formation of micelles above a critical, but low, concentration, called the *critical micelle concentration*. The hydrophobic 'tails' of the surface-active agent molecule are oriented towards the centre of the micelle that, as a consequence, is able to solubilise the disperse dye molecules within it. The dye then has a higher apparent solubility. The micelles, which carry negative charges on their surfaces, repel each other and consequently do not coalesce, therefore also preventing the dye molecules from aggregating. The hydrophobic chains of the surface-active agent are also adsorbed on the surface of the solid dye particles, thus further stabilising the suspension. When used in this way, the surface-active agents are referred to as *dispersing agents*.

3.5.3.2   *Basic Dyes*

Whereas most water-soluble dyes are anionic in nature (negatively charged), basic dyes are cationic (positively charged) and are held on to the fibre by the formation of salt links with anionic groups in the fibre. For this reason, basic dyes are often also referred to as *cationic dyes*. Basic dyes are mainly applied to acrylic fibres where they provide good lightfastness They are only occasionally employed for dyeing silk and wool, because they have poor lightfastness on these substrates. Most basic dyes are azo or methine types, such as C.I. Basic Red 18 (**3.62**) and C.I. Basic Violet 7 (**3.63**), respectively:

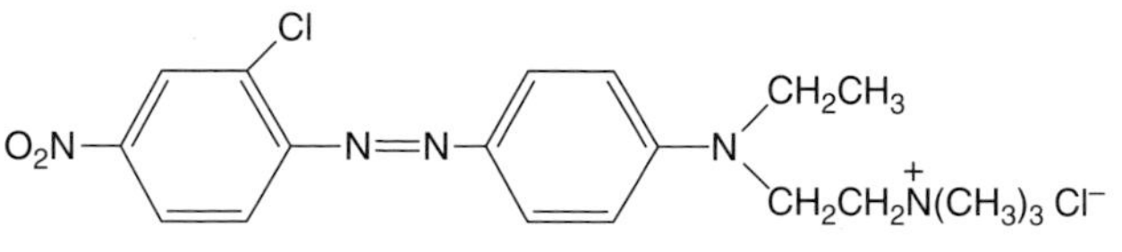

3.62

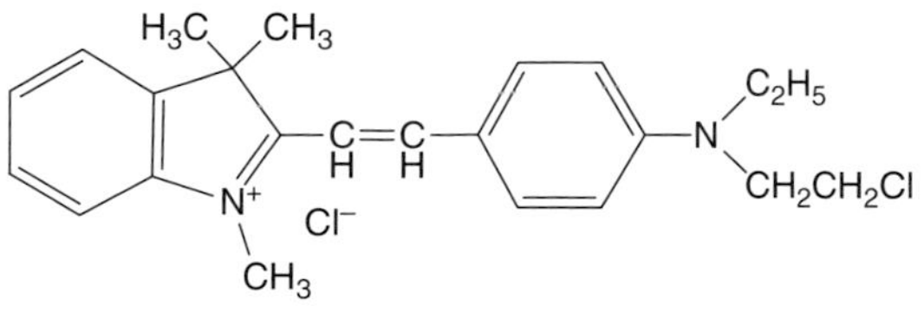

3.63

There are also a number of triarylmethane dyes such as C.I. Basic Violet 3 (**3.64**) and C.I. Basic Blue 1 (**3.65**):

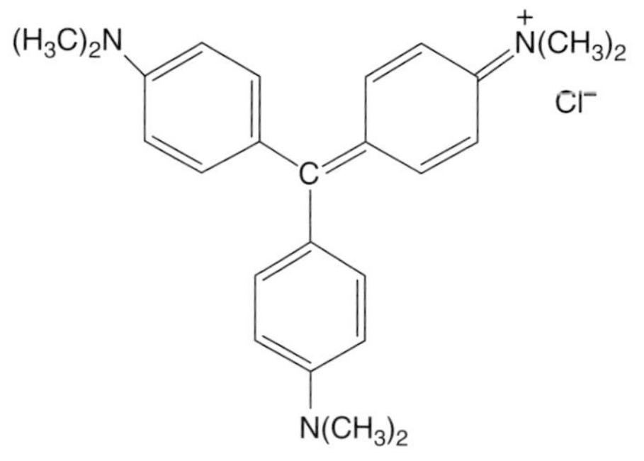

3.64

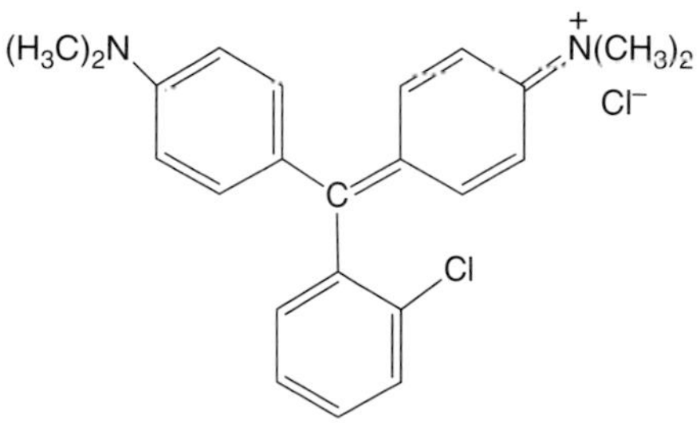

3.65

3.5.4   Pigments

The principal requirement for a substance to be classified as a pigment is that it must be insoluble in water or the solvents and the medium to which it is applied.

Some dye types, notably vat dyes that are water insoluble, can be used as pigments though they have to be produced with the required pigmentary physical properties. Vat dyes are considered to be dyes because prior to application to fibres, they are chemically reduced to a water-soluble form; then once dyebath exhaustion is complete, they are oxidised back to the water-insoluble 'parent' form. Examples of vat dyes applied in this way are Indanthrone (**3.37**) and Flavanthrone (**3.38**). In a similar manner, azoic dyes are insoluble once they are formed within the fibre, though they are considered to be dyes because the coupling components and diazo components are soluble in water at the point at which they are applied.

Pigments can be organic or inorganic but those most used in textile printing are organic. The vast majority are azo types and of these; they can be either metal-containing or metal-free. The metal-free azo types provide mainly yellow–orange–red hues, such as C.I. Pigment Yellow 1 (**3.66**) and C.I. Pigment Red 2 (**3.67**):

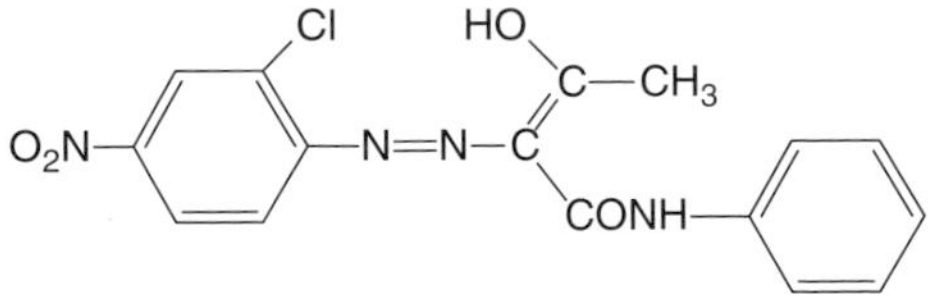

3.66 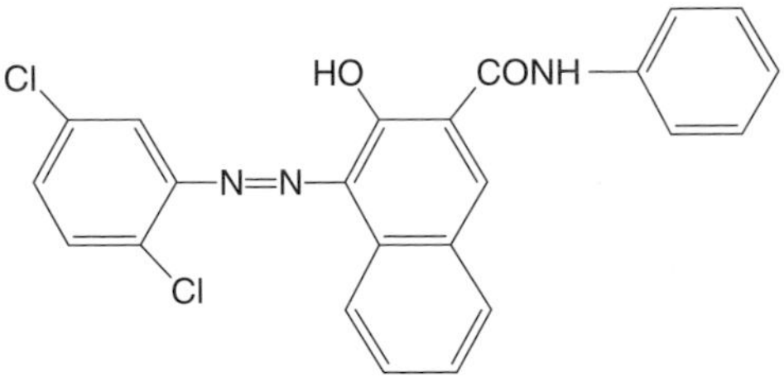   **3.67**

In metal-containing azo pigments, the sodium ion of the sodium salt form is replaced by a heavier metal, such as calcium, barium or manganese. An example of such a pigment is C.I. Pigment Red 48 (**3.68**), which is sold as the barium, calcium or manganese salts:

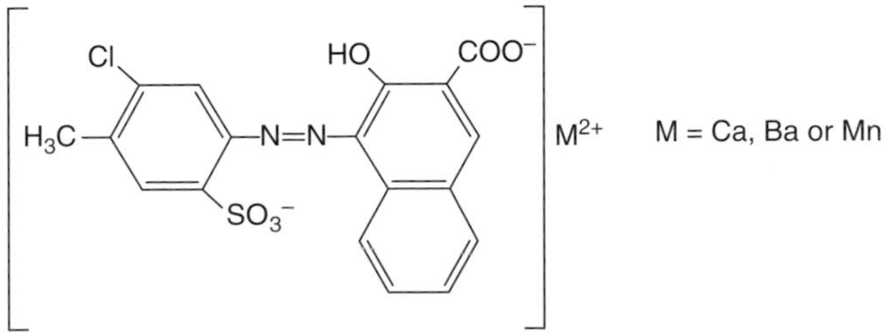

3.68

By far the most important metal-containing pigment in use is copper phthalocyanine (C.I. Pigment Blue 15, **3.69**). It has a brilliant blue colour together with excellent fastness to light, heat, solvents, acids and alkalis. A number of chlorinated copper phthalocyanines are also made, these having greener shades of blue, the greenness depending on the degree of chlorination. Yellower shades of green are obtained with brominated or a mixture of brominated and chlorinated copper phthalocyanines.

A particularly important range of red pigments are the diketopyrrolopyrrole (DPP) type, such as C.I. Pigment Red 254 (**3.70**), noted for their excellent technical performance in terms of fastness to light, solvent resistance and thermal stability:

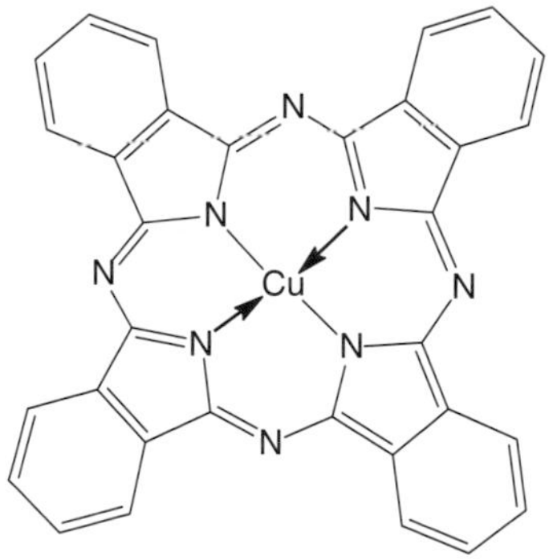

3.69 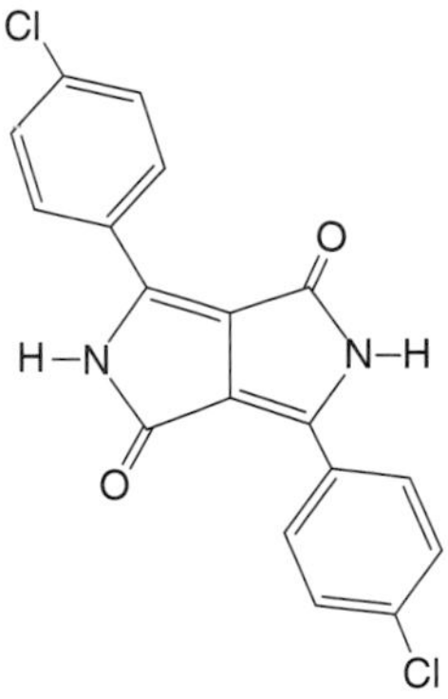   3.70

3.6 Commercial Naming of Dyes and Pigments

The brand name of a particular class of dye distinguishes between those originating from different manufacturers. Brand names also contain designatory letters because not all dyes of the same basic colour are exactly the same shade. Colours with a yellowish tint, such as yellowish-red, bear the letter G (German *gelb* = yellow). Similarly, those with a reddish tinge will be identified by R (*rot* = red) and those with a blue tinge by B (*blau* = blue). The designatory letters are sometimes preceded by a number that indicates the intensity of the deviation from the main colour. So, for example, a dye with the designatory letters Red 6B can be expected to be a considerably bluer shade of red than one labelled Red 2B.

3.7 Strength and Physical Form of Colorants

The dye powders sold by manufacturers may contain only about 30% colouring matter. When dyes are synthesised, their batch strength is much higher than the strength at which they are sold. During the final stage of manufacture, the batch of dye is adjusted to a standard strength by mixing the dye powder with a diluent, usually a neutral electrolyte. In some cases, such as with vat and disperse dyes, the dilution to standard strength is achieved using dispersing agents as well. A proportion of a second or even a third colour component may be added to ensure the shade exactly matches other batches of the same dye.

Dyes are marketed in various physical forms – fine powders, grains, pastes or liquids. The different forms are prepared not only for the convenience of handling but also for safety. During the handling of fine powders, during weighing, for example, a dust of the powder can form in the atmosphere, thereby creating a hazardous working environment. For this reason, de-dusting agents are often added to fine dye powders, but grains are often a preferred choice of physical form, even though they are bulkier to transport. An important consideration about the physical form is the ease with which the dye will dissolve in water or, in the case of disperse and vat dyes, disperse in water to form a consistent dispersion. Dyes are also sold in liquid form, but the challenge for the manufacturers is that of stability over long periods of storage. Water-soluble dyes may aggregate in solution (see Section 6.3) and the aggregates precipitate. Similarly when disperse dyes are sold in liquid form as concentrated dispersions, the dispersion may break down during prolonged storage.

References

[1] *Colour Index™ Online*, Society of Dyers and Colourists, Bradford, UK, www.colour-index.com (accessed 17 April 2017).
[2] J Shore, Ed., *Colorants and Auxiliaries, Vol 1 Colorants*, SDC, Bradford, 1990.
[3] R M Christie, *Colour Chemistry*, 2nd Edn., Royal Society of Chemistry, Cambridge, 2015.
[4] P Cowell, *Int. Dyer*, **200** No. 3 (2015) 36.

4

Industrial Coloration Methods

4.1 Introduction

In this chapter the processes by which dyes of the various application classes are applied to fibres are described. The second half of the chapter details the types of machinery used to dye materials in their different forms, such as loose stock, yarn and fabric.

4.2 Dye Application Processes

4.2.1 Wool Dyeing

4.2.1.1 Acid Dyes

Wool may be dyed with acid (anionic) dyes from an acidic dyebath because it is stable under acidic conditions (see Section 2.6.1.3). As noted in Chapter 3 (Section 3.5.1.1), acid dyes are classified as equalising acid, milling acid or supermilling acid.

Equalising Acid Dyes Equalising acid dyes are so called because of their ease of fibre penetration and migration ability, which facilitates the production of level-dyed material. Additionally they are characterised by their brightness of shade and fastness to light, though their fastness to wet treatments is poor. The application of equalising acid dyes requires a dyebath of pH 2.5–3.5, created by the presence of sulphuric acid or methanoic (formic) acid. Under these conditions of pH, hydrogen ions are adsorbed on the negatively charged $-COO^-$ groups of the salt links in the wool fibres (see structure **2.10** in Section 2.6.1.1), as shown in Scheme 4.1.

Electrical neutrality in the fibre is maintained by the adsorption of negative sulphate (SO_4^{2-}) ions. During the dyeing process, dye anions (e.g. represented by D^- for a monosulphonated anion) are adsorbed at the positively charged $-NH_3^+$ sites (Scheme 4.2).

The dye anions are adsorbed very rapidly initially and to decrease the rate of uptake at the $-NH_3^+$ sites a large amount of Glauber's salt (sodium sulphate, $Na_2SO_4.10H_2O$), around 5–10% on weight of fibre (owf), is added to the dyebath. The sulphate ions (SO_4^{2-}) are also attracted by electrical forces to the $-NH_3^+$ sites, but their affinity for the wool is considerably lower than that of dye anions. Nevertheless, they are present in such large numbers that they succeed in slowing down the uptake of the dye anions and thereby exert a levelling effect.

An Introduction to Textile Coloration: Principles and Practice, First Edition. Roger H. Wardman.
© 2018 John Wiley & Sons Ltd. Published 2018 by John Wiley & Sons Ltd.

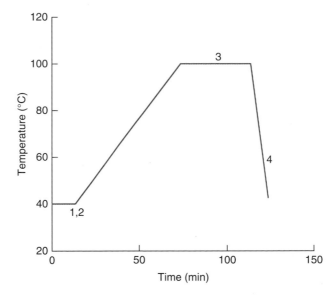

Scheme 4.1 Adsorption of hydrogen ions by protein fibres.

$$-NH_3{}^+ + D^- \leftrightarrow -NH_3\,{}^+D^-$$

Scheme 4.2 Adsorption of dye anions by protein fibres under acid conditions.

Figure 4.1 Dyeing profile for the application of acid dyes to wool. 1: Add $Na_2SO_4.10H_2O$, 5–10% owf. 2: Add dyes. 3: Boil. 4: Rinse.

Glauber's salt is used rather than sodium chloride because the chloride ion (Cl^-) has such low affinity for wool that about four times more would be required to produce the same levelling effect than Glauber's salt. The levelling effect of Glauber's salt is different for dyes of different degrees of sulphonation. Its levelling effect is greatest with dye anions of lowest affinity, that is, with those containing most sulphonate groups.

Levelness of equalising acid dyes is further improved by continued dyeing at the boil, even after the dyebath appears to have become exhausted. Therefore, even dyed material that may be unlevel initially can be corrected simply by further boiling due to the good migration properties of these dyes.

A typical dyeing profile for equalising acid dyes on wool is shown in Figure 4.1.

Milling Acid Dyes　Milling acid dyes have higher molecular weights and greater affinity for wool than equalising acid dyes. The pH of the dyebath is very important for the satisfactory application of these dyes because their higher affinity means that they are unable to migrate as easily during dyeing to improve levelness as equalising acid dyes. They are applied from an acidic dyebath but at a pH higher (around 4.5–7) than that used for equalising dyes. For this reason ethanoic (acetic) acid, rather than sulphuric acid, is used to obtain the required acidic conditions. Glauber's salt is not used in the application of milling acid dyes: as the affinity of the dye anion increases, Glauber's salt has little levelling effect and indeed can increase the rate of exhaustion and reduce levelling.

Supermilling Acid Dyes   Supermilling acid dyes have similar properties to milling dyes but are even more hydrophobic and possess higher affinities. Attraction to the fibre takes place mainly through non-polar forces, rather than by attraction at charged sites, such as $-NH_3^+$ sites. These dyes aggregate readily in water, and the presence of an electrolyte such as Glauber's salt would only increase aggregation. They do not migrate easily so any initial adsorption on to the fibre that is unlevel is difficult to remedy. For this reason these dyes are most often applied to loose stock or to yarns, where subsequent processing (carding, spinning) will hide any unlevelness.

Dyeing is usually commenced at pH 6.5–7.0 and the rate of dye uptake controlled by temperature. Dye uptake usually takes place slowly initially, up to a temperature of around 60 °C, but at higher temperatures, as the dye aggregates break down, it increases rapidly. Ammonium sulphate is added to the dyebath, which lowers the dyebath pH gradually as dyeing progresses.

An important feature of milling and supermilling acid dyes is their general lack of compatibility. For this reason and their relative expense, they are most often used in bright self-shades or binary combinations in recipes where one of the two dyes predominates.

4.2.1.2   Chrome Dyes

The principle of mordanting is to form a bridging link between the fibre and the dye. The mordants used are metal ions, the most common metal being chromium, though other metals such as aluminium, tin and copper have been used. The dyes used are usually the more substantive equalising acid dyes, and because they are usually applied using chromium, they are called *chrome dyes*. There are three methods by which the chrome can be attached to the acid dye: the chrome mordant method, the after-chrome method and the metachrome method.

In the chrome mordant method, the wool is treated first with a chromium compound and the acid dye applied afterwards. The process involves treating the wool with a hot solution of sodium dichromate ($Na_2Cr_2O_7$) and methanoic (formic) acid to give strongly acidic conditions. Chromate ions ($Cr_2O_7^{2-}$) are adsorbed at positively charged amino groups in the wool (Scheme 4.3).

The dichromate, in which the chromium is present as the anionic Cr(VI) state, is then reduced to the cationic chromic ion Cr(III), the overall reaction being represented by the equation (Scheme 4.4).

Strongly acidic conditions are required because the reduction reaction (Scheme 4.4) consumes acid, and if the pH rises as a result, the adsorption of dichromate decreases. The six electrons in Scheme 4.4 come from the cystine and amino acid residues in the wool,

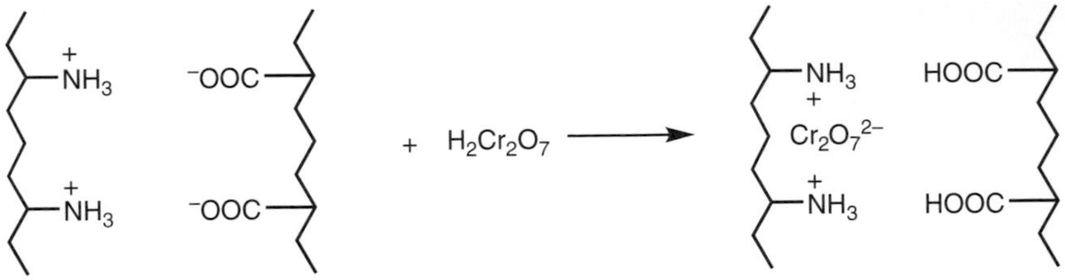

Scheme 4.3   Adsorption of dichromate ions by wool under acid conditions.

$$Cr_2O_7^{2-} + 14H^+ + 6e^- \rightarrow 2Cr^{3+} + 7H_2O$$

Scheme 4.4   Reduction of adsorbed dichromate ions by wool under acid conditions.

which are simultaneously oxidised. The reduction reaction to form the chromic ion (Cr^{3+}) is important because it is this cation that binds to wool and forms the dye–metal complex. It is not used directly because it has low substantivity for wool, whereas the dichromate has higher substantivity and also better migrating properties. Sometimes a reducing agent such as sodium thiosulphate is added to the mordanting bath in addition to the dichromate and acid to ensure complete reduction to the chromic ion.

Chroming is commenced at 60 °C and the bath raised to the boil at 1 °C/min and maintained at the boil for an hour. After thorough rinsing a fresh bath is set, containing the dye and methanoic (formic) acid. The temperature of this bath is raised to the boil and dyeing continued for another hour. The wool is then thoroughly rinsed to ensure the complete removal of residual dye and chrome. For machine-washable performance, after-treatment with ammonia may be used.

In the after-chrome method, the wool is first dyed with a suitable acid dye in the presence of ethanoic (acetic) acid (3% owf) at the boil to ensure the uptake is as level as possible. This is important since once the dichromate is added, the dye–metal complex formed will not migrate. Towards the end of this part of the process, methanoic (formic) acid, which is a stronger acid than ethanoic acid, can be added to complete the dye exhaustion. The dyebath is then cooled slightly to around 75 °C, sodium dichromate added and the bath temperature raised to the boil again. A reducing agent (again sodium thiosulphate) can be added also to ensure complete reduction of the dichromate to the chromic (Cr^{3+}) cation. This also helps to avoid damage to the wool due to the prolonged boiling.

During after-chroming, a marked change in the hue of a mordant dye takes place, with the final shades produced usually being fairly dull. However the lightfastness and wash fastness of the final dyed wool are excellent. For example, the samples shown in Figure 4.2 show the change in colour of the dye C.I. Mordant Blue 9 (**4.1**) as the chromic ion reacts with it after 5, 10, 15 and 20 minutes of the after-chroming part of the process. This change in colour can lead to difficulties in producing exactly the required colour:

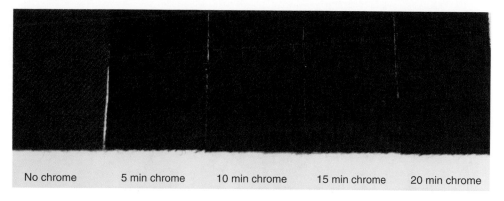

Figure 4.2 Change in shade of C.I. Mordant Blue 9 during after-chroming.

No chrome   5 min chrome   10 min chrome   15 min chrome   20 min chrome

4.1

One problem that occurs in this process is the preferential adsorption of the dichromate in the tips of weathered wool over that in the roots. Further, the reduction to the chromic ion occurs more efficiently in the tips, leading to what can be quite obvious unevenness of shade.

Although the dye–chromium complex formed is insoluble in water, it is possible to apply the dye and the chromate simultaneously in the process known as the metachrome method. However, certain conditions must be fulfilled, in that the dye must exhaust at pH 6–8.5, it is not salted out by the presence of the dichromate and it is not reduced by the dichromate to an insoluble complex during the process. Dyeing is carried out in the presence of sodium dichromate, ammonium sulphate and sodium sulphate at pH 6–7. Few dyes meet these needs, and at pH 6–7 the exhaustion of the dichromate is not particularly good, so there is a danger of residual dichromate being discharged to effluent. Consequently the metachrome process is now little used. For the reasons detailed in Section 3.5.1.2, use of the after-chrome method is also in significant decline.

4.2.1.3   Pre-metallised Dyes

It was explained in Section 3.5.1.3 that pre-metallised dyes can contain either one or two metal (chromium or cobalt) atoms in the molecule, giving what are called 1 : 1 or 2 : 1 pre-metallised dyes, respectively. The 1 : 1 types are soluble in water and are similar to levelling acid dyes in that they are applied from a strongly acidic dyebath (pH ~ 2). The 2 : 1 types are rather less soluble in water and are applied at higher pH values.

1 : 1 Pre-metallised Dyes Typical of the 1 : 1 pre-metallised dyes are the Neolan dyes, now produced by Huntsman. The wool is treated first in a blank bath, containing 6% owf sulphuric acid and 5% owf Glauber's salt at 50 °C. The dye is then added and the temperature raised to the boil and dyeing carried out for 1½–2 hours. Dyeing at the boil under these acidic conditions for such a prolonged period may cause damage to the wool fibre due to hydrolysis of the peptide bonds (see Scheme 2.5 in Section 2.6.1.3). To avoid this damage it is possible to apply these dyes with lower amounts of sulphuric acid by incorporating a levelling agent, such as a polyethylene oxide, in the dyebath. After dyeing and rinsing the fibre, it is necessary to neutralise the residual acid or buffer it using sodium ethanoate (acetate).

A modified range of 1 : 1 pre-metallised dyes, the Neolan P dyes, again produced by Huntsman, can be applied at pH 3.5, a little higher than for the Neolan dyes. In the molecules of these dyes, the chromic ion is coordinated with colourless hexafluorosilicate ligands (SiF_6^{2-}). They have higher affinity, so much less sulphuric acid is required, and indeed methanoic (formic) acid can be used instead. Dyeing is carried out in the presence of an amphoteric levelling agent. Since less acid is used, the neutralisation step after dyeing is not required.

The wet fastness of the 1 : 1 pre-metallised dyes is lower than that of after-chrome dyes but comparable with fast acid dyes. They are favoured for dyeing carbonised fabrics, as their use avoids the need for neutralisation after carbonising. Their lightfastness is good. They are still important for the coloration of loose wool and wool yarns for carpets.

2 : 1 Pre-metallised Dyes The 2 : 1 pre-metallised dyes, such as the Isolan dyes (DyStar), can be either weakly polar or strongly polar (see Section 3.5.1.3). The weakly polar types may be applied from a nearly neutral dyebath at a pH of 6–7 brought about using ammonium ethanoate (acetate). Ammonium ethanoate is used because, as the salt of a weak acid and weak base, it yields a solution with a pH of around neutral (see salt hydrolysis; Section 1.6.3). The more polar types are applied from dyebaths of pH in the range 4–6 using ethanoic (acetic) acid. The wool is usually treated in the dyebath at 60 °C with ammonium ethanoate and ethanoic acid for 10 minutes, and then the dye is added. The temperature of the dyebath is then raised to 98 °C and dyeing continued for about an hour.

The 2 : 1 complexes generally produce dyed fibres of higher wet fastness than the 1 : 1 complexes and without incurring fibre damage. However it can be difficult to produce adequate levelness with the strongly polar 2 : 1 dyes, so these are rarely used for dyeing fabrics. Instead they are used for loose stock, slubbings and yarns. They are often applied with dyebath additives to prevent the differential coloration of wool fibres from root to tip, a common phenomenon due to the slight changes in chemical composition caused by progressive weathering of wool fibre tips during growth. It is referred to as 'tippy dyeing' and the polar 2 : 1 complexes often exaggerate the effect when applied without levelling agents in the dyebath. The weakly polar types have better migration properties and do not yield skittery dyeings (an undesirable speckled effect arising from differences in colour between adjacent fibres or portions of the same fibre). Consequently they are used on piece goods as well as loose stock and yarn, but do not possess quite such good wet fastness properties.

There are many optimised dye ranges of metal complex and milling dyes available, such as the Lanaset range from Huntsman. These ranges consist of products that are single dyes or specific mixtures of dyes that have similar dyeing properties and are therefore very compatible and have a consistent application method.

4.2.1.4   *Reactive Dyes*

The number of ranges of reactive dyes for wool that have been commercially successful is comparatively low, in comparison with the number available for cellulosic fibres. Typical of the ranges marketed successfully are the Drimalan (Clariant), Lanasol (Huntsman) and Realan (DyStar), and the number of ranges is now increasing. The use of reactive dyes for wool is assuming a greater importance because of environmental concerns about the use of mordant and also pre-metallised dyes.

Reactive dyes are necessarily applied to cellulosic fibres under alkaline conditions, but alkaline conditions are usually inappropriate for wool because of the sensitivity of the disulphide cross-links (see Section 2.6.1.3). Fortunately, the chemical nature of the functional groups in wool (such as thiol –SH, amino –NH$_2$ and hydroxy –OH) enables reactive dyes to react with the fibres under slightly acidic conditions. Another difficulty is ensuring that the rate of adsorption of the dye into the fibre is greater than the rate of reaction with it. If this is not the case, then dye will react preferentially before migration has occurred, resulting in unlevel skittery dyeings.

Reactive dyes for wool typically have structures shown in Section 3.5.1.4 and react according to the schemes given in that section. However, the tendency to give skittery dyeings still exists with these reactive systems, so it is necessary to apply the dyes from dyebaths containing levelling agents. The most used type of levelling agent is amphoteric in nature (see Section 1.8.5.2). It is thought that these agents form a complex with the dye that, because of its large size, is more hydrophobic than the dye itself. The undamaged roots of wool fibres have lower hydrophilic character than the damaged tips, but the complex is not as sensitive to these differences as the free dye ions, so better levelness is achieved. Another feature of amphoteric agents is that they increase the electropositive nature of the fibre, thereby increasing the rate of adsorption of dye molecules. For the reason explained in the previous paragraph, migration and levelness is promoted.

The method of application of reactive dyes is very similar to that of acid dyes. Full shades are applied at pH 5.0–5.5, whilst pale–medium shades require a slightly less acidic bath, about pH 5.5–6.0, achieved in both cases using ammonium ethanoate (acetate) and ethanoic (acetic) acid. Dyeing is commenced at 40 °C and the temperature raised to the boil and dyeing at the boil continued for 60–90 minutes. During this time covalent bonding between the dye and the fibre will occur. Sometimes, during the heating up phase, the bath is held at 70 °C for 20 minutes to enable dye migration before fibre reaction takes place. The dyeing cycle is represented schematically in Figure 4.3.

After dyeing it is necessary to remove any unfixed dye, especially when heavy shades have been dyed. This is achieved by washing at 80 °C in an alkaline solution using dilute ammonia at pH 8, but sodium carbonate (or sodium carbonate plus sodium bicarbonate) provides a more economical process and avoids the discharge of ammoniacal nitrogen. Final washing in water-containing dilute ethanoic (acetic) acid ensures no alkali remains in the wool.

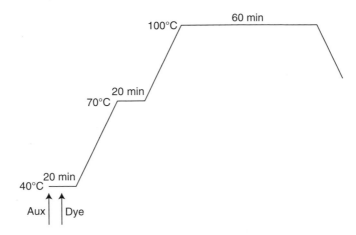

Figure 4.3 Dyeing cycle for application of reactive dyes to wool.

The alkaline after-treatment in reactive dyeing does not just remove unfixed dye. It also removes dyed soluble protein generated from the wool during the dyeing process. If this is not removed, it will cause staining of adjacent fabrics in wet fastness tests. The removal of dyed soluble protein is one of the critical factors in achieving good wet fastness.

Reactive dyes are primarily applied to shrink-resistant wool, wool that has been treated by the chlorine–Hercosett process (see Section 2.6.1.2). This resin has cationic character and is able to interact strongly with anionic dyes. There is therefore the potential for rapid unlevel uptake of reactive dyes, with reduced migration. To avoid this, the initial rate of strike is reduced by having a starting pH that is nearer to neutral than for untreated wool. An acid donor, a compound that gradually liberates acid as the dyebath temperature is raised, is included in the dyebath, so as dyeing proceeds the attraction of the dye for the fibre gradually increases.

Reactive dyes are becoming increasingly popular for wool fibres because they give bright shades over a wide gamut of hues and give very good wet fastness due to the fact that they form a covalent bond with the fibre. However, they are expensive and not all dye reacts with the fibre, some being lost to the competing hydrolysis reaction. Also their lightfastness in pale depths is not quite as good as can be achieved using pre-metallised dyes or chrome dyes.

4.2.1.5 Summary

The choice of the application class of dye to apply to wool is governed by technical consid-erations, such as the form in which the fibre is to be dyed: loose fibre ('loose stock'), top (combed wool in the form of an untwisted strand of parallel fibres), yarn or fabric. With each type of substrate, particular criteria have to be met in order to produce the required quality of colour, but also consideration has to be given to fibre/fabric wet treatments that are applied after dyeing, though they are not the responsibility of the dyer. These post-dyeing processes are usually more severe than those of normal aftercare used by the consumer.

In summary, the types of dyes used for the dyeing of wool, together with their charac-teristics, are shown in Table 4.1.

Table 4.1   Summary of dye types used on wool.

dye type	shades available	wash fastness	migrating power
Equalising acid	Bright, wide range	Low	Good
Milling acid	Bright, wide range	Good	Poor
Supermilling acid	Bright, wide range	High	Poor
Mordant	Dull, poor range	Excellent	Good
1 : 1 pre-metallised	Dull, medium range	Moderate	Good
2 : 1 pre-metallised	Dull, medium range	High	Poor
Reactive	Very bright, wide range	Excellent	Poor

4.2.2   Cellulosic Fibre Dyeing

4.2.2.1   *Introduction*

The widespread use of natural cellulosic fibres such as cotton and linen, and also of regenerated cellulosic fibres, means that there is a demand for dyes covering a wide range of shades and different standards of fastness properties. Not all end uses of goods require the same levels of fastness, for example:

- High lightfastness is required for high-quality curtaining, military and naval uniforms.
- High fastness to washing for goods that will be laundered often, such as towelling and underwear, where resistance of the dyes used to the chemicals present in detergents is also necessary.
- Fastness to bleaching is more important for articles that will be laundered under bleaching conditions to meet hygienic requirements.

There are also types of articles that do not require high fastness to light, washing or rubbing, and for these, a wide choice of dye application class is available. However it is important to remember that the dyes used must be suitable for the intended end use of the article during its working life, and therefore, fastness is an important consideration in the choice of dyes used.

This diversity of requirements has led to the development of several different coloration principles for cellulosic fibres:

(1)   Direct dyes – Dyes that are applied from a single bath and are attracted to the fibres by secondary forces of attraction

(2)   Reactive dyes – Dyes that are applied from a single bath and form a covalent bond with the fibres

(3)   Vat and sulphur dyes – Dyes that are water insoluble but are converted to a soluble form in the dyebath before being applied to the cellulosic fibres, after which the reverse process is carried out when the insoluble form is produced within the fibre.

4.2.2.2 *Direct Dyes*

Direct dyes (see Section 3.5.2.1) are anionic (the coloured dye ion is negatively charged), but cellulosic fibres acquire a negative surface charge when immersed in water. Consequently electrostatic repulsion occurs between the dye anions and the fibre (see Section 6.7.2). To overcome this repulsion, direct dyes are applied from aqueous solution containing an electrolyte, usually sodium chloride (NaCl), but sometimes Glauber's salt (sodium sulphate, Na_2SO_4). In practice the rate of dyebath exhaustion is controlled by both the addition of electrolyte and the regulation of dyebath temperature. Individual dyes vary in their response to electrolyte: the more sulphonate groups there are in the dye molecule, the greater is the effect of electrolyte and the greater the care needed to obtain level dyeings.

Direct dyes are classified into three classes according to the effect of changes in electrolyte concentration and dyebath temperature on their dyeing properties. The three groups of the SDC classification are the following:

(1)   Class A contains those dyes that can be applied easily, with electrolyte present from the start of dyeing. They are referred to as *self-levelling* dyes and are often used for the shading of faster dyes in hot dyebaths.
(2)   Class B contains those dyes for which the rate of addition of electrolyte throughout the dyeing process must be regulated in order to control dyeing. They are called *salt-controllable* dyes.
(3)   Class C contains those dyes for which regulation of both the rate of increase of dyebath temperature and the addition of electrolyte are essential for adequate control. Dyes in this class all possess very high substantivity, even in the presence of only small amounts of salt. Because of their marked additional dependence on the dyebath temperature, they are referred to as *temperature-controllable* dyes.

This scheme of classification is appropriate for the dyeing of both natural and regenerated cellulosic fibres, but since direct dyes are more substantive to regenerated cellulosic fibres than to cotton, there is a corresponding increase in the difficulty of producing level dyeings. A practical advantage may be taken of this difference by producing two-tone effects in the dyeing of cotton/viscose blends.

Typical dyeing profiles for the three classes of direct dyes are shown in Figure 4.4. When applying mixtures of dyes, such as in a *trichromatic dye recipe*, to avoid variations in shade as dyeing progresses, it is important to select dyes that have similar dyeing properties. This will ensure that as far as possible, the hue will build up on tone (see Figure 4.5), as dyebath exhaustion progresses. If it is not possible to select the three dyes from the same class, then the dyeing method used should be for the dye requiring most control.

The general method of application of direct dyes by a batch process is to commence dyeing at 40 °C and add a wetting agent to assist penetration and levelling. The salt is then added (depending on the class), followed by the dyes. The temperature of the dye liquor is then raised slowly to the boil, at about 2 °C/min, and boiling continued for 30–45 minutes, making salt additions where necessary for the class of dyes. The bath is then cooled to 60 °C and the fabric thoroughly rinsed.

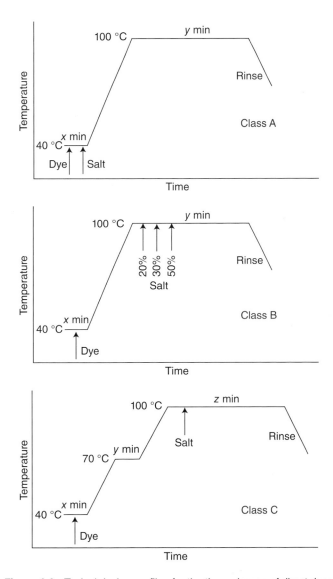

Figure 4.4   Typical dyeing profiles for the three classes of direct dyes.

Figure 4.5   On-tone build-up of a trichromatic mixture of direct dyes belonging to the same class.

Direct dyes are less suitable for continuous dyeing methods because prolonged batching or steaming is required to facilitate good migration and therefore levelness.

A pad/batch technique can be used in which the dyes are padded at 80 °C and the fabric rolled and batched. Careful control of fabric moisture content, padding liquor temperature and selection of dyes for mixture shades is necessary.

Fibres dyed with direct dyes can be after-treated to improve their fastness to wet treatments by increasing the molecular weight of the dye and rendering it less soluble in water. However, this after-treatment can cause a decrease in lightfastness. Different methods are available to improve the wet fastness, the most used being:

- Metal complex formation
- The use of cationic fixing agents

Formation of a Metal Complex In this method, the dyed fibres are treated at 70–80 °C with a copper salt to form a dye–metal complex. For the method to work though the dye needs to contain a o,o′-disubstituted azo groups of the type shown in structure **3.27** in Section 3.5.1.3, so it is not applicable to all direct dyes. Usually –OH or –OCH$_3$ groups are the substituents. Although there is an improvement in the lightfastness, the treatment gives a change in the hue, but the effect is not always permanent and de-metallisation can occur gradually during repeated domestic washing. Another disadvantage of this process is the use of copper salts, because discharge of these metal ions to effluent is subject to very strict tolerances, given their toxicity.

Cationic Fixatives Cationic fixing agents react with the sulphonate groups of the direct dye molecules to form a water-insoluble complex of higher molar mass than the dye ion alone. The fixing agent is usually a compound possessing a positive charge and a long aliphatic chain, for example, cetylpyridinium chloride, **4.2**. Whilst the wet fastness is improved, there can be a decrease in lightfastness:

$$N^+ \!-\! C_{16}H_{33} \quad Br^-$$

4.2

4.2.2.3 *Vat Dyes*

Dyeing with vat dyes is based on the principle of converting a water-insoluble pigment by chemical reduction (known as 'vatting') into a water-soluble ('leuco') form that has substantivity for cellulosic fibres. After it has diffused into the fibre, the colorant is oxidised back to the parent water-insoluble form (Scheme 4.5). The water-insoluble molecules are trapped within the fibre polymer matrix, and so wet fastness properties are excellent. In the case of indigo, reduction to the leuco form is shown in (Scheme 4.6)

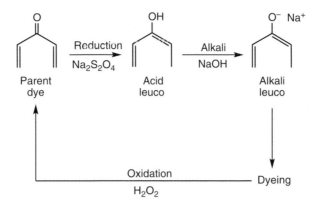

Scheme 4.5   Chemistry of vat dyeing – vatting to generate the soluble leuco form and subsequent oxidation back to the parent insoluble dye.

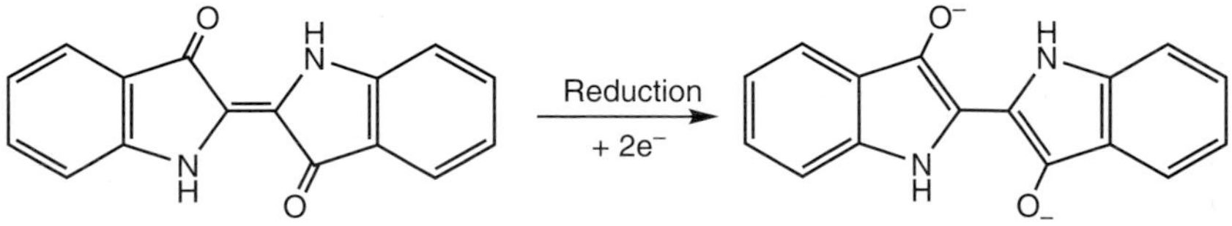

Scheme 4.6   Vatting of indigo.

In application, both the extent of reduction and the rate at which equilibrium between the reduced and oxidised forms is achieved are of practical significance. Vat dyes vary with the speed with which they undergo reduction. The most common reducing agent used for the reduction of vat dyes is sodium dithionite ($Na_2S_2O_4$), also known as sodium hydro-sulphite, or simply 'hydros'. This chemical is capable of reducing even the most stable of vat dyes. Any difficulties in vatting can be overcome by raising the vatting temperature, increasing the concentration of reducing agent or prolonging the vatting time. The vatted dye must be kept in a strongly alkaline solution because its leuco form is an insoluble acid. If, instead of being formed as its water-soluble sodium salt, it is formed as the free acid, it will not readily oxidise to the parent form. Variables such as pigment particle size and crystal form can affect the rate of reduction, but these are controlled by the dye manufacturer. Consequently the dyer needs to concentrate only on the temperature and concentration of reducing agent (sodium dithionite, $Na_2S_2O_4$).

The concentration of the reducing agent required depends on the number of reducible groups in the dye molecules, and allowances must be made for oxidation of the dithionite and adsorption by the fibre during dyeing. Oxidation of the dithionite (Scheme 4.7) causes its reducing power to be lost. Therefore a large excess of dithionite is essential.

It is important also that as little as possible of the leuco dye is exposed to air during the dyeing phase, since it may oxidise to the parent form prematurely creating deposits on the surface of the fibres.

$$Na_2S_2O_4 + 2NaOH + O_2 \rightarrow Na_2SO_3 + Na_2SO_4 + H_2O$$

Scheme 4.7 Oxidation of sodium dithionite during vat dyeing process.

The amount of alkali (sodium hydroxide (caustic soda), NaOH) used has to be sufficient to maintain a pH of 12–13, bearing in mind that some is consumed in the vatting process and is also adsorbed by cellulosic fibres. The presence of the sodium hydroxide causes the fibres to swell, which promotes dye exhaustion. In some cases neutral electrolyte (such as sodium chloride, NaCl) is required in the dyebath also to promote exhaustion of the leuco dye. However, the rate of vatting and the substantivities of vat dyes vary widely, so the concentration of caustic soda and sodium chloride and temperature of the dye liquor need to be adjusted to suit the dyes being applied.

Vat dyes can be grouped into three types:

(1)   Those for which the leuco form has high substantivity. No electrolyte is needed but a high concentration of caustic soda and a high vatting temperature (60 °C).
(2)   Those for which the leuco form has moderate substantivity, so some sodium chloride is required, but just a moderate concentration of caustic soda and a lower dyebath temperature (50 °C).
(3)   Those for which the leuco form has low substantivity, so a high concentration of sodium chloride is required, but just a low concentration of caustic soda and a low dyebath temperature (30–40 °C).

To decrease the use of reducing agents, pre-reduced forms of vat dyes are available, though there are some problems in their application. An example is pre-reduced indigo, sold as 20–40% pastes, which requires alternative reducing agents, including a catalytic hydrogenation process developed by DyStar that lowers the need for hydrosulphite in the final process.

After the exhaustion stage, the dyed material is rinsed to remove loose dye and any remaining reducing agent, alkali and electrolyte. The leuco form of the dye is then oxidised to the parent form using hydrogen peroxide and ethanoic (acetic) acid. Finally, a soaping treatment is given, which involves washing the material in detergent at the boil. This is a very important part of the overall operation, since it not only removes any loose particles of dye in the surface of the material but also develops the final crystal structure of the pigment particles and hence the final colour. The fastness properties are also improved by this operation.

Vat dyes are costly because they are difficult to synthesise and so their use is directed to higher-quality fabrics. Nevertheless, they are noted for their high fastness to light, very useful for the dyeing of fabric for awnings, curtains, upholstery, military and naval uniforms. High fastness to bleaching is another strong point of the anthraquinonoid group. These dyes are widely used in the manufacture of, for example, good-quality shirtings, tablecloths, towels, sportswear, high-quality overalls, fabrics for women's and children's clothing and tropical suitings. With careful selection, the use of vat dyes allows materials to be prepared with a guarantee against fading. A limitation, however, is that the range lacks scarlet, maroon and wine shades.

4.2.2.4 Reactive Dyes

There are two essential stages in the application of reactive dyes to cellulosic fibres:

(1) Diffusion of dye into the fibre, which is controlled by varying the dyeing time, the dyebath temperature and the salt concentration
(2) Reaction between the dye and the cellulose to form a covalent bond, which is controlled by the selection of an appropriate pH value

Reactive dye molecules have relatively low substantivity for cellulosic fibres, and this permits very easy levelling, but once they have reacted with the fibre, they cannot migrate further.

Once alkali is added to the dyebath, reaction of the dye with cellulose begins. At the same time, however, the dye begins to react with water (*hydrolysis*), and the hydrolysed dye becomes a nuisance as it cannot react with the cellulose. It retains its substantivity, but not its reactivity, and needs to be removed in the final washing off. This situation is usually expressed in terms of the 'efficiency' of the dyeing process, which is represented as a ratio of the amount of dye chemically combined ('fixed') to the amount of dye applied. This efficiency can never reach unity (100% efficient) because the presence of hydrolysed dye cannot be avoided. Indeed, in some cases the efficiency can be as low as 60–70%, which means a very significant amount of dye originally entered into the dyebath is discharged to effluent.

In a typical batchwise dyeing process, the material is treated in the dyebath containing the required amount of dye and electrolyte (usually sodium chloride), at a temperature of around 30 °C. The purpose of this part of the process is to transfer as much dye as possible from the aqueous phase to the fibre phase, the sodium chloride functioning in the same way as for the application of direct dyes, that is, in reducing the surface charge on the cellulosic fibres. Once adequate exhaustion has been achieved, the temperature of the dye liquor is raised and then alkali added to initiate chemical reaction. The alkali used is usually soda ash (sodium carbonate, Na_2CO_3). Since most of the dye is present within the fibre, the reaction should form the covalent bond between the dye and the fibre, though the competing hydrolysis reaction also occurs.

The reactivity of a particular dye increases with temperature and also with increasing pH of the dye liquor. A number of different fibre-reactive groups are described in Section 3.5.2.4, and their reactivities and temperatures of application are listed in Table 3.3 in that section. The dyeing profiles are customised according to the type of leaving group on the dye. Typical profiles for sulphatoethysulphone (SES) (cold dyeing type), monochlorotriazinyl (MCT) (hot dyeing type) and hetero-bifunctional dyes are illustrated in Figure 4.6.

For highly reactive dyes, that is, the cold dyeing types such as SES, the temperature can be set at between 30 and 60 °C, depending on the type of fabric and the strength of alkali. Typically, a pH of 10–10.5 and a dyeing temperature of 40 °C are used. Fixation is achieved using sodium carbonate as alkali. The pH is modified to 9.9–10.0 by using sodium bicarbonate when temperatures of 60 °C become necessary for certain fabric constructions.

For less reactive dyes, that is, the hot dyeing types such as MCT that require higher temperatures, the method is the same in principle, the only differences being the use of a dyeing temperature of between 65 and 90 °C and a more alkaline pH value of 10.5–11.0, again obtained using sodium carbonate.

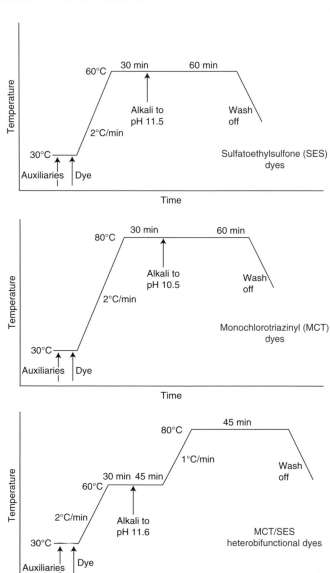

Figure 4.6 Typical dyeing profiles for reactive dyes on cellulosic fibres.

The final stage of the operation is that of rinsing, and it is vital that this is done effectively to remove all unfixed hydrolysed dye or any adsorbed dye that has not reacted. If not then, it will wash out in subsequent laundering by the consumer and possibly stain adjacent whites. It is also necessary to remove residual alkali and salt. Initial washing removes the alkali and salt and some of the unfixed dye, but it is necessary to boil in detergent to remove any remaining dye. This is especially important when heavy shades are dyed; indeed repeated soaping at the boil may be necessary.

Reactive dyes may be applied by both batch and continuous processes and there are certain issues to consider in each case.

Batch Dyeing   In batch dyeing, low liquor ratios favour exhaustion and fixation, and thus less electrolyte is needed. Increasing the electrolyte concentration increases exhaustion, but its effectiveness is reduced at high concentrations. Increasing the temperature reduces the substantivity but increases the rate of reaction and diffusion.

Semi-continuous Dyeing   For a semi-continuous process such as the pad/batch method, dyes of low substantivity and low reactivity are best. Dye and alkali are applied from the same bath, and the fabric is rolled on to a perforated or unperforated beam, covered with a plastic sheet to prevent drying out and then stored (batched) for long enough to allow dye diffusion and fixation to take place. The coloration process is then completed by washing the dyed fabric free from alkali and loose dye.

The pad/batch process is usually carried out at room temperature, and therefore the pH of the pad liquor is the most important variable in its control. Variations in the pH of the padding liquor are used to accommodate variations in the batching time needed for penetration and fixation. Flexibility is essential for various technical reasons. For example, the greater resistance to the penetration of aqueous solutions offered by a tightly woven fabric at room temperature means that a longer batching time is required, but this is accompanied by a greater danger of premature hydrolysis. Consequently, where longer batching times are used, the alkalinity is reduced.

The conditions used for cold-dyeing dyes are usually a 2 hour batching period for dye liquor at pH 10.5–11.0 (using sodium carbonate) and 24 hour batching, or even longer, for a pH of 10 or below using sodium bicarbonate.

The conditions used for the less reactive hot-dyeing dyes require a batching time of up to 48 hours using a stronger alkali and a pH as high as 12–13. Sodium hydroxide (caustic soda, NaOH) is the alkali used. These conditions lead to a greater loss of dye by hydrolysis, and the colour yield is usually lower than that obtained with cold dyeing types. Furthermore, the low colour yield cannot be counteracted by the use of the high-fixation dyes because their high substantivity and enhanced fixation lead to 'tailing'.

Speeding up the process is possible by batching at between 50 and 70 °C using a pad-roll machine. This gives improved penetration, diffusion and more rapid fixation but has associated technical problems. Variation in temperature across the width of the fabric due to cooling of the edges of the roll can bring uneven fixation across the width, a phenomenon called 'listing'. Dye may also migrate during the batching because of partial drying of the fabric from the edge. In general, therefore, the dyes chosen for application by the pad/batch process are those suited to cold batching.

If the fabric was originally batched on a perforated beam, it does not even need to be removed for washing since water can be pumped through it. Since the hollow beam is closed at one end, water is forced through the perforations into the bulk of the fabric; within the fabric the slight positive pressure forces direct replacement of the existing spent dye liquor held between the individual fibres with fresh water, thus flushing out the surplus dye and chemicals. This requires much less water than the conventional washing-off procedure, in which the fabric is passed through a large volume of water and which relies on the removal of loose dye and chemicals by water passing over the fabric surface.

Continuous Dyeing In continuous dyeing, the simplest way of applying a uniform distribution of dye on a fabric is by a padding method. The method is efficient in terms of dye consumption, levelling and water consumption, the liquor ratio being about 1 : 1. Although padding is a mechanically simple operation, care has to be taken to avoid loss of quality of the dyed fabric due to 'tailing'. The effect can become serious with dyes of high substantivity. Consequently dyes of lower substantivity, which do not exhaust too readily, are preferred for padding operations. It is only in continuous methods that dyes of low substantivity give high colour yield.

Padding operations are the first step for many of the continuous methods of application of reactive dyes in which the whole process is completed from start to finish without a break. After the initial padding stage, the method of fixation varies but, whichever sequence of operations is chosen, the aim is still to allow dye to diffuse quickly prior to fixation. There are different techniques for the fixation stage:

- Pad-dry
- Pad-dry-bake processes
- Pad-dry-steam process

Usually the levels of fixation achieved are higher than those obtained by batchwise methods.

The pad-dry process is really only suitable for dyes of high reactivity, such as the dichlorotriazinyl (DCT) types. The pad liquor contains the dye, sodium bicarbonate ($NaHCO_3$) and urea (see structure **1.42**), the function of urea being to assist in solubilising the dye in the limited amount of water present, to swell cellulosic fibres and to enable the fibres to retain water, thereby assisting dye diffusion during the drying stage. The rate of drying is low, the conditions being 2–5 minutes at 105 °C so that the fabric retains water (about 18%). During this drying stage, the sodium bicarbonate decomposes to sodium carbonate, a stronger alkali, giving a slightly higher pH and thus promoting fixation.

The pad-dry-bake process is used for dyes of lower reactivity, the final baking stage promoting fixation. A baking temperature of between 150 and 170 °C is used.

The pad-dry-steam process is a widely used continuous dyeing method. It involves padding the fabric (see Section 4.3.7) with a neutral solution of dye, drying and then padding with a solution containing alkali (in this case sodium hydroxide) and salt. The fabric is then steamed at 100–105 °C for 15–30 seconds in the case of DCT dyes but for 30–60 seconds for the less reactive MCT dyes.

Reactive dyes are relatively easy to apply to cellulosic materials and give bright shades with excellent lightfastness and wet fastness properties. However they do suffer from some disadvantages, relating especially to the effluent produced by their application:

- The main problem is the hydrolysed dye that ends up in the washing and rinsing water. This dye is not easily treated in sewage plants and de-colourising of waste water from the washing processes is desirable.
- The high concentrations of salt used also end up in effluent and this is also a cause of ecological concern.

To mitigate these problems, research in recent years has focussed on increasing the efficiency of fixation, so that as much as possible of the dye originally entered into the dyebath ends up chemically fixed to the fibre. Bifunctional reactive dyes, especially those containing two different reactive groups, have been commercialised, as have dyes that react with cellulose under neutral conditions. Manufacturers have also developed ranges that require less salt in exhaust dyeing processes, though the dyes used need to have higher substantivity, which in turn makes it more difficult to remove hydrolysed dye after dyeing.

Another problem with reactive dyes relates to the stability of the dye–fibre bond. The features of the reactive groupings that give the dyes their reactivity are still present after the dye–fibre bond has formed, so that if exposed to another strong nucleophile, such as a hydroxide ion, the bond may break. Thus, under severe alkaline conditions, rupture of the bond is possible.

4.2.2.5   Sulphur Dyes

Like vat dyes, sulphur dyes are reduced and applied as water-soluble leuco compounds that need to be kept under alkaline conditions, but sulphur dyes need only sodium sulphide to act as both alkali and reducing agent. A simplified version of the reduction reaction is represented in Scheme 4.8.

$$Ar-S-S-Ar' \quad \rightarrow \quad Ar-S^- \quad + \quad {}^-S-Ar'$$

Scheme 4.8   Reduction reaction of sulphur dyes.

The quantity of sodium sulphide required depends on the particular dye being applied around 2.5–5 g/l of the flakes (60%) but at least 20 g/l of the flakes for heavy depths. Sodium hydrosulphide is used as well, though this requires the presence of alkali. Another reducing agent used increasingly (to avoid the environmental problems associated with sodium sulphide) is glucose, though again alkali is required in addition. It can be used for the pre-reduced brands, either alone or with lower amounts of the sulphide.

After the dyebath is exhausted, the fabric is thoroughly rinsed prior to the oxidation stage. This is important because it removes loose surface colour that would otherwise be precipitated on the surface of the fabric, thereby reducing the rubbing fastness. Whilst indigo will oxidise in several passes through air (*skying*), sulphur dyes require a much more energetic oxidation process, achieved using a solution containing an oxidising agent. The oxidising agents used are:

- Hydrogen peroxide or peroxy compounds, though the wet fastness may be low with some dyes
- Sodium bromate at acid pH

Whilst sulphur dyes are a useful application class giving insoluble colorant in the fibre and consequent high wet fastness properties, they do suffer from disadvantages that arise from the chemicals required for their application:

- Since reducing chemicals are required, the effluent contains large amounts of sulphides, and the discharge of these to drain is not permitted due to their toxicological properties. The use of pre-reduced sulphur dyes that require less chemical for reduction means less sulphide is discharged to effluent.

- The presence of excess sulphur in the dyes arising from their manufacture. Again, this leads to effluent problems.
- The use of strong oxidising agents for the oxidation stage and their discharge to effluent streams.
- Poor rub fastness due to inadequate washing off prior to oxidation.

A process for the application of sulphur dyes to denim warp has been developed [1, 2], which is much more environmentally friendly than the traditional method. The new method involves impregnating the fibres with the dye solution, reducing agent and auxiliaries, followed by short air oxidation and then treatment in a bath containing a bi-reactive cationic fixing agent. This agent combines by ionic bonding with the dye and the fibre and also renders any unfixed dye, normally present in the fibre surface, insoluble. By combining this method with sizing, a considerable reduction in the use of water can be achieved.

4.2.2.6 Azoic Dyes

These dyes are formed in situ in the fibre by a *diazotisation* reaction. The process involves treating the fabric with a solution of a *coupling component*, then with a *diazo compound* (diazonium salt), so that an insoluble pigment is formed within the fibre matrix. The application of azoic dyes generally involves four stages:

(1) Impregnation of the fabric with the coupling component, usually a naphthol. The depth of shade of the final dyed material is governed entirely by the amount of the coupling component adsorbed by the substrate.
(2) Removal of the excess coupling component. If any coupling component remains on the surface of the fabric, insoluble pigment will form on the surface when the diazonium salt is applied, leading to poor fastness to rubbing.
(3) Treatment of the fabric at room temperature with a solution of the selected diazonium salt to form the dye by a coupling reaction. The impregnated fabric is worked in the solution of the diazonium salt for 20–30 minutes.
(4) The material is soaped at the boil for 30 minutes in a solution of detergent and sodium carbonate. As with the application of vat and sulphur dyes, soaping is important because it removes any mechanically held surface dye particles and develops the true shade of the dye.

The application process is complicated and time consuming, and there are difficulties associated with handling the component chemicals. Materials dyed with azoic dyes have excellent fastness to washing, very similar to that shown by vat dyes. However, their lightfastness, particularly of light shades, is less good. For these reasons, azoic dyeing is in decline.

4.2.3 Polyester Fibre Dyeing

The vast majority of polyester fibres are dyed using disperse dyes, which are substantially water insoluble. During manufacture, these dyes are milled to very small particle sizes, and a surface-active agent is incorporated, so that in water they form fine, stable dispersions, and so they are called disperse dyes.

These dyes have hydrophobic character and are capable of 'dissolving' in hydrophobic fibres. The fibre acts like an organic solvent, extracting the dye from water. This is analogous to the simple solvent extraction of organic compounds from water by shaking up the aqueous phase with an appropriate water-immiscible solvent. It is for this reason that the linear Nernst adsorption isotherms, described in Section 6.6.3, are obtained when disperse dyes are applied to hydrophobic fibres.

Disperse dye molecules contain polar groups, such as –OH, $–NO_2$, –CN (e.g. see structures **3.8** and **3.9**) so that although they are substantially water insoluble, they do have a small but important solubility. The transfer of dye molecules into the fibre during dyeing is a two-step mechanism, in which:

(1)   Individual dye molecules first dissolve from the small particles of the dispersion into the water to form a saturated solution
(2)   Transfer into the fibre then takes place because the dye is much more soluble in the fibre than the water

As the solution in water is no longer saturated, more dye molecules dissolve from the particles and again leave the water phase as they transfer to the fibre phase. This mechanism can be represented by the equation shown in Scheme 4.9.

Dye(solid in dispersion) ↔ Dye(in solution) ↔ Dye(in fibre)

Scheme 4.9   Mechanism of adsorption of disperse dyes into hydrophobic fibres.

A typical disperse dye will contain additives to improve its performance in use. The additives most commonly incorporated into commercial dye powders (or grains) are dispersing agent, humectants, de-foaming agent, wetting agent and biocide:

- Dispersing agents are surface-active agents (see Section 1.8.5.2) added to render the dye powder easily dispersible in water and to maintain the stability of the dispersion during dyeing. Also they improve the levelling properties of the dye.
- Humectants are added to maintain the moisture content of the dye and to prevent dusting of the powder. Urea (see structure **1.42**) is often used for this purpose.
- De-foaming agents are added to inhibit the formation of excessive foam during dyeing, since this may lead to unlevel dye uptake. These agents are usually polydimethylsiloxanes.
- Biocides are added to dye powders to prevent biodegradation during storage and are usually quaternary ammonium salts (see structure **1.45** in Section 1.8.3.7).
- A wetting agent is added to the formulation to improve the wetting out of the fabric. This is a type of surface-active agent described in Section 1.8.5.2.

Disperse dyes can be applied to most synthetic fibres using simple immersion techniques. Differences in dyeing properties between one fibre type and another are accommodated by changes in the dyeing temperature. Since neither ionic attachments nor covalent linkages are formed, the dyeing process is controlled by either:

- Accelerating dyeing by raising the temperature or
- Slowing it down by using a higher concentration of dispersing agent

The latter expedient assists levelling of the dye in the fibre and prevents the build-up of dye particles on the surface of the fibre filaments, a fault that otherwise leads to poor rubbing and wet fastness.

The dyeing of polyester presents a particular problem because the polymer chains are closely packed. The fibres do not swell, even as the temperature of the dye liquor is raised, so diffusion of dye molecules into the fibre matrix at temperatures up to the boil is so low that adequate dyebath exhaustion cannot be achieved. Dyes of lower molecular weight show slightly better diffusion than those of larger molecular weight, but their wet fastness is not good. Consequently the disperse dyes used are those with higher molecular weight, their problems of exhaustion being overcome by either dyeing in the presence of carriers or by high-temperature dyeing.

Section 1.8.5.3 gives examples of some carriers used for polyester dyeing. Their presence in the dyebath, usually as an emulsion, has the effect of lowering the glass transition temperature of the fibre, so the polymer becomes more flexible and swelling of the fibre filaments occurs. This change opens up space through which the dye molecules can diffuse; consequently even deep shades can be obtained at the boil. It is necessary to remove all of the carrier from the fibres after dyeing since they have objectionable smells and further, they are toxic. This has implications for the environment, and for these reasons carrier dyeing is not carried out now, the high-temperature dyeing method being favoured.

The high-temperature dyeing method involves dyeing at temperatures of about 130 °C using pressurised vessels. At this higher temperature, a temperature well above the *glass transition temperature*, T_g, of the fibre, the rate of dyeing is considerably increased and the use of dyes with high molecular weights is possible. Two advantages follow:

(1) At these high temperatures, migration is very good, so any irregularities in the fibre are covered.
(2) The use of dyes of high molecular weight means that the fastness properties, especially to light, washing and sublimation, are much better.

In a typical high-temperature dyeing operation, the dyes, dispersing agent and sufficient ethanoic (acetic) acid to give a pH of 4.5–5.5 (added to prevent hydrolysis of the fibre under alkaline conditions) are entered into the dyeing machine at about 60 °C. The temperature is raised at 1–2 °C/min to 130 °C and dyeing carried out at this temperature for 45–90 minutes, before allowing to cool.

There is a correlation between the dyeing behaviour of disperse dyes and fastness to heat treatments that allows dyes to be conveniently grouped into five main application groups – A, B, C, D and 'undesignated' – based upon their performance with polyester. Fastness to heat treatments is important in the context of polyester dyeing because it is related to the ease with which dyes will sublime out of the dyed material:

- 'Undesignated' dyes, which are unsuitable for use with polyester.
- Class A dyes, whilst they can be applied to polyester, are of limited value, giving poor fastness to heat.
- Class B dyes are excellent for use with polyester, particularly in covering variations in dyeing properties associated with textured yarns. Their fastness to heat is moderate.

- Class C dyes have all-round suitability for polyester in all dyeing methods. Their fast-ness to heat is not the best possible.
- Class D dyes arc used when maximum heat fastness is required.

As a general rule, the dyeing properties of disperse dyes become more difficult to deal with as the fastness to heat treatments (as measured by *sublimation temperature*) increases (Figure 4.7).

Sublimation temperature is the temperature at which the disperse dye passes directly from the solid to the gaseous state. This characteristic of disperse dyes, especially those with low molecular weights, enables them to be applied to polyester by sublimation (heat transfer) printing (see Section 5.9.4) and also by pad-Thermosol dyeing.

Just as sublimation temperature decreases as dyeing properties improve, so too does wash fastness. Wash fastness of a dyed polyester is crucially important because it is used extensively in cotton/polyester blends for the leisure market. Garments made of these blends are frequently washed, and there is the danger of cross-staining of the cotton component if dye leaches out of the polyester. There is therefore a trade-off between ease of dye application and wash fastness. Often class C dyes are preferred for high-temperature application because they give acceptable wash fastness and the dyeing performance gives greater flexibility of choice when dyeing mixtures, such as trichromatic recipes. The class D dyes are those of large molecular size (high molecular weight), which do not diffuse easily through the fibre and are difficult to apply level.

After the dyeing cycle is completed, the polyester material is given after-treatments to maximise the fastness properties. In particular, a thorough soaping is required to remove surface dye and auxiliaries from the material.

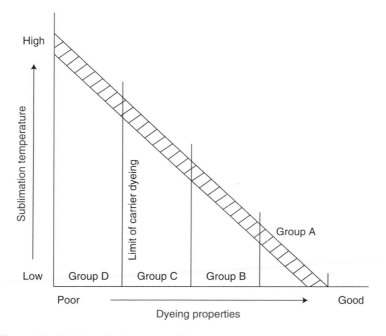

Figure 4.7 Relationship between sublimation temperature and dyeing properties.

However, another very important after-treatment is *reduction clearing*. This process involves treating the material with a solution containing 2 g/l reducing agent (sodium dithionite, $Na_2S_2O_4$), 2 g/l sodium hydroxide and 1 g/l of a non-ionic surfactant at 70 °C for 30 minutes. The purpose of reduction clearing is to remove any dye loosely bound in the surface of the polyester fibres by chemically reducing them. Azo disperse dyes are reduced to colourless amines, whilst anthraquinone-based dyes are reduced to their corresponding leuco forms (see Section 4.2.2.3), both of which are easily washed out. Because the temperature used is well below the glass transition temperature of the polyester, the fibres are in their hard, glassy state, so the reduction clearing solution cannot penetrate the fibres and destroy the dye in the interior of the fibres.

Another method of dyeing polyester is the Thermosol process. In this process, polyester fabric is passed through a pad bath containing the dye dispersion, ethanoic (acetic) acid to give pH 4.5–5 and a migration inhibitor (Figure 4.8).

After the dye has been padded on to the fabric, the fabric is dried. It is then fed directly into the stenter, in which the high temperature causes the dye to sublime, that is, change from the solid state into the gaseous state. In this gaseous state the individual dye molecules diffuse into the interior of the fibres.

An important requirement of this process is that the dyes must sublime at a suitable rate at the temperature used (~200 °C). If the rate of sublimation is too low (or the temperature in the stenter is too low), then the dye will not diffuse properly into the fibres. Conversely, if sublimation is too fast, more dye will exist in the gaseous phase than can be adsorbed by the fibre and will be lost. It is necessary therefore to carefully match the temperature in the stenter and the rate of passage of the fabric through it to the sublimation properties of the dyes being used.

The SDC classification is useful in selecting appropriate disperse dyes for the pad-Thermosol process. Dyes of classes B and C are best suited: class D dyes do not sublime readily enough, and class A dyes sublime easily and much of the dye vapour is lost.

The pad-Thermosol process is rarely used on 100% polyester fabrics because migration of dye occurs during drying, leading to unlevelness. It is, however, widely used for

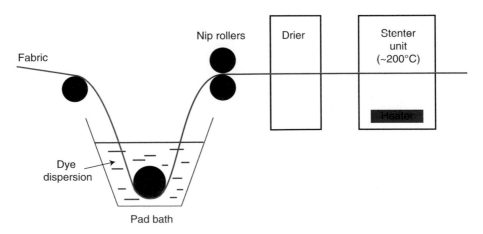

Figure 4.8   Schematic representation of the pad-Thermosol method.

dyeing woven polyester/cotton blends. Even though some dye dispersion is taken up by the cotton component of the blend in padding, during the heating stage, it transfers to the polyester. Dyes for the cotton component (e.g. reactive dyes) are also included in the padding bath, but because they are ionic, they do not sublime in the stenter. In this case the fabric is then fed through another pad bath, this one containing the chemicals required for fixation (in the case of reactive dyes, alkali and salt) and then through a steamer to allow diffusion and fixation of the dye. Next, a thorough soaping is carried out to remove hydrolysed dye.

One issue with polyester/cotton blends is that the reduction clearing process for the polyester component cannot be carried out because the reducing agent will destroy the dyes used for the cotton component. It is therefore necessary to select disperse dyes that do not stain cotton when domestic washing cycles are carried out.

Polyester microfibers (see Section 2.11) have become a significant part of the textile market, with garments made from such fibres being popular for the leisurewear and sportswear sectors. The dyeing of polyester microfibres is difficult because they require more dye to be applied to obtain a given depth than is necessary for fibres of 'regular' fineness and the rapid uptake of dye molecules means that care has to be taken to ensure good levelness. The wash fastness and lightfastness properties of dyed microfibers tend to be lower than for regular fibres.

4.2.4   Nylon Fibre Dyeing

4.2.4.1   Disperse Dyes

Dyeing nylon with disperse dyes has the advantage that their good migration properties cover irregularities and the problems of barré obtained when acid dyes are used are avoided. Usually the class B and C level-dyeing disperse dye types are used. Unfortunately the wet fastness properties are poor when heavy depths of shade are applied, so their use tends to be limited to pale shades for materials that will not be subjected to severe laundering. The dyeing process is straightforward. The temperature of the dyebath, which contains the required amount of dye dispersion and dispersing agent, is raised to the boil and dyeing continued at the boil for 45–60 minutes.

The use of disperse dyes for dyeing nylon is very limited and although practiced makes up only a very small percentage of the dyed nylon available.

4.2.4.2   Acid Dyes

Like wool, nylon contains amino groups that under acidic conditions become sites for acid dyes. The acid dyes used to dye nylon are classified into three groups:

Group 1 dyes are monosulphonated dyes with little substantivity for nylon under weakly acidic conditions but exhaust well under more strongly acidic conditions, such as pH 3–4. C.I. Acid Blue 25 (**3.22**) is an example of this type, and whilst it covers barré nylon well, the wash fastness is poor.

Group 2 dyes have greater substantivity than group 1 dyes and have better wash fastness
properties also. They exhaust at pH 3–5, though more care has to be taken in applying
them to ensure levelness. C.I. Acid Blue 41 is an example of a dye of this group.

Group 3 dyes have much higher molecular weights and therefore substantivity and are the
'neutral' dyeing type. They have poor levelling properties and do not cover barré, but
give good wash fastness. C.I. Acid Red 138 (**3.26**) is an example of a dye of this group.

Nylon fibres contain fewer amino groups (only about 30–50 mmol/kg) than wool (over
800 mmol/kg), and there is a limit to the amount of dye that can become attached to the
nylon fibre through electrostatic attraction. This is particularly the case with simple di- or
trisulphonated acid dyes of low molecular weight (see Section 6.7.2). However, with
increase in molecular size and/or reduction in the number of sulphonate groups in the dye
molecule, making it more hydrophobic, other forces of attraction such as hydrogen bonding and van der Waals forces play an important role. Indeed with such dyes, adsorption in
excess of the theoretical maximum number of amino groups occurs.

Compared with wool, the diffusion of acid dyes into nylon is slower, and the limited
number of charged sites in nylon can also cause problems in dyeing deep mixture
shades, where the individual dyes compete for available sites. Under these conditions
the faster-diffusing dyes may block entry of a second component, and the expected
shade is not achieved.

Another problem in dyeing nylon with acid dyes is the effect known as *barré*, which is
the formation of light and dark bars across a fabric. It occurs because of variations in the
amino and carboxyl end group content in the polymer during its synthesis or the oxidation
of amino groups during melt spinning of the fibres. Barré may also arise from variations in
crystallinity that can occur during melt spinning.

Barré can be reduced in dyeing nylon with acid dyes by incorporating auxiliaries in the
dyebath. These auxiliaries function in different ways. Some are anionic compounds that
compete with the dye molecules for the amino sites in the fibre, thereby reducing the rate
of strike by the dye. Others are cationic compounds that form complexes with the anionic
dye ions, and then the complexes gradually break down as the temperature of the dyebath
increases.

The dyeing process depends on the group to which the acid dyes being applied belong.
For dyes of groups 1 and 2, the fabric is treated in the dyebath, initially at 40 °C, first with
an anionic blocking agent and then with a cationic levelling agent at pH 3–5 using either
methanoic (formic) acid or ethanoic (acetic) acid. The temperature is then raised gradually
to the boil and boiling continued for 30–45 minutes. In the case of group 3 dyes, after treating the fabric with the anionic blocking agent, the temperature can be raised to the boil
quickly since adsorption only occurs when the dye–cationic auxiliary complex breaks
down as the temperature rises (Figure 4.9).

The wet fastness properties of dyes of groups 1 and 2 are not especially good, and it
is common to after-treat the dyed fabric with either natural or synthetic *tanning agents*
(syntans) to improve the fastness. The process with a natural tanning agent, called the
'full back-tan', involves treating the dyed nylon with a solution of tannic acid (2% owf)
with ethanoic (acetic) acid to give pH 4.0 for 20 minutes at 80 °C. Tannic acid is a highly
complex organic compound. Although it is acidic, it does not contain any carboxyl

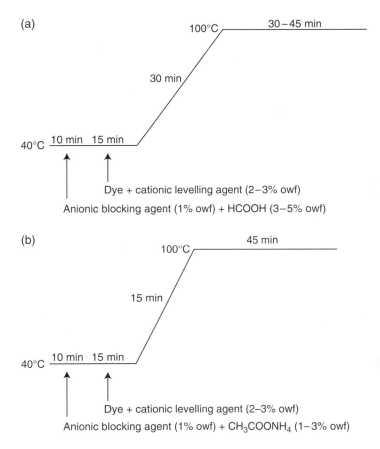

(a)

100°C   30–45 min

30 min

40°C   10 min   15 min

Dye + cationic levelling agent (2–3% owf)

Anionic blocking agent (1% owf) + HCOOH (3–5% owf)

(b)

45 min

100°C

15 min

40°C   10 min   15 min

Dye + cationic levelling agent (2–3% owf)

Anionic blocking agent (1% owf) + CH₃COONH₄ (1–3% owf)

Figure 4.9   Dyeing profiles of (a) groups 1 and 2 dyes and (b) group 3 dyes.

groups (–COOH); instead it is a type of polyphenol and its weak acidity is due to the many phenol groups in the molecule. The next stage of the process is to treat the fabric in a fresh bath of tartar emetic (2% owf) at pH 4.0 and 80 °C for 20 minutes. Tartar emetic is potassium antimonyl tartrate, $K_2Sb_2(C_4H_2O_6)_2$. The tannic acid and tartar emetic combine to form a water-insoluble complex as a film around the fibre filaments, thereby restricting diffusion of the dye out of the fibre during washing. Although the process significantly improves the wash fastness of acid-dyed nylon, it does have some disadvantages. Because it is a two-bath process, it is time consuming and expensive. Tannic acid discolours on exposure to light and back-tanning reduces the lightfastness of the dyed nylon.

The use of syntans instead of natural tannins is a much simpler alternative for improving the wash fastness of acid-dyed nylons, though they are not quite as effective. Syntans are polycondensates of sulphonated phenol–methanal or dihydroxydiphenyl sulphone–methanal of high molecular weight. They are applied at pH 3–5 and 40–50 °C for 20 minutes and can be added at the end of the dyeing cycle in the cooling dyebath. Syntans do not have the toxicity of tannic acid, but again they incur a reduction in lightfastness and cause slight changes in hue.

4.2.4.3 Reactive Dyes

Nylon can be dyed with reactive dyes, the reaction of the dye being with the amino groups in the fibre. However, as noted in Section 4.2.4.2, nylon contains only a modest amount of amino groups compared with wool, and covalent bonding is only possible for pale to medium depths of shade. If it is attempted to dye of heavy depths, much of the dye is present in the fibre as unfixed anionic dye, held only by weak van der Waals forces, so wash fastness is poor.

The reactive dyes used are anionic in nature (see Section 3.5.2.4), and it is necessary to apply them at pH 5.0–6.0 for them to be adsorbed by the fibre. However, acidic pH conditions are not ideal because protonation of the amine groups occurs (Scheme 4.10).

Dyes that react by either nucleophilic substitution or nucleophilic addition can be used: the reactions are shown in Schemes 4.11 and 4.12, respectively.

$$-NH_2 + H^+ \rightarrow -NH_3^+$$

Scheme 4.10 Protonation of amino end groups in nylon fibres under acid conditions.

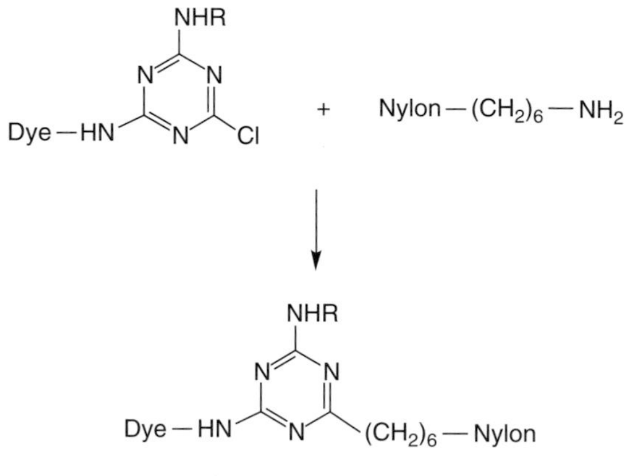

Scheme 4.11 Reaction of chlorotriazinyl reactive dyes with nylon fibres by nucleophilic substitution.

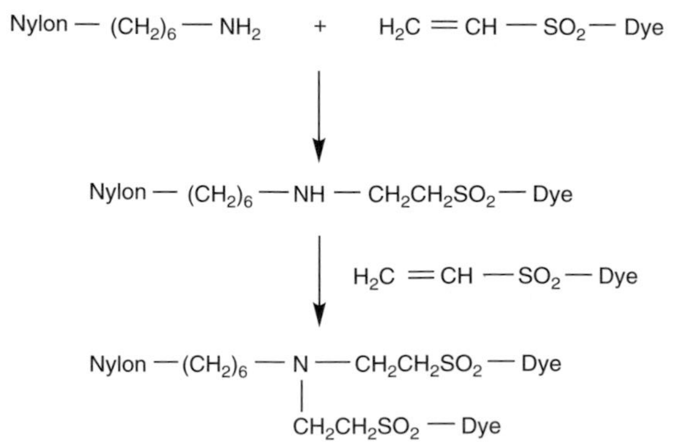

Scheme 4.12 Reaction of vinyl sulphone reactive dyes with polyamide fibres by nucleophilic addition.

4.2.5   Acrylic Fibre Dyeing

4.2.5.1   Basic (Cationic) Dyes

In Section 2.8.2.2 it was pointed out that commercial acrylic fibres are not made of 100% pure polyacrylonitrile because the polymer is difficult to convert to the fibre form. The pure polymer has a glass transition temperature (T_g) of 105 °C, so dye adsorption at the boil is very low indeed. The commercially available acrylic fibres are copolymers, whose T_g values are in the region 55–60 °C, so they can be dyed easily at temperatures below the boil.

The fibres contain sulphate ($-SO_4^-$) and sulphonate ($-SO_3^-$) groups at the ends of the polymer chains, and these negatively charged groups give them substantivity for dyes carrying positive charges – basic (cationic) dyes. Adsorption of the dye at the negatively charged sites in the fibre occurs by a simple ion-exchange mechanism (Scheme 4.13).

$$\text{Fibre–SO}_4^-\text{Na}^+ + \text{Dye}^+\text{Cl}^- \rightarrow \text{Fibre–SO}_4^-\text{Dye}^+ + \text{Na}^+\text{Cl}^-$$

Scheme 4.13   Ion-exchange mechanism of dyeing acrylic fibres with basic (cationic) dyes.

Some acrylic fibre types also contain weakly dissociating –COOH groups, whilst some contain only these groups and no sulphonate or sulphate groups. Consequently the pH at which acrylic fibres are dyed has to be carefully chosen, since at low pH values the carboxyl groups are unlikely to be dissociated, and if this is the case, they do not act as sites for the dye cations.

The different types of acrylic fibre on the market therefore have dyeing properties that are influenced by the type and number of anionic groups along and at the ends of the polymer chains. This leads to different saturation levels and affinity for basic dyes, so the dyeing process has to be optimised for each type of acrylic fibre.

Dyeing is usually carried out at pH 4.0–5.5 using a buffer solution of ethanoic (acetic) acid and sodium ethanoate (acetate). The dyes are highly substantive and do not migrate easily once adsorbed into the fibre. It is necessary to retard the rate of adsorption to ensure level dyeing, and for this reason retarding agents are also added to the dyebath. These agents are commonly quaternary ammonium compounds (see structure **1.44** in Section 1.8.3.7) and are adsorbed preferentially at the anionic sites in the fibre. During dyeing, they are gradually replaced by the more substantive dye cations.

In order to ensure level dyeing, careful control of temperature is required. Whilst exhaustion of the dye is slow at temperatures below T_g of the fibre, it occurs rapidly at temperatures just above the T_g, as polymer chain flexibility suddenly increases. Once the dyebath temperature has reached 70 °C, the rate of heating is reduced to about 0.5 °C per minute until 100 °C is reached. Dyeing is then continued at the boil for approximately 1 hour. Care has to be taken in cooling the dyebath at the end of the dyeing process since the thermoplastic nature of acrylic fibres means that creases can be set into the yarns or fabrics if it is cooled too quickly. Usually the material is cooled slowly to 50–60 °C to avoid this problem.

In addition to the problems in obtaining level dyeing due to the chemical and physical properties of acrylic fibres, difficulties also arise when dyeing mixtures of basic dyes. The rate of uptake of a given dye can be lower if another dye is present than when it is applied alone at the same concentration. There are only a limited number of dye sites available in the fibre surface, and the dye that is adsorbed quickest retards the uptake of the other dye. Therefore, the compatibility of basic dyes in admixture has to be determined, but measures such as the time of half dyeing or exhaustion curves (see Section 6.5) made on individual dyes are not helpful.

An alternative method is to establish a compatibility value, K, for a dye from a series of combination dyeings. The dye under test is applied in combination with a set of five blue or five yellow dyes, the particular set chosen to contrast the most with the hue of the dye being examined. The dyes in each of the two sets of five form a scale of compatibility ratings ranging from 1 (the most rapid dyeing) to 5 (the slowest dyeing). The dye from the set of five (1–5) that builds up on tone with the dye under test is its compatibility value. For example, if the dye under test is an orange hue, it will be tested against the set of five blues hues. If it builds up on tone with blue dye number 4, its compatibility rating is 4. It is then reasonably safe to assume that the test dye can be applied successfully in mixtures with other dyes of compatibility rating 4.

Basic dyes have high tinctorial strength and are bright, with very good lightfastness properties. Thus, despite the difficulties in obtaining level dyeing of the correct shade outlined earlier, they are the most commonly used application class.

4.2.5.2   Disperse Dyes

Whilst dyeing with disperse dyes was carried out after the introduction of acrylic fibres, it is not now the norm for acrylic fibres to be dyed with them. Disperse dyes are generally unsatisfactory because they do not build up well and are duller than basic dyes and their wet fastness properties are not especially good. They have good levelling properties, but if they are used at all, it is only for dyeing pale shades and for applications where good wet fastness is not a requirement.

4.2.6   Polypropylene Fibre Dyeing

Polypropylene fibres are very hydrophobic and have little substantivity for the dyes of the various application classes, even disperse dyes. Consequently, they are solution dyed, a process called *mass pigmentation*, using organic pigments.

4.2.7   Dyeing Fibre Blends

4.2.7.1   Wool Fibre Blends

The most common blends of wool are wool/nylon, wool/polyester and wool/acrylic. In these blends, the function of the wool is to provide warmth, softness, moisture absorbency and anti-static properties, whilst the synthetic component present provides characteristics such as abrasion resistance and durability.

Wool/Nylon   The most common wool/nylon blend is the 80% wool/20% nylon blend used in the manufacture of carpets. The nylon enhances the wear properties and wool provides the warmth and resilience to the carpet pile. Reference has already been made (see Section 4.2.4.2) to the difference in amine end group content between wool and nylon, and this difference makes it difficult, but not impossible, to achieve heavy depths of shade with blends of these fibres.

Wool/nylon blends are typically dyed to a solid shade with equalising acid, acid milling or 2:1 dye: metal–complex dyes. The 1:1 dye: metal–complex types have only a low substantivity for nylon, so the 2:1 types are used for obtaining medium to heavy depths of shade. The dyeing process is essentially the same as for 100% wool, with the exception that a *retarding agent* is often required to control the *partition* of dye between the two fibre types. Partition is the difference in dye distribution between the two fibre types in a blend. Poor partition means that the two fibres dye to different depths of shade to an extent that is visible. The difference in dye distribution depends on factors such as the dye type, the depth of shade being applied, the pH and the blend ratio of the wool and nylon. Another difficulty in the application of anionic dyes is that they have different substantivities for nylon 6 and nylon 6.6. These two types of nylon require quite different levels of reserving agent when dyeing in their blends with wool.

The rate of dyeing of nylon, particularly at 60–80 °C, is higher than that of wool, so there is preferential uptake on nylon, in the absence of a retarding agent. As the applied depth is increased, a critical value is reached at which a solid shade is obtained. This critical depth is specific for the dye and is higher for monosulphonated dyes than for disulphonated dyes. Application of dye at higher depths leads to preferential uptake by the wool, and there is no auxiliary that can prevent this from occurring with acid dyes. When dyeing pale shades, the preferential uptake of dye by nylon can be overcome by dyeing a mixture of mono- and disulphonated dyes of similar hue. They are applied using methanoic (formic) acid instead of sulphuric acid, the slightly higher pH causing less damage to the nylon. Levelling agents are used to prevent uptake of dye by the nylon component in the early stages of dyeing.

Acid milling and 2:1 metal–complex dyes offer improved fastness at the expense of poorer levelling qualities. Levelling agents optimise the partition between the two fibre types and is maintained as the temperature rises, due to the poor migration properties of the dyes.

Acid milling dyes are often used as a single component to brighten shades based on 2:1 metal–complex dyes. 2:1 metal–complex dyes are used where high fastness to light and washing is required. The saturation limit of these dyes on nylon decreases with an increasing level of sulphonation of the dye, and disulphonated dyes tend to dye wool preferentially. Monosulphonated dyes display a more balanced distribution between the wool and nylon that can be regulated by retarding agents. The usual conditions of dyeing are pH 5–6 using ethanoic (acetic) acid and ammonium ethanoate (acetate) at the boil.

Wool/Polyester   This is another popular blend, mainly used for outerwear, such as woven suitings and dresses. Whilst the wool again provides softness and warmth, the polyester gives the fabrics abrasion resistance, crease recovery and increased strength. The polyester component also provides durable pleating properties, though it is necessary for it to be present in the greater proportion for it to have this effect, one reason for the 55/45 polyester/wool blend being the most popular. Other blend ratios are produced for different end uses, not all of which are intimate blends. Some fabrics are woven with the warp and weft having different compositions, such as a 55/45 polyester/wool warp and a pure wool weft.

Whilst the polyester component has to be dyed with disperse dyes, the wool component can be dyed using acid milling or 2:1 metal–complex dyes. Fortunately all these dye classes can be satisfactorily applied at slightly acidic pH values, but there is a difference in temperature requirements. Disperse dyes usually have to be applied to polyester at 130 °C, a temperature that will cause considerable degradation of the wool component. Another problem is that of cross-staining of disperse dyes on to the wool component. Dyeing processes for wool/polyester blends must therefore consider these two issues. One way around these problems is to dye each fibre type separately with their appropriate dye application classes before blending. However, many yarns are intimate mixtures, as of course are the fabrics.

The dyeing of the blended yarns or fabrics can be carried out using either one-bath or two-bath methods:

- In the one-bath method, dyeing is typically carried out in a slightly acidic dyebath (pH 5.0–5.5) in the temperature range 95–120 °C, the temperature used depending on the depth of shade. Nevertheless, because these temperatures are lower than that normally used to dye polyester, low medium energy disperse dyes (classes B and C – see Section 4.2.3) have to be used. Pale shades may be dyed at 95 °C, and medium depths at 105 °C, but deep shades, navies and blacks require a temperature of 120 °C, for which a wool protective agent is required. The one-bath method is much used because it is economical in terms of time, water and energy.
- In the two-bath method, the polyester component is dyed first, again in the temperature range 95–120 °C, followed by a mild reduction clear at 45–50 °C using sodium dithionite (hydrosulphite), ammonia and a non-ionic detergent. The wool component is then dyed, this time at the boil with a 2 : 1 metal–complex dye at pH 5.5–6.0, in the presence of a levelling agent. The advantage of the two-bath method is that any staining of the wool by the disperse dye is removed by the reduction clear. Consequently a large number of disperse dyes can be chosen to obtain the required fastness.

Wool/Acrylic The blends tend to be used for knitted sweaters, skirts and outerwear. Typically they are 50/50 or 80/20 acrylic/wool and are noted for their good dimensional stability, wear resistance and pleat retention.

They are usually dyed by a two-stage process, in which basic (cationic) dyes are applied at the boil to dye the acrylic component first, cooling to 60 °C, then adding anionic dyes and dyeing the wool component at the boil. 1 : 1 metal–complex and cationic dyes can be applied at pH 2–3, whilst 2 : 1 metal–complex, acid milling and cationic dyes are applied at pH 6–7.

Reactive dyes such as α-bromoacrylamido and vinyl sulphone types with cationic dyes can be used to achieve bright shades and are applied at pH 4–5. For medium to full depths, an alkali after-treatment is necessary to fix the reactive dyes to the wool.

4.2.7.2 Cotton Fibre Blends

Polyester/Cotton In these blends, the cotton is responsible for good moisture absorption properties (and hence comfort in wear), anti-static properties and reduced propensity to pilling. The polyester confers good tensile strength, abrasion resistance, dimensional stability and easy-care properties to the goods. To improve crease resistance effectively, around 50% or more of polyester in the blend is required, though the use of durable-press finishes on cotton means that lower proportions of polyester can be used. However, if the proportion of polyester in the blend is too low, the strength of the yarns is weaker than that of either fibre individually. Consequently, the most common blend ratios produced are 50/50 and 65/35 polyester/cotton.

The polyester component is always dyed using disperse dyes, whilst the cotton may be dyed using direct, reactive, vat, sulphur and azoic dyes. Of these, use of the vat, sulphur and azoic dye types is uncommon due to the environmental hazards associated with the chemistry of their application. Also the wet fastness properties given by direct dyes are not particularly good.

For these reasons the most popular dye combination used for polyester/cotton blends is the disperse/reactive dye system. As with the use of disperse dyes in dyeing polyester in blends with wool, the cross-staining of them on to cotton can present a challenge. The disperse dye is only loosely bound to cotton, so without a suitable wash-off procedure, the dyeing can exhibit poor wet fastness. Another factor to consider is that if the cotton is to be treated with a cross-linking agent that has to be cured at high temperature, it is necessary to use high energy disperse dyes (class D dyes – see Section 4.2.3) that have high sublimation temperatures.

Dyeing can be carried out by either a batch (a one-bath or two-bath process) or a continuous process. In two-bath batch dyeing there are two methods that can be employed:

(1)   The polyester component is dyed first using appropriate disperse dyes at the usual application temperature of 120–130 °C. Reduction clearing is then carried out to remove residual disperse dye from the polyester fibre surface and from the cotton. The cotton component is then dyed with reactive dyes, followed by thorough washing off to remove unreacted hydrolysed dye.

(2)   The reactive dye is applied first; then after rinsing to remove the alkali and salt, the disperse dyeing is carried out at 130 °C. Performing the dyeing operation this way round allows the hydrolysed reactive dyed to be removed from the fibres during the high-temperature dyeing of the polyester component. However it also means that the reduction clearing operation is not possible because the hydrosulphite may chemically degrade the reactive dye bonded to the cellulose.

Overall, these methods are time consuming and so rapid one-bath dyeing procedures have been developed. In the one-bath process, the reactive and disperse dyes are both present initially in the dyebath at pH 6.5. Dyeing is commenced by adding salt and raising the temperature to 80 °C to promote exhaustion of the reactive dye on to the cotton. The dyebath temperature is then raised to 130 °C, so the disperse dye dyes the polyester, then cooled and alkali added, to initiate fixation of the reactive dye to the cotton. After dyeing at the required temperature for the type of reactive dye being used, the material is washed and scoured. If this method is carried out, the reactive dyes must be stable at 130 °C and the disperse dye must not be sensitive to alkali or salt.

When direct dyes are used instead of reactive dyes, in combination with disperse dyes, there is less concern about the cross-staining of the cotton by disperse dyes, since the wet fastness of direct dyes is only modest anyway. For this reason direct dyes are not used for producing heavy depths of shade.

In the one-bath method in which the disperse dye and the direct dye are together in the dyebath initially, the polyester component of the blend is dyed first by raising the temperature of the dye liquor to 130 °C. At this temperature, adsorption of the direct dye on to cotton occurs and levelling is very good. The bath is then cooled to 90 °C and alkali added to promote exhaustion of the direct dye. The bath temperature is reduced further, which also encourages exhaustion of the direct dye, and then finally the material is thoroughly washed.

The use of vat dyes in combination with disperse dyes gives dyed polyester/cotton blends of very good wet fastness. In the one-bath method, both dyes are initially present in the dyebath, and the temperature raised to 130 °C to dye the polyester component. During this phase the cotton becomes impregnated with vat dye, so when the temperature is then reduced to 60–70 °C and alkali and sodium dithionite added, it is reduced to the leuco form more or less in situ. After dyeing to allow the substantive leuco form to diffuse into the cotton fibre, during which time any disperse dye in the surface of the polyester is simultaneously reduction cleared by the alkaline dithionite conditions, the oxidation and soaping phases are carried out.

The dyeing of polyester/cotton blends is also widely carried out on a continuous basis, though it is only viable when long lengths of fabric, typically over 10 000 m, are to be dyed. A high degree of control is required to ensure uniform dyeing, as is careful control of drying after padding so that migration of the dyes is avoided. The pad-Thermosol process for the continuous dyeing of polyester/cotton blends with disperse/reactive dyes is described in Section 4.2.3. If vat dyes are used instead of reactive dyes, the second pad bath contains the chemicals required for the reduction of the vat dyes, that is, alkali and sodium hydrosulphite.

4.3 Dyeing Machinery

4.3.1 Introduction

Coloration may be carried out at any stage in the manufacture of textile goods, and dyeing machines are available for dyeing textiles in the form of loose stock, tow, slubbing, sliver, yarn or fabric. Fabric may be dyed in all its forms, such as woven, non-woven and knitted fabrics or as hosiery or garments, so there is a wide variety of dyeing machines available.

Whilst it is usual for synthetic fibres to be dyed using conventional exhaust dyeing procedures, other options exist. Significant quantities of filaments of man-made fibres are coloured by a process called *mass pigmentation*, in which heat-stable pigment particles are dispersed in the molten polymer prior to extrusion. This is a particularly useful method for olefin fibres, such as polypropylene, that are too hydrophobic to be coloured from an aqueous dyebath, but it may also be used for polyester and polyamide fibres.

Another variation, one used for acrylic fibres, is to bring the freshly extruded filament into contact with a solution of basic dye before the filament is allowed to dry out. This process is referred to as *producer dyeing* and is used in the production of acrylic filaments in tow form (untwisted aligned fibres or filaments collected together in the form of a loose strand).

There are advantages and disadvantages to dyeing textile materials at the various stages of manufacture. Dyeing textiles early in the manufacturing sequence, for example, loose stock, has the advantage that any unlevelness that may arise with dyes having poor migration properties can be randomised during subsequent carding and spinning operations and becomes unnoticeable in the final form of the product. However dyeing so early in the sequence means that there has to be a firm demand for the colour being dyed. Conversely, dyeing in the later stages of manufacture, such as fabric, means that manufacturers can avoid overproduction of less popular colours and respond quickly to repeat orders for popular colours. Stock holding is then reduced and delivery times are shortened.

Dyeing can be carried out at the garment stage. The problem here though is that considerable care has to be taken to ensure that levelness of shade in the fabric or garment is achieved. If this means that dyes of better migration properties are used, there could be a trade-off in the wet fastness properties of the final dyed product. Also, garment dyeing is restricted to garments of simple construction.

All machines in a textile dyehouse have comprehensive control systems linked to management and workflow systems that help to maintain the productivity, energy efficiency and complete management of the flow of goods through the production facility.

In any dyeing process it is necessary to transfer the dye molecules effectively from the dye liquor to the fibre surface and then allow them to diffuse to the interior of the fibre structure. To achieve efficient transfer, the dye liquor and the textile material have to be moved relative to each other, so agitation of the system is necessary. Dyeing machines accomplish this requirement in one of three ways:

(a)   By moving the textile material through the stationary dye liquor
(b)   By pumping the dye liquor through the stationary textile
(c)   By moving the textile material and dye liquor simultaneously

Dyeing machines have been developed on all these principles, additionally providing the option of batchwise, semi-continuous or fully continuous processing.

There are a number of fundamental requirements of dyeing machine construction. For example, they are usually constructed of stainless steel so that they can withstand the chemical and physical conditions used in not only dyeing operations but also in preparatory processes such as desizing, scouring and bleaching. The latter process involves the use of hydrogen peroxide, and if iron or copper components in the machine come into contact with it, free radicals are formed, which will attack and degrade cellulosic fibres. Stainless steel has the advantage that it is inert to most process conditions and is easy to clean.

There are different methods of heating and cooling of dye liquors, the main requirement being to maintain a uniform temperature throughout the whole of the machine. It is usual to use high-pressure steam, passed through closed coiled pipes to heat the liquors, a method that avoids any contamination of the dye liquor by heavy metal ions from the boiler or pipework. The coils are designed to have a large surface area, so that rapid heat transfer to the surrounding liquor occurs. The coils can also serve for cooling dye liquors after dyeing by passing cold water through them instead of steam.

The machines designed for batchwise dyeing are limited in the amount of goods they can accommodate at any one time, but if the quantity to be dyed exceeds the capacity of the machine, the load has to be split into manageable batches. Equipment for batchwise dyeing is constructed to ensure that the goods are in constant contact with the dye liquor. The *liquor ratio* of the various batchwise dyeing machines ranges from 5 : 1 to 30 : 1, or even higher. Liquor ratio is the amount of water required in relation to the mass of textile, expressed as a ratio. For example, a liquor ratio of 10 : 1 means 10× the volume of liquor to the mass of textile, so if 100 kg of textile is to be dyed, then 1000 l of water is used. The dyebath composition is always formulated with a specific liquor ratio in mind.

4.3.2 Dyeing Loose Fibre

The dyeing of staple fibre (loose stock) or continuous tow is referred to as either *loose stock* or *pack dyeing*. The tow is usually formed into packages convenient for handling and dyed in machines similar to yarn package machines. The hot dye liquor is circulated by the use of pumps through the fibrous mass packed into a large annular basket or cage. This method is not widely used because the quality of the tow after dyeing is often disappointing.

The machines traditionally used for dyeing fibre in loose stock form are the conical pan and pear-shaped types. The fibres are packed relatively loosely so the dye liquor can be pumped through them at relatively low pressures, ensuring adequate penetration without causing undue mechanical damage. In the conical pan machines (Figure 4.10), the fibres are held in an inner container with a perforated bottom and covered at the top with a perforated lid. The dye liquor, which is heated by steam coils at the base of the machine, is pumped upwards through the fibre mass, which has the effect of pushing the fibres against the conical walls of the inner vessel. The liquor flows out through the top plate and falls back to the base of the machine.

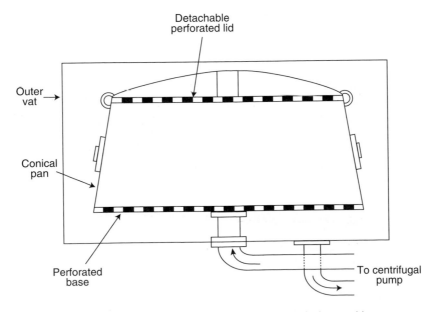

Figure 4.10 Schematic diagram of a conical pan dyeing machine.

The method has traditionally been used mostly for wool, rather than cellulosic or synthetic fibres.

4.3.3 Top Dyeing

A top is made from a sliver, a thick band of carded and combed but untwisted fibres that lie parallel to each other, by cross-winding it into a ball about 40 cm in diameter, securing it with strings and then flattening it somewhat into a more cylindrical shape. Bump tops, as they are called, weigh about 10 kg and are loaded for dyeing into cylindrical cages about 1 m high, located on a carrier. The carrier (Figure 4.11) is hoisted into the dyeing machine for the dyeing cycle to be carried out. Machine capacity varies, some accommodating 45 bumps (a total of 450 kg), whilst machines, just big enough to take two bumps (20 kg), enable smaller batches to be dyed. Examples of dyed woollen bump tops are shown in Figure 4.12. The method is popular for dyeing wool fibres because dyes of high wet fastness properties, which tend to have poor migration characteristics, can be used.

Figure 4.11 Nine cages, each containing five bump tops, ready to be hoist into the dyeing machine (Source: Photograph courtesy of Bulmer and Lumb Group, UK).

Figure 4.12 Bump tops after dyeing (Source: Photograph courtesy of Bulmer and Lumb Group, UK).

4.3.4 Yarn Dyeing

Yarn is dyed either as hanks or wound on to compact reels known as *packages*, mounted on hollow spindles.

4.3.4.1 Package Dyeing

Packages can take a variety of forms, with names such as *cheeses*, *cones* or *cakes*. The most commonly used type is the cheese, which is parallel sided and formed by winding the yarn on to cylindrically shaped formers. Some formers are made of stainless steel wire that interlaces with a helix of coarser wire (springs) constructed in such a way that they are compressible, allowing the packages to be compressed in the dyeing machine, at the same time avoiding any significant increase in the diameter of the cheese. Another type of former is rigid plastic with a network structure that allows dye liquor to pass through (Figure 4.13), though a disadvantage of plastic formers is that they can become stained by dye, especially if the dye liquors are concentrated.

Cones, as the name suggests, are conical-shaped packages and are used for yarns that are difficult to unwind after dyeing, such as yarns that swell when wet. The conical shape facilitates the unwinding.

Cakes are made in a similar way to cheeses, but the former is removed after winding, which causes the innermost yarns of the package to fall into the void left by the former. During dyeing the whole package becomes more relaxed and liquor flow through it much easier. This type of package is used for dyeing high-bulk acrylic yarns.

Figure 4.13 Cheese packages on rigid plastic formers, after dyeing. In the background are dyed wool tops (Source: Photograph courtesy of Bulmer and Lumb Group, UK).

Whatever the type of package, it is important that winding the yarn round the former is carried out in as uniform a manner as possible to ensure that the flow of dye liquor through the package is regular, with no channelling. In this way unlevel dyeing and damage to sections of yarn on the package will be avoided. Packages can be prepared directly from spinning machines, so that a separate winding operation is unnecessary. Staple yarns, for example, are often delivered to dyehouses on plastic formers, ready for dyeing.

Uniform movement of dye liquor is essential for the production of level dyeing, and several factors can impede the flow if they are not taken into account when the package is being wound. In general permeability is low for regenerated cellulosic yarns, higher for staple yarns and highest for synthetic fibre packages. Provision must also be made for the effects of fibre swelling or yarn shrinkage, as these may cause a change in porosity as the dyeing process proceeds. Other factors can be controlled more directly.

The package will be required to withstand frequent reversal of the forced flow of hot dye liquor without disruption of the orderly arrangement of the wound yarn. The porosity of the package must also remain uniform, because dye liquor will be able to flow more rapidly through less densely packed regions, and this may lead to uneven dyeing. Obviously the use of a high tension during winding will produce a denser package.

With non-textured yarns permeability increases with increasing tex, but the finer the filaments, the more readily the permeability is reduced due to flattening of the yarn. The same difficulty may occur with low-twist yarns. Packages of textured yarns and yarns with a rounded compact structure imparted by a high degree of twist are more permeable. Care is needed with non-textured fibres since their dimensional instability can also cause problems. Non-textured filament nylon, for example, is usually wound on to collapsible bobbins and relaxation induced by steaming. The increased tensions produced are then released before dyeing by rewinding on to fresh formers.

The packages of yarn, each weighing around 1.5 kg, are mounted on vertical hollow spindles, approximately 70 mm in diameter and 1250 mm high, perforated so that dye liquor can flow through them. The spindles are screwed in a concentric arrangement into holes set in a frame with a circular hollow base and a centre pillar with an eye at the top for lifting. At the base of the frame is an inlet for the dye liquor leaving a pump. About 8–10 packages are loaded on to each spindle, and the number of spindles depends on the capacity of the individual dyeing machine. Figure 4.14 shows packages that have been dyed in a machine of capacity 900 kg.

The pump forces the dye liquor up the hollow spindles and through the packages. The flow is reversed from time to time. The frame holding the spindles and the packages is contained in a cylindrical vessel with a dome-shaped lid (Figure 4.15). The vessel can be sealed to enable the pressurised dyeing of polyester at temperatures of 130 °C. The liquor ratio is around 10:1, and an expansion tank at the side of the vessel accommodates the

Figure 4.14 Packages of wool yarn on the spindles of a package dyeing machine of 900 kg capacity (Source: Photograph courtesy of Bulmer and Lumb Group, UK).

Figure 4.15 Loading a package dyeing machine (Source: Photographs kindly supplied by Thies GmbH & Co. KG, Germany).

increasing volume as the temperature rises. It is also used for making additions to the dye liquor, for example, of auxiliaries or more dye, during dyeing.

Package dyeing machines are available either as vertical loading, as shown in Figure 4.15, or as horizontally loading machines that do not require heavy lifting gear to load the packages (Figure 4.16).

Ribbons and tapes can also be dyed using package dyeing machines.

After the dyeing process is completed, the packages are dried either by using a radio-frequency machine or a pressure drying machine. In radio-frequency drying the packages are placed on a polypropylene conveyor belt (Figure 4.17), which carries them through the drier. The power rating used depends on the fibre type: for wool and cotton it is about 50 kW, whilst for a more hydrophobic fibre such as polyester, it is about 40 kW. Care also has to be taken with acrylic fibres with their low T_g value, so a power of about 30 kW is used for them. In pressure drying the complete frame holding the packages on their spindles can be lowered into the machine and warm air forced through in the same way as the dye liquor was during the dyeing cycle.

4.3.4.2 Beam Dyeing for Yarns

Beam dyeing machines are very similar to package dyeing machines but are used to dye yarn that is wound on to a beam ready for its use on the weaving loom as the warp. The yarn is wound carefully onto a specially perforated hollow beam. The dye liquor is forced through the perforated beam and through the yarn. As in the package machine the flow of

Figure 4.16 Horizontal loading package dyeing machine (Source: Photographs kindly supplied by Thies GmbH & Co. KG, Germany).

Figure 4.17 A radio-frequency drying machine (Source: Photograph courtesy of Bulmer and Lumb Group, UK).

liquor can be either in or out or changed between the two to even out the different pressures. The arrangement is similar to the horizontal package machine in that the beam is loaded horizontally into a vessel. Further details are given in Section 4.3.5.3.

4.3.4.3 Hank Dyeing

Hank dyeing machines are relatively simple in their construction, though a number of different designs are available, including:

(1) Hussong type
(2) Cabinet style
(3) Hank spray type
(4) Space dyeing

Hussong Type   The Hussong machine comprises a rectangular bath in which the hanks of yarn are suspended from horizontal 'sticks' in the dye liquor (Figure 4.18). The dye liquor is gently circulated around the yarns using a suitably placed pump and overflow arrangement. Uniform dyeing is ensured by reversing the direction of flow from time to time. The liquor flow tends to raise the hanks off the sticks to a certain extent, which is useful in that it prevents a resist on the yarn where the liquor would be otherwise unable to penetrate, though bunching of the yarns at the top of the machine can impede liquor flow through them. To prevent this some machines, called 'double stick' machines, also have sticks at the bottom to prevent the hanks from rising too far. However, allowance has to be made for the shrinkage of some types of yarn, especially high-bulk yarns, when wet.

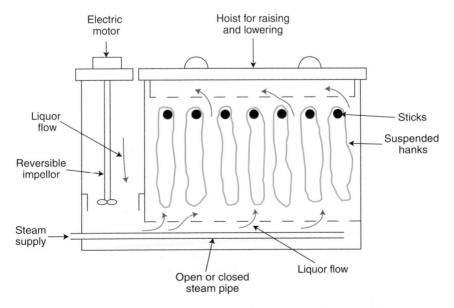

Figure 4.18   Schematic diagram of a Hussong hank dyeing machine.

Hank dyeing is much less common than package dyeing but is still used for dyeing lofty yarns such as wool and high-bulk acrylic yarns for knitting and carpet manufacture. The hanks can weigh up to 10 kg and either have tie bands fastened round them or are wrapped in a stockinette made of heat-set polyester to prevent tangling as the dye liquor flows through the machine. Figure 4.19 shows a small machine in which the container housing the sticks, with the hanks suspended from them, raised out of the main tank.

Whilst hank dyeing is fairly straightforward, the machinery is not especially expensive and reproducible results can be achieved easily, it has the disadvantage of being labour intensive because of the need to reel the yarn into hanks; then after dyeing to wind, it back again on to cones. Loading and unloading of the hanks from the sticks requires considerable care and also takes time. It is for reasons such as these that package dyeing is preferred.

Cabinet Type   The cabinet-style hank dyeing machine is most suitable for dyeing delicate yarns such as wool, wool and nylon, wool and acrylic, shrink-resistant wool,

Figure 4.19   Small-scale hank dyeing machine.

wool and silk, mercerised cotton, cashmere and linen. In these machines the liquor flow is controlled to be in the direction of the yarn, causing less chance of felting or disruption. It is therefore a suitable method of dyeing fancy yarns such as chenille. The yarn is held both at the top and bottom on sticks to further reduce the movement of the yarn and maintain the feel of the yarn.

The machines are modular and can work efficiently at 50% capacity when it is not necessary to increase the liquor ratio because of the method of delivery of the liquor to the fibre. They also offer flexibility in being able to be used to dye narrow fabrics and even socks, tights and other lightweight materials in special drawers that can be inserted into the cabinet. The totally enclosed nature of the machine provides for a better working environment than with a Hussong-type machine. If the means for pressurisation is available, a temperature of 110 °C can be achieved.

Spray Dyeing Machines   In spray dyeing machines the yarn is held on rotating arms. The rotating arms have the capability of spraying dye into the fibre both from the outside and the inside of the hank. Below the hanks is a collection tray incorporating a heating element to heat the liquor before recirculation to the spray arms. The spray pressure and flow rate can be varied to accommodate different yarn types to prevent any yarn distortion or damage. The machines are designed to be modular so that varying loads can be dyed. They can also be used to dye very delicate yarns.

Both cabinet machines and spray dyeing machines are designed with removable arms, so future batches can be prepared before the previous dyeing cycle is finished and loaded quickly to reduce the time the machine is not being used.

Space Dyeing   The space dyeing of hanks can be performed in one of two ways:

(1)   Multiple strands of yarn sprayed with colour at set intervals
(2)   Multiple strands of yarn continuously printed using rollers as set intervals

After the colour is applied, the dye is fixed in a steam oven and washed off before drying. This technique is used to create design effects in knits. Further processing can include doubling of the yarn. Space dyeing is also carried out when the fibre is in the form of a top or sliver and the effect of the different colours is then spun into the yarn. Similar techniques are also possible on packages of yarns. Other techniques to get the same result are available, such as the knit–de-knit process.

4.3.5   Fabric Dyeing

4.3.5.1   Winch Dyeing

In winch dyeing machines, the oldest kind of equipment used for dyeing fabric, the fabric is dyed in rope form, gathered together across its width. The rope of fabric is looped over the winch reel, which lifts the fabric and allows it to drop into the dye liquor. A free-running jockey acts as a support for the fabric as it is pulled forwards.

As the fabric drops into the dye liquor, it becomes folded over in a concertina fashion. The folding and opening action, which persists for as long as the winch reel is turning, keeps the dye liquor in motion.

The method is used for fabrics such as woollens, loosely woven cottons, synthetic fibre fabrics and most knitted fabrics. As well as being used for dyeing, winches have a role to play in both the preparation of fabrics and the finishing of woollen fabrics. All these can withstand creasing during the dyeing process. The winch machine operates at fairly long liquor ratios, though shallow draft machines are more efficient in this respect. The warp threads of the fabric remain under tension throughout dyeing, and as a result of this and of the mechanical action used to propel the fabric through the dye liquor, the yarns in woven fabric often develop a crimp. With knitted goods there is an increase in loop length. Both effects lead to a fuller thicker fabric with a more resilient handle and improved crease recovery.

Variations in the design details of machinery have been introduced to suit the characteristics of different fabrics. The two main types of winch are the deep-draught and shallow-draught machines. Wool and heavy types are best suited to the deep-draught winch, where fabric piles up on the sloping back and pushes forwards the fabric in front. For filament viscose, acetate or any fabrics that crease whilst wet, the shallow-draught winch is preferred because the lack of depth and flat base reduce the compressive pressure of the falling fabric. The design of the winch reel governs the movement of the fabric, which when wet grips the reel because of the gravitational pull and the friction between the wheel and the wet fabric. The reels are usually made of stainless steel.

With the deep-draught machine, a circular or slightly elliptical reel is used to lift the fabric over the jockey roller, so that it falls straight into the dyebath from the reel with only a little plaiting action. Elliptical winch reels are better for other fabrics and provide greater plaiting. Mechanical action on more delicate fabrics is reduced by the use of a larger elliptical wheel. The presence of an adjustable horizontal bar (the 'gate') in the middle of the bath restrains the fabric at the back of the machine until it is pulled past by the revolving winch reel. This action minimises creasing by straightening the warp direction folds.

Winch machines can either be open or enclosed, though most are enclosed for reasons of health and safety and to reduce heat loss, thereby reducing costs and also helping to maintain a constant temperature. They can only operate at temperatures up to 100 °C, or close to that temperature, depending on the surrounding conditions.

Figure 4.20 shows a machine with many lengths (ropes) of fabric arranged side by side, in parallel. In other machine types, a very long continuous length can be dyed in a spiral configuration.

4.3.5.2 *Jig Dyeing*

The jig machine (Figure 4.21) is one of the oldest ways of dyeing fabric in open width. In this machine a batch of fabric is rolled backwards and forwards from one roller to another through the dye liquor, all the time being kept in open width which avoids any running marks forming. The direction of movement of the fabric is automatically reversed as the machine reaches the end of the fabric roll.

The duration of the dyeing process is monitored by the number of *passes* (or '*ends*') through the liquor. Machines open to the atmosphere can accommodate a roll of 500–1000 m

Figure 4.20 Winch dyeing machine for multiple lengths of fabric (Source: Photograph courtesy of DP Dyers, Holmfirth, UK).

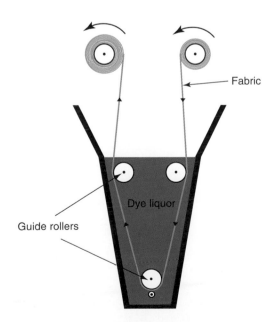

Figure 4.21 Schematic diagram of a jig dyeing machine.

in length, but more modern enclosed machines can operate with a roll of 5000 m. The rollers vary in size but can be up to 2 m wide.

An enclosing lid helps to reduce heat losses and consequent temperature differences between the edge and the centre of the roll. Such differences lead to *listing*, an uneven dyeing caused by a reduction in the dye uptake at the edges of the fabric.

In addition to listing another uneven dyeing effect that can arise is *tailing*. Tailing is a gradual reduction in depth along the length of the fabric, caused by a reduction in dye uptake between the two ends of the fabric.

After passing through the dye liquor, the fabric towards the end of the roll will soon pass through it again as the direction of travel is reversed, before the dye from the previous pass has had sufficient time to be adsorbed into the fibres. In comparison, the fabric in the centre of the roll will pass through the trough at more regular intervals. The consequence of this difference is a slight gradual decrease in depth of shade towards each end of the roll of the fabric. As the fabric passes from one roller to the other, through the dye liquor, it experiences a warp tension that can stretch it lengthways. To overcome this both of the rollers are driven, but careful control is necessary since they rotate at gradually differing speeds as the fabric transfers from one to the other. It is only when there is an equal amount of fabric on each roller that they revolve at the same speed.

Pressurised jigs are available for the more difficult hydrophobic fibres such as polyester, and these machines can operate at 130 °C. An example of a pressurised jig is shown in Figure 4.22. Depending on the fabric construction, beam dyeing (see Section 4.3.5.3) is another option for high-temperature dyeing.

Figure 4.22   High-pressure jig dyeing machine (Source: Photograph kindly supplied by Thies GmbH).

In jig dyeing low liquor ratios can be used, down to 1.5:1, which offers advantages over other techniques. This is useful for applying those dyes that have only medium exhaustion properties. The fact that a given section of the fabric spends most of the time tightly wound on one of the rollers and is only immersed in the dye liquor from time to time means that it is well suited for applying vat dyes on to cellulose, because premature exposure to air oxygen is minimal. The method is also well suited to the dyeing of fabrics that are readily creased, such as taffetas, poplins, suitings and satins, but less well for knitted goods because they distort very easily.

4.3.5.3   Beam Dyeing of Fabric

In beam dyeing, fabric is wound on to a rigid perforated beam and dyed in an enclosed vessel. This enables the dye liquor to be forced through the substrate in a manner analogous to yarn package dyeing. The use of reversible pumps enables the flow to be in either the 'in-to-out' or the 'out-to-in' direction, but the former is favoured because the latter flow can cause compression of the fabric and reduce the porosity of the goods. The beam dyeing method is good for dyeing delicate woven or knitted fabrics that may, in other types of machine, become creased due to the way the fabric is manipulated in the liquor. The fabric remains stationary in an open-width configuration and is supported throughout the dyeing process. It is also best for fabrics that are not too tightly woven or knitted, so that the dye liquor can flow relatively easily through the many layers on the beam. It is also essential that in winding the fabric on to the roll any creases are avoided. Beam dyeing machines usually operate at liquor ratios of about 10:1, so they are efficient in terms of water and energy usage.

Before winding the fabric, the beam barrel is wrapped with a few layers of cotton fabric that provides a good bed for the roll of fabric. This is necessary since otherwise the perforations in the beam show as overdyed marks on the inner layers of the fabric. When all the fabric has been wound on to the beam, a short length of calico or polypropylene net, slightly wider than the fabric, is stitched on and wrapped round the roll. The function of this net is to provide extra stability, especially restraining the centre portion of the roll from ballooning under the pressure of the liquor flow.

To obtain uniform liquor flow during dyeing, the porosity of the fabric must remain consistent. In this respect it is essential to wind the fabric on to the beam with an even pressure. The fabric must be dimensionally stable. Any dimensional instability of the fabric will lead to unlevel dyeing because stretching allows localised increases in the porosity within the roll wherein liquor may flow more rapidly.

If shrinkage occurs during dyeing, an increase in the internal mechanical pressure in the roll brings about flattening of the fibres. With thermoplastic fibres this results in deterioration of the appearance of the fabric surface. Fabrics made of synthetic fibres, particularly nylon, are given either a heat-setting treatment by passing them through a stenter or a scouring treatment before dyeing to minimise such difficulties.

A pressure beam dyeing machine is shown in Figure 4.23. A small expansion tank (not shown in Figure 4.23) is also fitted to enable additives to be introduced at the requisite point of the dyeing process.

The diameter of the beams around which the fabric is wound varies according to the size of the dyeing machine, but is about 50 cm, fitting into machines of up to 200 cm in internal

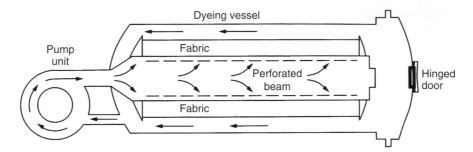

Figure 4.23 Sectional diagram of a high-temperature beam dyeing machine.

diameter. Their lengths vary also, from around 2 m for woven fabrics to 4.5 m for warp-knitted fabrics. The first 20 cm or so in from each end of a beam is not perforated: the perforations (small holes of about 5 mm diameter) start thereafter, enabling the dye liquor to flow through it. One end of the beam is closed, whilst the other end has a fitting that aligns with the liquor delivery pipe when the machine is closed. The beam is usually loaded into the machine through this end, from a cradle.

4.3.5.4   Jet Dyeing

In jet dyeing machines a strong jet of dye liquor is pumped out from an annular ring through which a rope of fabric passes in a tube called a *venturi*. This venturi tube has a constriction, so the force of the dye liquor passing through it pulls the fabric with it from the front to the back of the machine. Thereafter the fabric rope moves slowly in folds round the machine and then passes through the jet again, a cycle similar to that of a winch dyeing machine. The jet has a dual purpose in that it provides both a gentle transport system for a fabric and also to fully immerse the fabric in liquor as it passes through it.

In all types of jet machines there are two principle phases of operation:

(1)   The active phase in which the fabric moves at speed, passing through the jet and picking up fresh dye liquor
(2)   The passive phase in which the fabric moves slowly around the system back to the feed-in to the jets

Jet dyeing machines are unique because both the dye and the fabric are in motion, whereas in other types of machine either the fabric moves in stationary dye liquor, or fabric is stationary and the dye liquor moves through it.

The design of the jet dyeing machine with its venturi means that very effective agitation between the fabric rope and the dye liquor is maintained, giving a fast rate of dyeing and good levelness. Although this design can create creases longitudinally in the fabric, the high degree of turbulence causes the fabric to balloon out and the creases disappear after the fabric leaves the jet. However, the rapid flow of the dye liquor can lead to a high degree of foaming when the machines are not fully flooded. The machines operate at low liquor ratios of about 10:1, so as with beam dyeing, exhaustion is good and water and energy consumption efficient.

Jet dyeing machines were initially designed specifically for dyeing knitted textured polyester, and indeed they were originally designed to operate at high temperature for this purpose. Jet dyeing machines through their various designs and transportation systems provide a great amount of versatility and are used for many woven and knitted fabrics. Figure 4.24 shows a jet dyeing machine being unloaded after the dyeing cycle has completed.

The strong liquor flow in early machines gave somewhat rough treatment to fabrics, so they were not suited for dyeing delicate fabrics. The development of fully flooded machines, in which the rope is always kept fully immersed in dye liquor, overcame this issue. After the fabric passes through the venturi, it passes to a much wider storage tube, where the speed is much slower. Thereafter it is drawn through a narrow tube back towards the venturi.

Another variation is the soft-flow machine, in which the fabric rope and the dye liquor move only slowly, giving a gentler mechanical action. The main dye vessel is tubular in shape and contains the bulk of the fabric. A circular driven reel gently lifts the rope from the dye liquor and feeds it into the vessel down, which it progresses to a narrow transportation tube that delivers the fabric back to the reel. Some machines use a jet to transport the fabric that provides greater speeds, but this does sacrifice some of the gentle action. The transportation tube can be located above or below the main vessel and is filled with dye liquor. The fabric rope experiences no tension, so the machine causes less creasing of delicate fabrics, though the liquor ratio is slightly higher at 15 : 1. Figure 4.25 shows the principle of the soft-flow type of machine.

Figure 4.24 Jet dyeing machines (Source: Photograph kindly supplied by Thiess GmbH).

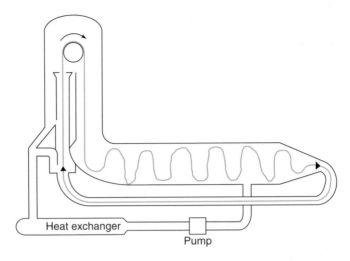

Figure 4.25 Principle of the soft-flow jet dyeing machine.

There are several versions of both jet dyeing and soft-flow machines. Some designs have two or more tubes, mounted side by side, which share a common dye liquor. The speed of the rollers, the circumference of the tube and the force of the jets can be altered to accommodate different fabrics and reduce potential issues of fabric creasing and any stretching of the fabric, thereby making the machines versatile in the range of fabrics that can be processed.

4.3.6 Garment Dyeing

Garments of relatively simple construction, such as womens' hosiery and socks, are typical examples of textile goods that can be dyed in garment form, and the dyeing of fully fashioned knitted wool garments is equally long established. It is common for more elaborate garments to be dyed too, offering the retailer scope for a quick response to the shades that prove most popular with customers, together with a reduction in the level of stock holding. The conventional approach of dyeing the fabric, followed by making up, is associated with a delay of at least 2–3 weeks between the preparation of the fabric for coloration and its appearance as a garment in the retail store. The time between request and delivery may be reduced to 4 or 5 days, however, by holding a stock of undyed garments. Even after making allowance for the manufacture of the fabric and the making-up time, the garment dyeing approach is much faster.

The interest in garment dyeing has been stimulated by the growth in the demand for casual wear and leisurewear. Although garments made from most fibre types can be dyed successfully, the main growth has been with cotton goods. The drawback of garment dyeing is its greater expense. Consequently success is dependent on reduction of the overall cost through retailing methods and attention to design features that are compatible with this method of production.

Adequate attention to fabric preparation is essential for satisfactory dyeing, and the fabric must be thoroughly scoured to remove contaminants such as spinning oils. Some garments may shrink during dyeing, and one strategy is to make up garments oversize to allow for it. With wool garments some form of shrink-resistant treatment is usually applied before dyeing, and with all knitted goods pre-relaxation treatments are equally important for a good-quality end product. Careful selection of stitch type, stitch density, thread tension and thread type helps to reduce problems such as the development of seam pucker due to differential relaxation of the fabric and sewing thread during dyeing.

Other problems are entirely the responsibility of the dyer. For example, fabrics of different construction in the same garment must appear to be of the same intensity of colour. The compacted fabric enclosed by the seam must be as well penetrated and dyed to the same depth as the rest of the garment. This calls for dyes with good migration properties and a dyeing method that will give good penetration of the dye liquor. For this reason the hot-dyeing reactive dyes are often used with cotton garments in preference to cold dyeing types, since the higher dyeing temperature makes for more efficient penetration. For goods that will be subjected to hand washing, however, dyes will probably be selected using the more levelling 'A' and 'B' classification of direct dyes (see Section 4.2.2.2), which, whilst possessing only modest wash fastness, are generally suitable for hand washing.

The sewing thread is also of concern. Usually it must remain inconspicuous, and this entails the use of dyes with a similar substantivity on both the sewing thread and the fabric. Occasionally stitching is deliberately used to produce decorative effects, in which case differential dyeing of the thread and the fabric may be a desirable feature.

Knitted wool garments present similar problems of penetration, and in addition care is needed to avoid fibre damage, particularly when long dyeing times or low pH values are used. Again, dyes are chosen to match the requirements of the attached aftercare label. Thus, equalising acid dyes are used for garments that will be hand washed, whereas metal complex and milling acid dyes provide better wet fastness and deeper shades for more robust articles. On the other hand, reactive dyes are generally the choice for wool fabrics finished to machine-washable standards.

Articles such as buttons or zip fasteners may present additional hazards to the dyer. The presence of metal ions in the dyebath from buttons or zips can influence the shade of the dye. Conversely, dyebath auxiliaries such as acid, alkali or reducing agent (for vat dyes) may do irreversible damage to the appearance of such trimmings.

The machines used for dyeing garments have to be constructed so that they do not cause damage to the garments. The surfaces that come into contact with the garments must be perfectly smooth and the fluid flow must be gentle. Different types of machines have been developed, with most designed to operate at atmospheric pressure. However, high-temperature machines are available for dyeing garments containing texturised polyester.

4.3.6.1   Side-Paddle Machines

These machines are very simple in construction. The dye liquor is circulated in an oval-shaped vessel, with an oval-shaped island, by a paddle arrangement (Figure 4.26). There are several types of machines differing in the position of the paddles, but in all of them the action is gentle. The axle of the paddle is mounted just above the level of the dye liquor, so the blades of the paddle enter the dye liquor and the motor is geared to rotate them at slow speed.

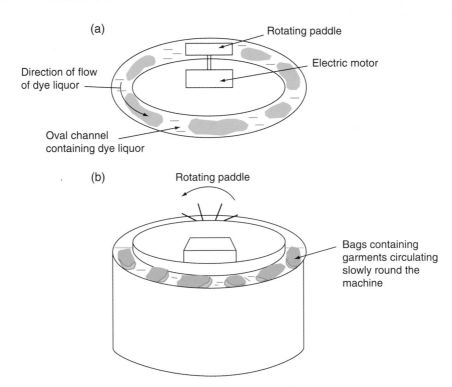

Figure 4.26 Schematic diagram of a side-paddle garment dyeing machine. (a) Top view. (b) Side view.

It is common practice to hold the garments in open mesh bags made of polyester or polypropylene so they do not become tangled. A few garments can be placed in one bag, as long as they are not compressed, so the dye liquor is able to move freely amongst them. The only point at which the liquid is forced through the garment structures in any way is when the bags containing them pass under the paddle. Thereafter the bags make their way round the channel in the gentle flow of the dye liquor. The lack of any forceful movement of dye liquor through the fabric creates predictable penetration problems with thick seams and with garments made from heavy fabrics. The liquor ratio in side-paddle machines is relatively high, at about 30 : 1, so dyebath exhaustion is not always as complete as desirable from both energy cost and effluent standpoints.

The operation of side-paddle machines is labour intensive, especially the unloading phase, which is a laborious, time-consuming task. It is, of course, necessary firstly to drain the machine of the exhausted dye liquor after dyeing and then refill it with cold water and allow the goods to circulate to remove unfixed dye. After removing the bags, the garments then have to be hydro-extracted than dried.

4.3.6.2 Rotating Drum Machines

This type of machine comprises a perforated cylindrical drum that is mounted horizontally in a vessel that just encloses it and contains the dye liquor (Figure 4.27). The garments are placed in the drum, which is then rotated slowly (about 3–4 rev/min) in the dye liquor, the

Figure 4.27 Rotary drum garment dyeing machine (Source: Photograph courtesy of Tonello S.r.l., Sarcedo, Italy).

direction of rotation being reversed frequently. As the drum rotates, the garments are carried up its inside wall and then fall back under gravity, a motion that gives frequent agitation of the garment in the dye liquor. These machines operate at lower liquor ratios (about 15:1) than side-paddle machines, so they are more efficient in terms of energy and water consumption. One of the latest types of rotating drum machine to be marketed operates with a liquor ratio of just 5:1. There are different designs of drums: some have just one compartment and are called 'open pocket' types, others have two compartments ('D pockets') and some three compartments ('Y pockets').

After the dyeing process is completed, the drum can run a centrifuge cycle that removes much of the exhaust dye liquor. This means that fewer rinsing cycles are then necessary and also that the goods leaving the machine are only damp, rather than very wet, which reduces drying time. Some machines have the facility to offload the garments on to a conveyor belt for onward transfer to a tumble drying machine, this reducing the amount of man handling of the goods. Traditionally, rotating drum machines have mainly been used for dyeing hosiery but increasingly are also used for socks and other small garment types.

4.3.7 Continuous Dyeing

When there is a large volume of goods, it is usually better to process them continuously without the inconvenience of separating the job into batches and repeating the process several times. In this situation processing is arranged as a continuous uninterrupted sequence of events from start to finish. The goods move at a constant rate throughout and, on completion of dyeing, are taken up mechanically at the far end of the equipment. The process is run until the whole length of textile has passed through.

Although most continuous coloration is carried out on fabric, equipment is also available for dyeing fibre in the form of loose stock, tops and tow. Whatever the form of the textile, a continuous dyeing range comprises the stages of impregnation followed by fixation, washing and drying. Various sequential operations are used for the dyeing of fabric. An initial padding stage is common to all sequences. It involves immersion of the fabric in the dye liquor that is contained in a trough of minimal volume, kept constantly replenished from a stock tank. A liquor ratio as low as 1 : 1 may be used (see Section 4.2.2.4). Next, the fabric passes in open width through a 'nip'. The nip is the padding mangle, in which heavy rollers (called 'bowls') are pneumatically pressed closely together along their length at high pressure. The rollers, about 2 m long and 35 cm in diameter and having a rubber surface, are rotated in opposite directions to carry the fabric through the system at constant speed, squeezing out any superfluous dye liquor (Figure 4.28a). Heavier fabrics are passed through two consecutive troughs and a second nip using a three-bowl mangle (Figure 4.28b).

The fabric, now uniformly penetrated with dye and the appropriate dyebath auxiliaries, usually passes directly to a steamer or dry heater for fixation. In some cases however, where evaporation of the water in the fixation chamber causes unwanted directional migration of the dye, the fixation step follows a preliminary drying operation. After fixing, the fabric passes from the fixing chamber to the washing range and is dried either on heated

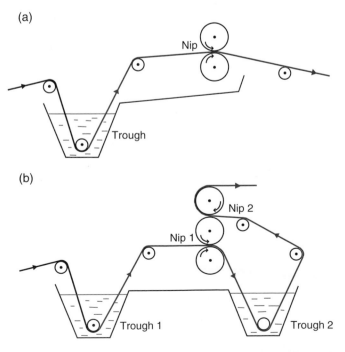

Figure 4.28   Principle of pad mangle operation: (a) two-bowl mangle, (b) three-bowl mangle.

cylinders or by some other means such as radio-frequency heating. Finally it is wound on to a beam ready for transportation. The sequence is similar for all processes, but of course the dyes, additives and operating temperatures vary from one situation to another.

4.4 Supercritical Fluid Dyeing

Globally, dyeing and finishing processes consume millions of gallons of water, mostly due to multistep processes that are largely inefficient. After the processes are completed, much of this water becomes polluted with unfixed dyestuff and chemicals such as auxiliaries and detergents. In response, dye manufacturers have developed dyestuffs and application processes that aim to increase the exhaustion of dye on to the fibre and also reduce substantially the need for auxiliary chemicals. Examples of such developments are reactive dyes that can be applied without the need for salt and can be applied at low temperatures. For their part, dyeing machinery manufacturers have focussed on the development of machines that significantly reduce water usage.

The ideal situation is to be able to dye fibres without the need for a water-based process, and this can be achieved by dyeing from supercritical fluids [3]. The most useful compound for this application method is carbon dioxide (CO_2), mainly because it is cheap, non-toxic and non-flammable. Supercritical carbon dioxide ($scCO_2$) is formed at high temperature and pressure, typically around 40–120 °C and over 100 bar (Figure 4.29). In the supercritical state it behaves like a gas as well as a liquid. The gaseous properties enable it to diffuse into polymers, giving a swelling and plasticising action, whilst its liquid properties enable it to dissolve disperse dyes.

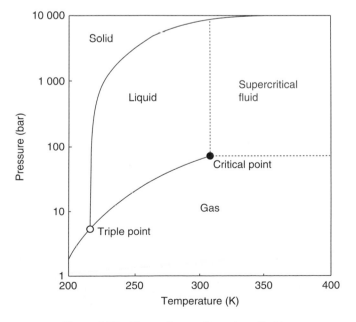

Figure 4.29 Phase diagram for carbon dioxide.

One unfortunate property of $scCO_2$ however is that it has a low polarity, so it can only easily be used for the application of the non-ionic disperse dyes to hydrophobic fibres such as polyester and nylon: it is not suitable for the application of ionic dyes, such as acid, direct or reactive dyes, which means it is not easily adapted to the dyeing of natural fibres. Nevertheless, given the huge popularity of polyester, dyeing in $scCO_2$ offers significant savings in water consumption. The dye formulations do not require auxiliaries; indeed pure dyes are used. The dyeing times are shorter than that by conventional aqueous methods, and no drying of the material is required after dyeing. However conventional machinery is required for any preparation and finishing processes that need to be undertaken either before or after dyeing.

The cost of supercritical dyeing machines is more expensive than conventional machines for aqueous dyeing, but the operating costs are lower and the quality of the dyed polyester is comparable in terms of colour yields and fastness properties.

During the dyeing of polyester, in addition to dissolving the disperse dye, the small CO_2 molecules are able to diffuse into the internal free volume of the amorphous regions of the fibre and cause plasticisation by increasing the segmental mobility of the polymer molecules, in much the same way as carriers do when present in aqueous dyeing processes. The action reduces the glass transition temperature (T_g) of the fibre by some 20–30 °C, and the increase in free volume causes the polymer matrix to swell. Under these conditions the disperse dye molecules are able to diffuse easily into the fibre. Indeed the diffusion coefficients of disperse dyes applied in $scCO_2$ are between 10 and 1000 times higher than the same dyes applied in an aqueous dyeing process. When the system is depressurised at the end of the supercritical dyeing process, the CO_2 molecules are released by the polyester. Some 95% of the CO_2 is recycled in a closed-loop system. One problem resulting from the plasticisation of polyester by the CO_2 molecules is that of the diffusion of oligomer to the fibre surface. Oligomers are very small cyclic trimers formed as bi-products during the polymerisation reactions to form polyester. They are slightly soluble in $scCO_2$ and their migration to the fibre surface can reduce the rub fastness properties of the dyed fibre.

The components of a typical $scCO_2$ dyeing machine are a tank containing CO_2, a compression pump, an autoclave, a dye container and a circulation pump. The material to be dyed, either as yarn wound on to bobbins or fabric wound on to a beam, is entered into the autoclave; then the system is pressurised and heated to the required settings for the dye. The $scCO_2$/dissolved dye is pumped through the autoclave by the circulating pump. After dyeing is complete, the system is partially depressurised and washed through with clean $scCO_2$ to remove unfixed dye. An example of a commercial machine is shown in Figure 4.30.

Besides the issue with oligomer, there are other problems with dyeing in $scCO_2$. Uneven distribution of dye through the material, for example, through a package or through multi-layers of fabric, may occur due to non-uniform flow of the $scCO_2$, or fluctuations in the temperature or pressure in the autoclave. The actual flow rate of the $scCO_2$ can also exert an influence, and in general reversing the direction of flow is helpful in avoiding unlevelness. The incomplete dissolution of the dye in the $scCO_2$ can cause dye spots to form on the material. The dyeing of mixtures is not straightforward: the uptake and colour yields of individual dyes can be different when they are dyed in admixture than when dyed alone.

Figure 4.30 DyeCoo scCO$_2$ beam dyeing machines installed at Yeh Group factory in Thailand (Source: Reproduced by kind permission of the Yeh Group).

Also the optimum conditions of temperature and pressure of the scCO$_2$ can be different for different dyes, so finding compatible dyes is a challenge.

Supercritical dyeing of other synthetic fibres, such as nylon and polypropylene, has not been quite as successful as with polyester and is not commercially available, mainly because of lower dye sorption and poorer fastness properties. However the use of special reactive dyes enables nylon fibres to be dyed, though this has not been done commercially. Also, despite much research, the dyeing of natural fibres is not successful because of the insolubility of ionic dyes in scCO$_2$. Although disperse dyes have been tried with these fibres, they inherently lack substantivity and have proved unsuitable. It is possible that reactive disperse dyes may be a possibility for natural fibres, but as yet there is no commercially viable process using them.

Despite the difficulties mentioned a number of companies have developed scCO$_2$ dyeing machines, and dye manufacturing companies have developed specially selected disperse dyes. For example, the Yeh Group in their Tong Siang factory in Thailand are an early adopter and pioneer in the commercialisation of this technology. Supercritical dyeing of polyester with disperse dyes is therefore now possible industrially and is likely to become a more dominant method of dyeing polyester in the future.

References

[1] W Cowell, *Int. Dyer*, **200** No. 3 (2015) 36.
[2] I Holme, *Int. Dyer*, **200** No. 5 (2015) 16.
[3] M Banchero, *Color. Technol.*, **129** (2013) 2.

Suggested Further Reading

A D Broadbent, *Basic Principles of Textile Coloration*, Society of Dyers and Colourists, Bradford, 2001.

C Hawkyard, Ed., *Synthetic Fibre Dyeing*, Society of Dyers and Colourists, Bradford, 2004.

D M Lewis and J A Rippon, Eds., *The Coloration of Wool and Other Keratin Fibres*, John Wiley & Sons, Ltd, Chichester, 2013.

J Shore, Ed., *Cellulosics Dyeing*, Society of Dyers and Colourists, Bradford, 1995.

J Shore, *Blends Dyeing*, Society of Dyers and Colourists, Bradford, 1998.

D H Wyles, Functional Design of Coloration Machines in *Engineering in Textile Coloration*, Ed. C Duckworth, Society of Dyers and Colourists, Bradford, 1983.

5

Textile Printing

5.1 Introduction

The word 'printing' is derived from Latin, meaning 'pressing', and implies the process in which pressure is applied during the printing process. Printing involves the production of multiple copies of images or designs of various complexities and commonly with a range of colours. The printing process differs from the dyeing methods of colour application in that it involves the *localised* dyeing or pigmentation in a way dictated by the image. This localised application of colour must be carried out with precision since each colour of the image is applied via a screen and registration of the print is paramount to the success of the operation. On an industrial scale, printing is used to provide repeat copies of artistic designs on textiles in a short period of time. However, the time taken depends upon the complexity of the design and the colorants, either dyes or pigments, used to colour the image.

From a historical perspective, printing may have originated in India, Egypt or China at around 3000 BC. Wooden block printing centres existed in India around the first century and became prevalent in Europe in the fourteenth century. The use of stencils to make multicoloured prints began to emerge from Japan in the seventeenth century. Nixon introduced an engraved printing technique in Dublin in 1752. In 1783 copper roller printing became firmly established with several patents for the process being submitted by Bell. In 1850 the first silk-screen printing methods started to emerge in France. Rapid success of this technique followed and in 1976 screen printing had taken over from roller printing as the most important industrial process, utilising both flat and rotary screens. The 1990s saw the introduction of ink jet printing (based on digital technology), and this technology soon began to impose a serious challenge to previous printing techniques.

Currently worldwide annual output of printed textiles is in the region of 20 billion metres. Although ink jet printing has not superseded rotary screen printing in terms of machine running speed, ink jet lends itself to more complex, intricate designs and allows for shorter runs of cloth, making it a competitive process. There are also several other significant advantages of ink jet printing compared with either flat or rotary screen printing. For example, ink jet printing does not require the costly operation of producing a screen for each colour to be printed, and there is less waste of printing paste.

However, in order to make comparisons of the various printing methods, it is necessary first to describe each in detail, encompassing print paste formulation, colorants, screen manufacture, printing styles and printing methods.

An Introduction to Textile Coloration: Principles and Practice, First Edition. Roger H. Wardman.
© 2018 John Wiley & Sons Ltd. Published 2018 by John Wiley & Sons Ltd.

5.2 Print Paste Formulation

The print paste is the medium by which the colorant (dye or pigment) is applied to the textile substrate. Many print works will have their own unique formulations of print paste that have been developed over a number of years and in latter times were closely guarded secrets, known only to the master printer. In general, the formulation depends upon a number of factors, including the dye or pigments being printed, the substrate and the fastness requirements of the print. The print paste must exhibit specific properties in order to achieve a successful print; the ingredients of the print paste and their functions are listed as follows:

• Dye or pigment (provides colour)
• Solvent (always water with textiles, the medium in which the other ingredients are dissolved/dispersed)
• Thickening agent (to increase the viscosity of the paste)
• Solution assistants (to assist the dissolving of the ingredients)
• Chemicals required to aid fixation of dyes (e.g. acids or alkalis)
• Reducing or oxidising agents (required by specific dyes for water solubilisation, also to prevent the reduction of dyes during steaming)
• Humectants (assist in swelling the textile fibre during steaming)

The above list comprises the general ingredients of a generic print paste, and as stated earlier, the various print works will have variations on the above. Taking into consideration the colorant, the print paste formulation is different when using a dye compared with a pigment and vice versa. Whereas dye-based printing pastes will facilitate the localised diffusion of the dye into the textile substrate, the pigment printing paste will focus upon the adhesion of the pigment to the substrate surface and the film forming over the pigment to assist in improved fastness properties of the print.

A *dye*-based print paste has the following ingredients:

• Colorant – Dye.
• Thickener – To provide print paste viscosity.
• Humectant – To maintain moisture content during printing and fixation (steaming). Most commonly used is urea (see structure **1.42** in Section 1.8.3.7).
• Acid/Alkali – Depends upon dye class being used. Alkali used for reactive dyes to facilitate covalent fixation to the substrate.
• Oxidising agent – Prevents 'frosting' of prints or reduction of dye during steaming. Compounds such as sodium *m*-nitrobenzene sulphonate.
• Water – Solvent for the ingredients.

A *pigment*-based print paste has the following ingredients:

• Suitable pigment dispersion
• Bonding and cross-linking agent (the binder) – To adhere the pigment to the surface of the substrate and form a transparent film over the pigment

- Thickener system – To provide print paste viscosity
- Softening agents – To modify the handle of the print
- Evaporation inhibiters – To prevent screen clogging
- Foam suppressers – To minimise foam build-up during printing
- Wash-off assistants – To prevent the paste drying out on the screen and to ease screen washing after printing

5.3 Thickeners

Print paste thickeners are colourless materials that form viscous pastes with water. The function of the thickener is to ensure that the print paste has satisfactory *viscoelastic* properties. The term 'viscoelastic' relates to viscosity and how viscosity changes during the application of the print paste to the fabric.

The print paste must be thick (viscous) enough to hold the colour in place on the fabric, but the viscosity of the print must reduce enough, under the *shear forces* applied, to flow through the print screen during application. The notion of a shear force can be understood by imagining a print paste sandwiched between two flat plates. If the top plate is moved whilst the lower plate kept stationary, the print paste experiences a shear force. The faster the top plate is moved, the greater the applied shear force. Most print pastes show a type of behaviour called *shear thinning*, which means that their viscosity reduces as the shear force increases. It is also a requirement of the print paste that once the shear force is removed, it returns to its original viscosity. Then when the print paste is deposited at the exact locality on the fabric where it is required, it stays there and does not bleed across the fabric.

The requirements of a thickener may vary depending upon the dye class being printed and the substrate. However the thickener should have the following properties:

- Compatibility with all the other components of the print paste
- Easy to dissolve in water
- Stable during the printing and fixation stages
- Capable of being removed in the final print washing stage

Thickeners may be divided chemically into three main types depending upon their chemical constitution: natural products, modified natural products and synthetic products.

5.3.1 Natural Products

Natural products are sugar-based carbohydrates and are categorised as polysaccharides. They are obtained from various natural sources:

- Starch-based products, which include British gums, obtained from plant seeds
- Alginates, obtained from seaweed
- Xanthans obtained from microorganisms
- Cellulose ethers, obtained from cellulose pulps

5.3.1.1 Starch-Based Thickeners

Starch contains both amylose, a linear chain polymer of α-glucose (1,4 linkages), and amylopectin, a branched polymer of α-glucose (1,6 linkages). Amylose has considerable hydrogen bonding between the polymer chains, making it difficult to dissolve in cold water. However the pastes produced have high viscosity with high shear thinning properties. Conversely, amylopectin is readily water soluble, producing low viscosity print pastes with no shear thinning. These different properties influence the character of a specific print paste depending on the ratio of the two within the print paste. Most starches contain about 20–30% amylose.

British gum thickeners are starch based, though the starch has been roasted at 135–190 °C for 10–24 hours. This produces random hydrolysis of the 1,4 bonds and formation of 1,6 linkages, increasing the branching of the polymer. This is shown chemically in Scheme 5.1.

The conversion to British gum results in increased water solubility of the thickener due to the change in amylose/amylopectin ratio. British gums have a high stability to alkaline conditions and are ideal for printing vat dyes.

5.3.1.2 Alginates

Alginates are obtained from the alginic acid found in brown seaweed. A copolymer is formed between β-D-mannuronic acid (**5.1**) and α-L-guluronic acid (**5.2**), linked in 1,4-positions in a ratio of 1.5 : 1:

5.1   **5.2**

Alginates are especially important when printing with reactive dyes on cellulosic substrates. The carboxylate ion present in the alkali print paste prevents the approach of the dye anion, and since neither monomer contains a primary alcohol ($-CH_2OH$) group, there is little chance of the reactive dye reacting with the print paste instead of reacting with the substrate. The viscosity of pastes is controlled by the incorporation of calcium ions (Ca^{2+}) into the print paste to encourage cross-linking, thereby increasing paste viscosity.

5.3.1.3 Xanthans

These soluble polysaccharides contain mannose, glucose and glucuronic acid. The print pastes produced have a high viscosity at low solid content and are economic for producing pastes of high viscosity. They are stable to electrolyte and rise in temperature and have a good working pH range of 1–11, making them suitable for the application of a wide range of dye application classes.

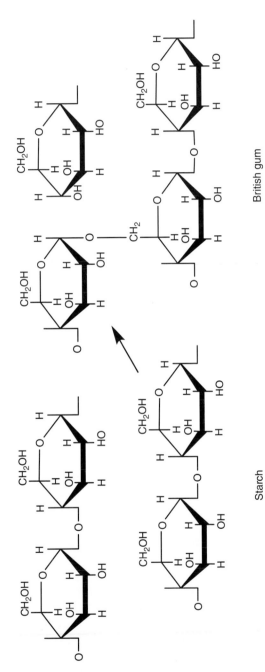

Scheme 5.1 Conversion of starch to British gum showing increased branching of the polymer chains following conversion.

5.3.2 Modified Natural Products

5.3.2.1 Carboxymethyl Cellulose

Carboxymethyl cellulose (CMC) is a cellulose derivative with carboxymethyl groups bound to some of the hydroxyl groups on the glucopyranose of the cellulose backbone (**5.3**);

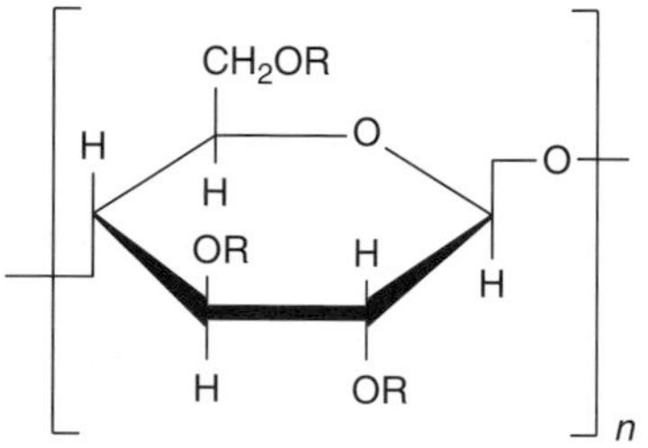

$R = H$ or CH_2COOH

5.3

CMC is synthesised by the alkali-catalysed reaction of cellulose with chloroacetic acid to modify the hydroxyl group(s) on the backbone ring. Various formulations of the CMC can be in the form of both high and low solids, so this enables the production of a wide range of print pastes. The pastes tend to have good stability during printing and good stability to highly alkaline pastes, allowing printing of both vat and reactive dyes to cellulosic substrates.

5.3.3 Synthetic Products

5.3.3.1 Emulsions

Emulsions in textile printing are classed as oil-in-water dispersions, whereby the 'oil' is white spirit (a turpentine substitute) and is called the *dispersed phase* and is typically 60% of the dispersion. The water is termed the *continuous phase* and accounts typically for 40% of the dispersion.

The viscosity of the dispersion is governed by the ratio of white spirit to water. Increasing the amount of oil increases the viscosity of the dispersion. The viscosity will also depend upon the size of the dispersed droplets – the smaller these are, the higher the viscosity of the dispersion.

The low water content within a dispersion may not allow for the complete dissolving of dyes and other auxiliaries within the dispersion, so emulsions are predominantly used for the printing of pigments to textile substrates. If emulsions are required for printing dyes, the print paste needs to incorporate a 'film former', a small quantity of natural or synthetic thickener. Incorporating the film former increases the viscosity of the continuous phase of the dispersion, in turn improving emulsion stability, since the amount of white spirit in the dispersion may be reduced.

Other synthetic emulsions have been evaluated for textile printing, with limited success achieved with polyvinyl alcohols for pigment printing. Emulsions for dye-based printing have tended to be copolymers of acrylates. Examples of these are the use of low cost copolymers of methacrylic acid and ethyl acrylate that produce low viscosity dispersions in water.

5.4   Binders

Binders are used specifically for the printing of pigments on to textile substrates. The role of the binder is to adhere the pigment to the substrate surface and create a protective film over the pigment. The following are the requirements of a binder:

*   Capable of application by dry/cure to form a film, trapping the pigment on the surface of the substrate
*   Provide adhesion of the pigment to the substrate surface
*   Able to cross-link in order to form a protective film
*   Possess 'elastic' properties to allow for bending and stretching of the substrate after printing

The cross-linking ability of the binder is paramount to the printing operation. Poor film formation or adhesion will result in the pigment peeling or flaking off from the substrate, resulting in loss of the printed image and a print of poor fastness.

Cross-linking may be achieved by self-cross-linking or with the use of suitable cross-linking agents incorporated within the print paste. Initiation of cross-linking may be brought about by temperature or pH change. Free radical polymerisation is commonly the mechanism to initiate the film forming process. Components within the binder generate free radicals that initiate the process. During drying the film is formed from the dispersed phase in two stages: the dispersed solids coagulate to form a gel layer of tightly packed spheres. Gel particles (spheres) flow together to form a continuous film. Recently ultraviolet light has been used to initiate the cross-linking process. The cross-linking reaction should only occur during processing and not during the binder/print paste storage, since this would result in an unusable print paste. The cross-links need to be covalent bonds to ensure a strong film is formed, which is resistant to hydrolysing agents used in aftercare processes, such as washing detergents and (during wear of the garment) to bodily fluids such as sweat.

Typically, binders are composed of unsaturated monomers such as methylolacrylamide or methylolmethacrylamide, acrylic acids, acrylamide, methacrylic acid and vinyl esters/ethers.

5.5   Pigments and Dyes

The choice of whether to use pigments or dyes as the colorant of the print will depend upon a number of factors, such as the complexity of the image to be printed, whether the required colour can be achieved with stock colorants, cost of printing, fastness requirements of the print and substrate type. The chemical structures of dyes and pigments have been covered in Chapter 3, so only specific properties relating to textile printing will be outlined within this chapter.

In terms of textile printing, dyes will diffuse into the textile substrate during steaming, form intermolecular bonds between the dye and the substrate and, in the case of reactive dyes, form covalent bonds with the substrate.

In contrast, pigments are insoluble in water and cannot be applied to textiles in the same manner as dyes. They have little substantivity for the substrate, that is, they are not retained by textile fibres and need to be adhered to the textile using binders, as discussed in Section 5.4. They are attached only on the surface of the fibres as a coating.

5.5.1   Pigments

Pigments are the most important group of colorants for textile printing. As stated previously they are adhered to the surface of textile substrates and are essentially trapped on the substrate surface by a binder. Pigments are relatively inexpensive and can be applied to almost all textile substrates provided that a suitable binder is available. This means that a wide range of textile substrates, for example, protein, cellulosic and synthetic, may be printed using just one system.

The pigment dispersion is key to the success of colour in pigment printing. The supplier formulates pigments into dispersions, and the pigments require milling to a specific particle size before being incorporated into the dispersion. Milling is carried out until an optimum particle size is achieved, which is usually in the region of 0.03—0.5 μm. If the pigment is not milled enough and the particle size is therefore too large, the resulting prints will appear flat and dull. Conversely, if the milling continues so that the pigment size is too small, such that the particle size is less than the wavelength of visible light, there will be a loss in covering power and colour intensity of the pigment.

There are significant advantages of printing textile substrates with pigments, since the process is relatively simple and productivity, especially for rotary screen printing, is quite fast. There is no need for the substantial wash-off processes associated with dye-based printing systems. Pigment printing is applicable to all textile fibres, and the only drawback may be the construction of the textile material. Since there is little or no wash-off after curing, considerable saving may be made in water, energy and labour costs. Savings in colorant cost, compared with dye-based systems, may be made since less colour is required for a pigment system because the colour sits upon the surface of the substrate rather than diffusing into the substrate for a dye-based print. Pigment printing has less of an environmental footprint compared with dye-based systems since there is little or no wash-off; therefore there are lower effluent loads. However there is still the need to wash pigment from the printing screens as is the need to wash dye-based print paste from screens, so there is still some environmental impact. Also any volatile chemicals leaving the substrate during the drying and curing stages need to be taken into account when viewing the process from an environmental perspective.

There are some disadvantages to pigment printing of textiles. Since the pigment is on the surface of the textile and covered with a film, the final handle of the textile may be compromised, especially if there are large areas of print. To overcome this problem, some printers apply a softener to the fabrics in post-curing stages. There have also been developments in the creation of 'soft' binder systems to reduce the impact on fabric handle. Problems with fastness, especially rubbing (abrasion) fastness, can arise, due to poor adhesion of the binder to the textile substrate or incomplete cross-linking of the binder during the curing stage. These defects may result in the lifting off of part of the binder together with the pigment, and this will flake off under any small rubbing action, for example, in domestic washing.

Careful selection of binder is required to suit the textile substrate being printed. For knitted structures, binders with high elastic properties are required so that the printed areas may stretch with the unprinted areas of the substrate. Failure to stretch will result in cracking of the binder, and this will begin to lose adhesion to the substrate and flake off from the surface.

Although, as stated previously, pigment printing has a lower environmental impact compared with dye-based printing systems, care needs to be taken when choosing thickeners and binders for pigment systems. Thickener choice, especially emulsion types, is important since a greater number of solvents are coming under increased legislation with regard to discharge limits to the environment. White spirit itself is becoming less environmentally acceptable and manufacturers are constantly looking to use more environmentally acceptable solvents. The list of acceptable alternative solvents is diminishing, therefore increasing costs on the few that remain that may not necessarily be readily obtained through simple fractionation processes of crude products.

5.5.2 Dyes

Dye-based textile printing is still popular since the benefits from dyes have many advantages compared with pigment printing systems. The physical form of dyes may be powder, grain or liquid, although for the powder and grain formulations, good water solubility is desirable since the thickener will take up a large portion of the available water in the paste. Dyes, during the steaming stage, will diffuse into the textile substrate, creating a print with good abrasion resistance and a soft handle compared with pigment-based printing systems. Wash fastness can also be significantly better for dye-based prints depending upon the dye class being used for the substrate.

Since reactive dyes form covalent bonds with the substrate, these prints will have superior wash fastness properties if the washing-off processes have been conducted correctly. Lightfastness properties of dye-based prints may be superior to pigment-based prints because light can damage and degrade the film protecting the pigment and cause loss of pigment from the substrate surface, whereas this does not occur on dye-based prints.

There are some disadvantages of using dye-based print systems. The lengthy steaming process to fix prints needs to be carefully controlled to maximise colour yield, and washing-off treatments are required to remove unfixed dye. Printing auxiliaries are required in order to obtain optimum fastness properties of the print. These processes are costly in terms of labour, time, energy and water and have an environmental impact if effluent is not treated correctly before discharge to sewers or open water. Care must also be taken if powdered dyes are used since there can be dusting issues when weighing out significant quantities of dye and also during the dissolving of the dyes.

The dye classes and/or pigments that are applicable for each substrate have been discussed at length in Chapter 3. It is safe to assume that the appropriate dye class to apply to a fibre, discussed in Chapter 4, is also applicable to printing.

5.6 Printing Screens

Apart from transfer printing and ink jet printing, print screens are required in order to create the desired image on the textile substrate. Each colour of the image/design requires a separate screen, for example, a three-colour design requires three screens, one per colour. Print designs are becoming more complex, and it is not unusual for 20 colours to be used in the design. This can be quite costly to set up since a screen is required for each colour and setting up the registration of the design (*print registration*) can be quite time consuming.

Print registration is the term used to describe aligning the distance between the screens so that the colours of the design appear in the correct position during printing as the substrate moves beneath the printing screens. This is illustrated schematically in Figure 5.1.

5.6.1 Flat Screens

Flat screens were traditionally made from a silk mesh that was stretched over a wooden frame. However, as designs became larger, the number of dye classes increased and designs became more complex, the wooden frames were replaced by alloy frames that were significantly lighter, making handling easier. They were also stronger and could support a larger surface area of mesh for the design. The woven mesh was also replaced with either polyamide or polyester, since these are stronger than the silk filaments and, in the case of the polyester mesh, does not adsorb dye during the printing process. Polyester mesh screens are easier to clean since polyester is hydrophobic whilst the majority of dyes are hydrophilic. On washing polyester screens after printing a dark shade, there is less chance of staining the substrate when a lighter colour is applied using the same screen. This may not be the case for polyamide mesh screens.

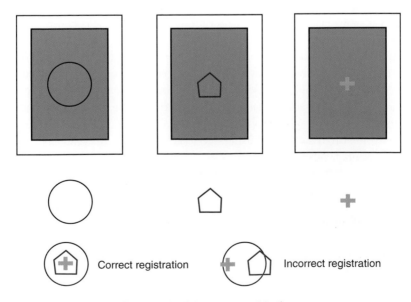

Figure 5.1 Print screen registration.

Since the design is put onto the woven mesh of the screen, the edges of printed areas appear to be serrated. This is particularly noticeable on designs with curves and swirls and designs requiring a coarse mesh gauge. The gauge of mesh used for the screen is dictated by the design being printed. Fine patterns such as small flowers require a fine mesh size since the amount of colour in a given area of the design is small and must be controlled to maintain the integrity of the design. Fine mesh gauges allow small quantities of low viscosity pastes through them, so obtaining medium to heavy shades with such frames can be difficult. Coarse mesh gauges are used for large blocks of colour, but as mesh gauge increases, so does the propensity of serrated image edge.

Each colour of the design is separated out and a screen is manufactured corresponding to each colour and the position of the colour in the design. The woven mesh is stretched across the screen, fastened into place and degreased to make sure it is completely clean. The screen is placed vertically in a jig and then coated with a light-sensitive polymer from the bottom of the screen to the top. The polymer is sensitive to ultraviolet light, so the coating is usually carried out under 'yellow' light that has the ultraviolet component filtered out. The polymer is allowed to dry out on the screen. A 'positive' of the repeat design (pattern) is fixed to a horizontal light box. The light box is essentially a glass-topped table that has a series of lights underneath the glass. The screen is laid flat on the light box so that the face of the mesh that will come into contact with the substrate is in contact with the pattern laid out on the light box. The lights are switched on so that the screen is exposed to the light. Where the pattern has been laid on the light box, no light will reach the screen, but all other areas will be exposed to the light. The polymer on the exposed mesh cross-links so that it is no longer water soluble. The screens are then soaked or washed to remove the water-soluble (unexposed) polymer, thereby enabling the dye to pass through the mesh during printing. No dye will penetrate through the other areas of the mesh since the cross-linked polymer will remain in these areas. Following washing, the screens are dried and examined for any defects or pinholes in the cross-linked polymer that may require filling in so that print paste may not pass through during printing. The screen is now ready for use in the printing machine.

Modern print screen manufacture set-ups include the use of ink jet printers to apply the light-sensitive polymer to areas that need to be impervious to print paste during printing. This allows for small reductions in cost since the whole screen is not covered in polymer and will not require washing following exposure to light. Other set-ups include the use of lasers to 'burn out' the required design on a screen that has been coated in polymer. Again this may be carried out digitally, so this will not require a 'positive' image being made for each colour of the design.

5.6.2 Rotary Screens

Lacquered rotary screens are produced in much the same way as flat screens, the only difference being that the screens are cylindrical, not flat. The mesh has hexagonal holes and is nickel coated so that it can endure the printing process. Again the design may be laser cut along the cylindrical roller for ease of screen manufacture.

Galvano rotary screens are thin inflatable nickel tubes, coated with a light-sensitive polymer. A full-length negative of the design is wrapped around the tube followed by a negative

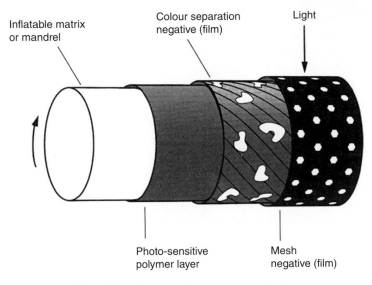

Inflatable matrix
or mandrel

Colour separation
negative (film)

Light

Photo-sensitive
polymer layer

Mesh
negative (film)

Figure 5.2 Galvano rotary screen construction.

of the mesh pattern. The screen is exposed to light, washed and dried. Only the non-pattern areas and areas corresponding to the supporting mesh in the pattern have no polymer. The polymer acts as an insulator upon subsequent nickel plating, nickel only building up in areas where there is no polymer. This results in a thin nickel sheet with holes only in areas of the pattern. Figure 5.2 shows the different layers in the manufacture of a galvano screen.

5.6.3 Engraved Rollers

Engraved screens are heavy copper rollers whereby the required design is etched or engraved into the copper to produce a 'cell' for the print paste. During printing the cell is filled with print paste, and as the screen rotates and the cell comes into contact with the textile substrate, the print paste flows out of the cell onto the substrate in the desired area of the design.

The engraving of the copper may be conducted by various operations. Chemical etching is carried out by coating the copper roller in a photosensitive polymer and dried. A positive image of the design is wrapped around the roller, and the roller is exposed to light. The polymer becomes insoluble and acid resistant, whereas an unexposed polymer is washed from the roller. Etching is carried out in a bath of ferric chloride solution and continues until V-shaped grooves have almost met each other in the area where there is no polymer on the roller. The etching is observed as grooves, and the number of grooves per inch determines the depth of the engraving and the amount of print paste held in the cell (Figure 5.3).

The depth of the etching is also important when considering the type of substrate being printed. An etching scale of 55 gives an engraving depth of 0.1 mm, which may be suitable for smooth-faced synthetic substrates. An etching scale of 35 gives an engraving depth of 0.20 mm, which is more suitable for cellulosic substrates. The relationship between cell depth and depth of shade on the textile substrate is shown in Figure 5.4.

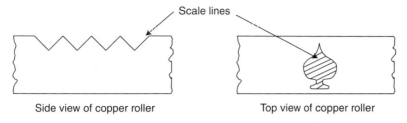

Figure 5.3 Etched grooves on a copper roller.

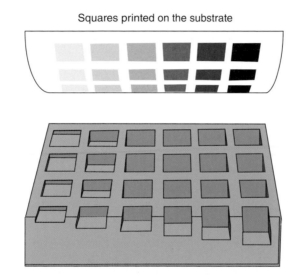

Figure 5.4 Engraved cells and depth of shade on a copper roller.

5.7 Stages of Printing

The printing process can be broken down into three stages:

- Transport
- Fixation
- Wash-off

Each stage is unique to the substrate being printed, the dye application class or pigment being used and the type of design being printed.

5.7.1 Transport

This is the stage that describes how the colour is applied to the textile substrate. This may be either the dye being forced through the mesh of a flat or rotary screen or deposited from the cell of a copper roller. The dye is transported to the fabric surface, and upon drying the paste forms a thin film containing the dye. This is not to be confused with pigment printing, since thickeners in dye-based print pastes form a film containing the dye, although the film does not cross-link as it does in pigment printing.

5.7.2 Fixation (Dye-Based Prints)

Fixation is the movement of the dye from the film to the interior of the substrate by diffusion. For this to occur, energy and moisture are required, and the size of the dye molecule, the chemical and physical structure of the substrate and the temperature/duration of the fixation process govern the time needed for fixation to take place. The dried print is subjected to steam to facilitate the diffusion of the dye into the substrate. For some dye application classes, such as reactive dyes, steam and alkali are required for the dye to diffuse and form covalent bonds with the substrate.

The type of steam used during the fixation process is important since this can influence the duration of the fixation process. Steaming may be conducted by using:

- Saturated steam
- Dry saturated steam
- Superheated steam

Saturated steam is steam that contains droplets of water and the temperature is typically 104 °C. The use of saturated steam does have disadvantages however, since condensation may build up within the steam, especially near cold metal surfaces. The condensation has a tendency to drip onto the printed fabric, causing the dye to bleed either into areas of the substrate where no dye is required or onto another colour of the design.

Dry saturated steam is steam that does not contain droplets of liquid water. Atmospheric steam is rarely used since most industrial boilers are pressurised. Table 5.1 shows the guidelines for the temperature and pressure of dry saturated steam.

Superheated steam is dry saturated steam that has been heated to a higher temperature (between 130 and 160 °C) for the pressure involved. Superheated steam is a gas that does not condense until it has given up its superheat; therefore steam pipes are not transporting 'water'. This also means that smaller-bore steam pipes may be employed in order to reduce cost and reduce heat loss due to the smaller surface area of the steam pipe.

Regardless of the type of steam employed for the fixation process, the steam must provide sufficient 'water' to swell the thickener film so that movement (diffusion) of the dye is possible. The 'water' allows for dispersion and diffusion of the dye, providing a liquid medium for the dye to diffuse to the fibre surface. This process can be envisaged as 'localised dyeing'. Raising the temperature of the dye/fibre system accelerates the whole 'dyeing' process to achieve fixation in a reasonable time (the duration of the steaming process). The steam also provides 'water' to swell the fibre to allow penetration (diffusion) of the dye into the fibre. The presence of urea in the print paste, to act as a humectant (see Section 1.8.3.7), enables the paste to hold on to the water from the steam that keeps the dye in solution. In the case of reactive dyes, the reaction with the fibre then takes place effectively and promotes colour yield.

Table 5.1 Guidelines for the temperature and pressure of dry saturated steam.

absolute (per kPa)	pressure (lb/in^2)	temperature (°C)
135	5	108
170	10	115
308	30	134

The most common type of steaming unit found in print works is the Festoon steamer. This type of steamer allows for a large quantity of fabric to be steamed in a short space due to the manner by which the fabric passes through the steamer.

5.7.3 Wash-Off (Dye-Based Prints)

Following steaming it is essential to wash off the printed steamed fabric in order to remove thickener, ancillary chemicals and unfixed dye. The wash-off provides the printed textile substrate with the full wet fastness performance of the print. Washing off is usually conducted in a series of bowls in a continuous line, similar to that employed in continuous dyeing as described in Chapter 4. The number and temperature of the washing bowls are determined by the dye application class being printed and depth of shade of the print. Usually the first bowl is cold water to remove a significant quantity of unfixed dye, without the dye migrating onto the unprinted areas of the fabric. The temperature of the bowls is steadily increased to 60–80 °C for the soaping stage and then decreased during the rinsing stages. In the final bowl a softener or other finishing chemicals may be applied to the fabric, followed by drying.

5.7.4 Pigment Prints

The printing of pigments does not require the steaming and lengthy wash-off stages employed in the printing of dye-based systems. Fabric is printed with the print paste containing the pigments, binder and ancillary chemicals. This is followed by drying/curing where the printed fabric is dried to facilitate film formation and adhesion of the binder containing the trapped pigment to the surface of the substrate. As stated previously, a curing stage follows drying in order to cross-link the binder to strengthen the film formation to increase the fastness properties of the print. There is no specific wash-off stage for pigment prints; however depending on the end use of the fabrics, specific finishes may be applied after the curing stage.

5.8 Printing Styles

There are many different styles of printing depending on the print design and the final appearance of the printed fabric. Some prints are produced on a white (undyed) background; other print designs require the print to have a coloured background. The coloured background may be applied to the fabric before dyeing using either a batch or continuous dyeing process. There are also print designs that are printed on blend fabrics containing two different fibre types. Certain printing techniques can be employed to remove (burn out) one of the fibres in the blend to create an almost three-dimensional print.

5.8.1 Direct Printing

Direct printing is probably the simplest printing technique, in which colour is applied directly onto the fabric, followed by fixation. This style of printing produces a coloured print on a white background with each colour of the design requiring a separate screen.

5.8.2 Discharge Printing

In discharge printing the background colour of the print is firstly applied to the fabric prior to printing. This background colour can be applied by exhaust batch dyeing or by continuous dyeing processes. A print paste that, instead of containing dye, contains a reducing agent is then applied to the local areas of the fabric required by the design. It is essential to the printing style that the background colour is produced from dyestuffs that can be 'discharged', that is, destroyed by chemical reduction in order for the pattern to be displayed after printing of the discharge paste. This process is illustrated in Figure 5.5.

The chemical reduction leaves a white image on the coloured background. This technique allows for deep, rich background shades to be achieved. Common discharging chemicals are based on:

- Formaldehyde sulphoxylates such as sodium formaldehyde sulphoxylate (C.I. Reducing Agent 2)
- Thiourea dioxide (C.I. Reducing Agent 11)

There are certain disadvantages to this style of printing. The choice of reducing agent depends upon the fibre being printed and the dyes used for the background. The background dyes need to be relatively easy to discharge, so they tend to be azo-based colours. However, dyes with specific structural characteristics are more easily dischargeable than others. In general, monoazo disperse dyes based on azobenzene typified by structure **5.4** are most easily discharged:

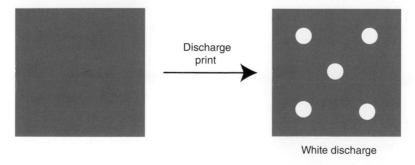

White discharge

Figure 5.5 The discharging of the background colour during discharge printing.

5.4

Dischargeability is dependent upon the substituents in the ortho- position to the azo group, particularly in the R_1 and R_2 positions. Substituents in the R_3 position have less influence. The dye C.I. Disperse Orange 5 (**5.5**) is easily discharged:

(5.5)

Of the two dyes, **5.6** and **5.7**, dye **5.6** is the easier to discharge; dyes with two nitro groups and one chlorine atom are generally more difficult to discharge:

(5.6)   **(5.7)**

It is possible to create an 'illuminated discharge' by incorporating a dye into the discharging paste. The dye or pigment in the paste must be resistant to the discharge chemical present also, so that it remains on the fabric when the illuminated discharge paste is printed. The illuminated discharge paste destroys the background colour of the print, but the 'illuminating' colour remains. This effect is illustrated in Figure 5.6.

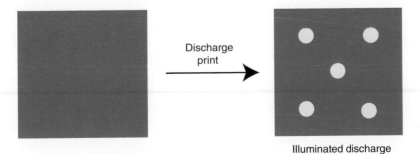

Illuminated discharge

Figure 5.6 Example of an illuminated discharge print.

Structure **5.8** is a typical example of an illuminating colour:

5.8

The above discharge printing styles have focused upon discharging the background dye. However printed effects may be produced whereby the substrate may be partially destroyed. This type of print, termed 'devore' or 'burn-out', is applicable to fabrics constructed of a blend of two or more different fibre types, typically cotton/polyester or wool/polyester. The partial destruction of the substrate removes one of the fibre types of the blend in the printed area. This type of printing forms a stable 'lace' type of effect since the non-burned-out portion of the substrate maintains the fabric construction. Therefore it is important that the fibre remaining after the printing process is a substantial proportion of the fabric blend and is contained in both the warp and weft of the substrate. Typical chemicals used to destroy one of the fibre types in the blend are acid generators such as sodium hydrogen sulphate, which destroys cellulose in cotton/polyester blends, or alkalis such as sodium hydroxide, which destroys wool in a wool/polyester blend.

5.8.3   Resist Printing

Resist printing, also known as reserve printing, involves printing the white fabric first with a paste known as a *resist paste*. The resist paste prevents the dye from being taken up in a subsequent dyeing process. The final result resembles much the same as that achieved using the discharge printing route. However, the resist printing route allows the use of a wider range of dyes that can be used in discharge printing. There are two types of resist printing technique depending upon how the resist is achieved:

- Mechanical resist, which is achieved by using materials such as resin, clay or wax (African batik prints). These form a physical barrier between the fabric and the dye and are used mainly for coarse decorative styles.
- Chemical resist printing, where the fabric is printed with the resist paste followed by overdyeing, either by batch or continuous methods. The resist paste prevents fixation or development of the ground colour by chemically reacting either with the dye or with the reagents necessary for the fixation of the dye. For example, the use of an acid will prevent the fixation of a reactive dye to a cellulosic substrate. The resist agent is applied prior to dyeing and can be used to achieve a white or coloured resist if a dye is incorporated within the resist paste. The dye must not be susceptible to the chemicals used in the resist paste in order to produce an illuminating colour.

5.9   Printing Methods

So far this chapter has discussed the chemical aspects of printing and the types of prints that can be achieved by altering print paste formulations. This section will concentrate of the types of printing methods in terms of the machinery used to deliver the print paste and the mode by which paste is delivered to the substrate, be it flat screen, rotary screen, copper roller, ink jet printing or another mechanism.

5.9.1   Flat Screen Printing

As the name suggests, the screens for this printing method are flat as opposed to circular as in rotary screen printing. The screen is a woven mesh, made from either polyester or polyamide. The mesh is stretched over a rectangular frame, originally made from wood, but now made from metal alloy to reduce weight and increase durability. Screen preparation has been covered in Section 5.6.1.

The operation of flat screen printing can be described in a number of steps. The substrate to be printed is attached to a continuous washable rubber blanket of the printing machine using an adhesive. The screens are placed on the fabric against registration stops to prevent the movement of the screen during the action of printing. A separate screen is used for each colour to be printed. The print paste is applied against the inside edge of the print frame and is forced through the open holes in the design area of the screen by means of a rubber squeegee moving across the frame, taking the print paste with it. The screens are then lifted off the fabric and the fabric is transported forward one repeat unit of the design by moving the blanket forward. The process is repeated again and again throughout the length of the substrate to be printed.

Print paste needs to be regularly applied to the screens as the process progresses to avoid any loss of the design. When the fabric reaches the end of the printing machine, it is taken off the blanket and dried prior to any further fixation processes. The blanket moves to the underside of the machine to be washed in order to remove any print paste that has penetrated through the substrate and to avoid staining of subsequent substrate on the blanket. Following washing the blanket is re-gummed, so it will adhere to the next portion of the substrate ready for printing. A schematic of a flat screen printing machine is shown in Figure 5.7, and a photograph of an industrial flat screen printing process is shown in Figure 5.8.

5.9.2   Rotary Screen Printing

As described previously rotary screen printing uses cylindrical screens as opposed to flat screens. Again, a separate screen is required for each colour of the design being printed. More complex designs require the application of many different colours, and typical rotary screen printing machines have the capacity for up to 20 screens. The screens rotate in contact with the substrate and the print paste is fed from inside the screens. The paste is forced from out of the inside of the screen by means of a metal squeegee blade. Again the fabric is adhered to a continuous washable rubber blanket,

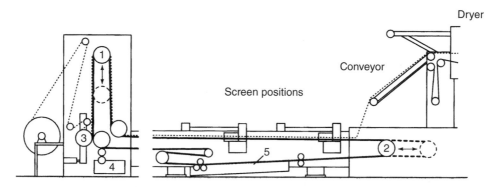

Figure 5.7 Schematic of a flat screen printing machine. Rollers 1 and 2 maintain the motion of the blanket. Roller 3 is a pressure roller. Roller 4 applies adhesive to the blanket. Roller 5 is the blanket washer.

Figure 5.8 Commercial flat screen printing (Source: Photograph courtesy of Standfast and Barracks Ltd., Lancaster, UK).

although in rotary screen printing the fabric and substrate run continually through the machine as opposed to a start stop motion in flat screen printing. The printed fabric is again taken off the end of the machine and dried whilst the rubber blanket is washed and re-gummed. A diagram of a rotary screen with squeegee blade is shown in Figure 5.9. A magnetic roller can also be used.

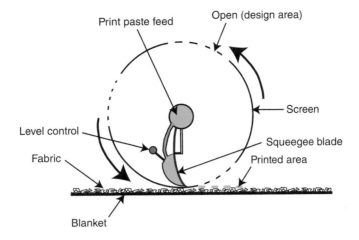

Figure 5.9 Cross-sectional diagram of a rotary screen.

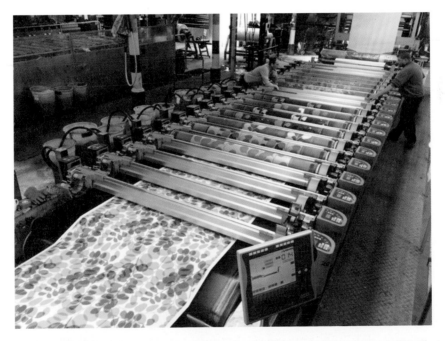

Figure 5.10 Commercial rotary screen printing (Source: Photograph courtesy of Standfast and Barracks Ltd., Lancaster, UK).

Figure 5.10 shows a typical rotary screen printing machine in operation. The design being printed is not especially complex, so not all screen locations are required.

5.9.3 Copper Roller Printing

Copper roller printing utilises engraved rollers as described in Section 5.6.3. Print paste is applied to the outside of the roller via a colour box so that it fills the cells of the engraved area. A colour doctor blade in contact with the roller surface removes

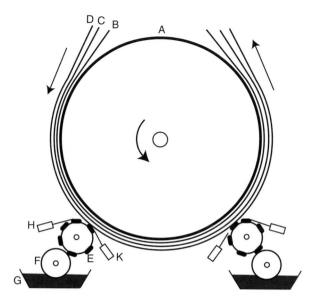

Figure 5.11 Diagram of a two-colour roller printer. A, pressure bowl (impression cylinder); B, endless printing blanket; C, back grey; D, fabric being printed; E, engraved printing cylinder; F, furnishing roller; G, colour box; H, colour doctor; K, lint doctor.

unwanted colour from the non-engraved areas of the roller. The substrate is moved through the printing machine on a gummed rubber blanket as in flat and rotary screen printing. As the paste-filled cell part of the roller comes into contact with the substrate, the print paste is transferred from the cell onto the substrate. Since this is a contact printing mechanism, and a heavy roller is in contact with the substrate, it is possible that small fibres from the substrate are transferred onto the printing roller. A lint doctor blade, in contact with the roller surface, removes these fibres since they may cause problems during printing if not removed. The problems that may arise are cross-contamination into non-printed areas, lifting of the colour doctor blade, thereby depositing paste onto non-design areas and contamination of the engraved cells that could result in loss of print paste. A schematic diagram of a two-colour copper roller printing mechanism is shown in Figure 5.11.

The complexity of producing copper rollers engraved with the required design for each colour to be printed means that to be cost effective, roller printing is only suitable when long lengths of fabrics need to be printed. For this reason it is a printing method very much in decline.

5.9.4   Heat Transfer Printing

This process is also a contact printing system though it does not use wet viscous pastes. The design is firstly printed on a special type of paper using sublimable disperse dyes. The printed paper can be stored until it is required. Prior to printing, the paper is inspected for any faults in the printed design before it is loaded onto the printing machine. The fabric is placed face down onto the printed paper (face up) and the two squeezed together as they

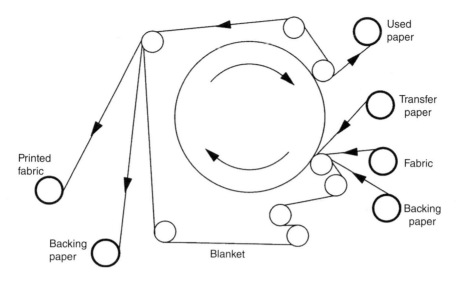

Figure 5.12 Continuous transfer printing.

pass through heated rollers (similar to calendars). The design is transferred from the print paper onto the textile substrate by heat and pressure.

During the transfer process, the heat causes the disperse dye to sublime. The dye vaporises and migrates into the fibre. This process is conducted on polyester or substrates with a high proportion of polyester in the blend. A diagram illustrating the operation of transfer printing is shown in Figure 5.12.

5.9.5 Ink Jet Printing

Ink jet printing is a non-contact printing system because the print head does not touch the substrate surface. The dyes to be printed are in the form of inks rather than pastes, and the viscosity of the inks is low compared with the pastes used in the printing systems previously described. The inks themselves have a high purity in terms of dye content, and only a few other additives are present in their formulation. This is in contrast to the commercial dyes used for flat and rotary screen printing where the dye content can be as little as 50%. For ink jet printing there is no need for the manufacture of printing screens since the design is a digital image.

Ink jet printing involves the deposition of small jets (as ink droplets) of a coloured ink on the substrate. The jet of ink is controlled by a computer as required by the digital image. The quality of the image being printed is defined by the resolution of printing by 'dots per inch' (dpi).

Rather than the traditional primary colours red, green and blue used in dyeing, ink jet printing is based upon the colours cyan, yellow, magenta and black, known as CYMK. Although there are four main colours, the colour palette is quite restrictive, but 'spot' colours can be added by the printer since print heads offer the capability of dispensing 6, 8 or even 12 colours during printing. These 'spot' colours tend to be orange or green but can be colours specified by the client if their design is based upon a specific shade. The use of spot colours allows other aesthetics to be added to the print, such as glitter or pearlescent effect.

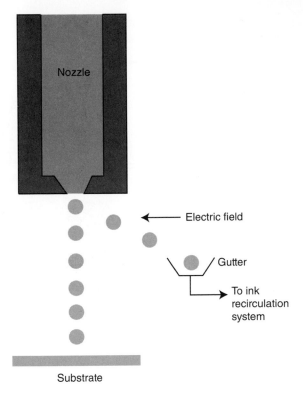

Figure 5.13 Continuous ink jet printing system.

Ink jet printing on textiles was originally used for printing designs on carpets; however it has become much more popular for printing on textiles for apparel wear. As print runs per design become shorter, ink jet printing is fast become a commercially viable method of printing textiles, since the cost of expensive inks is offset by the elimination of the need for screens per colourway of a design as is required in flat and rotary screen printing.

5.9.5.1 *Continuous Ink Jet Technology*

In this ink jet system a continuous stream of ink is fired out of the print head nozzle. An electrical charge is then used to direct individual droplets to either the substrate or a collection gutter where the ink is recirculated back to the ink reservoir. The disadvantage of this system is that the charging of droplets requires extreme control for the system to work satisfactorily. A diagram showing the principle of operation of this system is shown in Figure 5.13.

5.9.5.2 *Thermal Ink Jet Printing*

This type of system is known as 'drop on demand' since the ink is fired out of the printing plate nozzle when required by the digital design. This is made possible by a small heating element inside the print head that ejects the ink. The element heats up, rapidly causing a small amount of ink to vaporise, creating a bubble within the print head chamber; this forces the ink out of the nozzle. This is shown schematically in Figure 5.14.

5.9.5.3 *Piezo Ink Jet Printing*

This system is also a 'drop-on-demand' system, but the mode of ink ejection is different to that of a thermal ink jet printing head. The piezo ink jet head utilises a diaphragm that is attached to a ceramic plate. As an electrical charge is applied to the plate, the material expands and contracts. This movement forces the diaphragm to move in and out, generating pressure on the ink, forcing it out of the nozzle onto the substrate. A schematic diagram of a piezo print head is given in Figure 5.15.

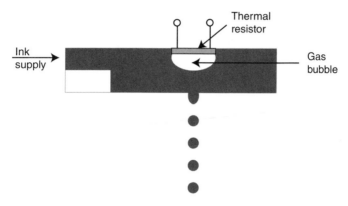

Figure 5.14 Diagram of a thermal ink jet printing head.

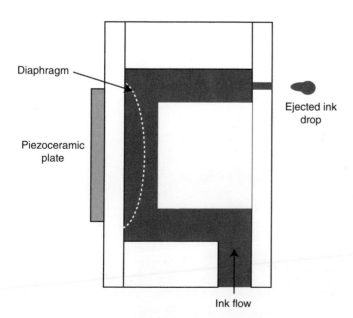

Figure 5.15 Diagram of a piezo ink jet printing head.

5.9.6   Comparisons between Ink Jet Printing and Screen Printing

Since the evolution of ink jet printing, it is envisaged that it will eventually overtake conventional printing systems such as flat and rotary screen printing as the most commercial form of textile printing. Ink jet printing does not require screen manufacture or machine set-up for print registration. Also there are no costs involved in making up print pastes since the inks are usually supplied in replaceable cartridges.

When ink jet printing first emerged, its main drawback was its low printing speed, and this, together with the high cost of inks, prevented its ready adoption by printers. However, over the last 5 years, there have been significant developments in ink jet printing technology that have led to a considerable increase in printing speed. This, together with the trend towards printing short runs, has made ink jet printing more commercially viable compared with more traditional textile printing systems. Currently, flat screen printing speeds tend to be in the region of 900–1800 m²/h depending upon complexity of the design being printed and the number of colours in the design. Rotary printing speeds are in the region of 3000–7200 m²/h, again depending upon complexity of the design being printed and the number of colours in the design. Ink jet printing speeds vary depending upon the printing machine. The Stork Sphere has a printing speed of 555 m²/h, hardly making it competitive to flat screen printing speeds; however the cost of screen manufacture and registration needs to be factored into the equation when comparing the cost of printing. The Xennia Osiris is capable of printing speeds of 2880 m²/h, making it faster than flat screen printing and not far short of rotary screen printing speeds. The MS LaRio is capable of printing speeds of up to 3,500 m²/h, making it faster than flat screen printing and comparable with the slowest of rotary screen printing machines.

It is not hard to imagine that within the next few years, developments in the technology in ink jet printing machines will produce machines capable of the same speeds as rotary screen printing machines. There will be the added benefit of being able to print smaller print runs

Figure 5.16   Digital ink jet printing (Source: Photograph courtesy of Standfast and Barracks Ltd., Lancaster, UK).

without the need to produce printing screens and with reduced downtime between designs, so ink jet printing will be a favourable option in the future for textile printing. Figure 5.16 shows a commercial ink jet printing set-up, in which the contrast in the cleanliness of operation with that of flat screen (Figure 5.8) and rotary screen printing (Figure 5.10) is obvious.

Suggested Further Reading

T L Dawson and B Glover, Eds., *Textile Ink Jet Printing*, Society of Dyers and Colourists, Bradford, 2004.

L W C Miles, *Textile Printing*, 2nd Edn., Society of Dyers and Colourists, Bradford, 1994.

6

Theoretical Aspects of Dyeing

6.1 Introduction

When a textile material is placed in a solution of dye, the dye molecules spontaneously transfer from the aqueous phase into the fibre phase. What is of importance to the dyer is that the dyes in the dyebath transfer to the fibres in a reasonable period of time, become uniformly distributed through the fibres and result in level dyeing. Other important considerations are that in the case of applying mixtures of dyes, the dyes build up on-tone and when they are finally fixed in the fibres, they do not wash out in subsequent wet processes such as washing. A dye that meets all of these requirements is said to have *substantivity* for a fibre, but different dyes can have very different substantivities on the same type of fibre. In order to be able to compare the substantivities of different dyes, it is necessary to make quantitative studies of the uptake of dyes by fibres, and such studies involve principles of physical chemistry.

There are two types of study that can be carried out:

(1) The kinetics of dyeing processes, which are concerned with the factors that influence the speed with which the uptake of dyes by fibres takes place.
(2) Studies of dyeing processes that have reached equilibrium between the dye diffused into the fibre and the dye remaining in solution. This aspect involves the application of the thermodynamic principles established in Chapter 1, and these studies involve dyeing for times considerably longer than are commercially viable. However the measurements made indicate the true ability of a fibre to take up a dye, expressed as the *affinity*, which is a quantitative expression of substantivity.

Studies of these two aspects make it possible to describe quantitatively the influence of process variables such as temperature, concentration and the presence of auxiliary chemicals on dyeing behaviour. An understanding of the behaviour of dyes leads to the formulation of numerical models that describe various dye/fibre combinations. These models in turn provide the information necessary to establish groups of dyes that are compatible with each other when applied to a fibre in admixture and to ascertain the optimum conditions for a dyeing process.

An Introduction to Textile Coloration: Principles and Practice, First Edition. Roger H. Wardman.
© 2018 John Wiley & Sons Ltd. Published 2018 by John Wiley & Sons Ltd.

6.2   Kinetic Aspects of Dyeing

The transfer of dye molecules from the bulk dyebath solution into the fibre involves three stages:

(A)   Transfer of the dye molecules from the bulk aqueous solution to the surface of the fibre. This stage is influenced by the solubility of the dye and the flow rates of the dye liquor through the fibre, yarn or fabric in the dyeing machine.

(B)   Adsorption of the dye molecules onto the fibre surface. The word 'adsorption', rather than 'absorption', is used here because this stage is concerned with the dye becoming attached to the surface of the fibres. The factors of importance governing adsorption are the nature of the surface and the dye/fibre interactions occurring, temperature, pH, the presence of auxiliary chemicals and dye concentration.

(C)   Diffusion of the dye molecules from the fibre surface into the interior of the fibre. The rate at which this process occurs depends on factors such as the molecular size and shape of the dye, the structure of the fibre (e.g. how crystalline it is) and temperature.

Of these three stages, stage (B) is very rapid in comparison with the other two stages, and stage (A) is only fast if there is efficient movement of the dye liquor around the individual fibres. These stages are represented diagrammatically in Figure 6.1.

For the transfer of dye molecules to the fibre surface in stage (A), it is important that the dye liquor moves efficiently, relative to the fibres, in a process known as *convective diffusion*. Dyeing machines vary in their operation in that some circulate the dye liquor around the stationary fibres, others move the fibres in the stationary dye liquor, whilst others move both dye liquor and fibres at the same time. Convective diffusion is made up of the *macroscopic motion* of the dye liquor and *molecular diffusion*. If there is no movement of the dye

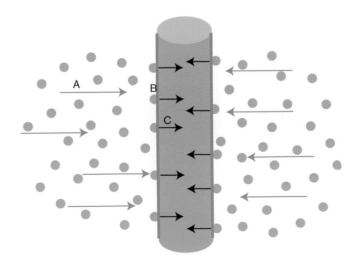

Figure 6.1   Transfer of dye molecules from dye solution into a fibre. A: Transfer of molecules to the fibre surface. B: Adsorption of dye molecules at the fibre surface. C: Diffusion of dye molecules into the fibre matrix.

liquor relative to the fibres at all, then transfer of the dye molecules to the fibre surface will rely totally on molecular diffusion, which is a slow process. However, whatever the speed of movement of the dye liquor, there is always a layer of liquid adjacent to the fibre surface that is stationary through which the dye must pass by molecular diffusion. This layer is called the *diffusional boundary layer* (see Figure 6.2). Because it is necessary for dye molecules to diffuse across this layer, stage (A) can influence the rate of dyeing. It is necessary therefore that this layer of stationary liquor is as thin as possible, which is achieved by effective stirring of the dye liquor (or movement of the fibres in the dye liquor).

As the dye molecules reach the fibre surface and are adsorbed onto it, the concentration of dye at the surface builds up. All naturally occurring processes involve levelling out differences in concentration between regions, so diffusion from the fibre surface into its interior occurs. Initially dye builds up at the fibre surface and the fibre is 'ring-dyed' ('A' in Figure 6.3). Gradually, as the dyeing operation proceeds, the concentration of dye increases towards the centre of the fibre ('B' in Figure 6.3) so that in an ideal situation, at the end of the process, there is a uniform distribution of dye throughout the polymer matrix of the fibre ('C' in Figure 6.3).

The diffusion of dye molecules from the surface of a fibre to its interior is a slow process and is influenced by the fact that the polymer structure is very non-uniform in terms of its ease of access. Some parts are highly crystalline, whilst other regions are more amorphous with voids through which dye molecules can pass more easily (see Section 2.2). There are

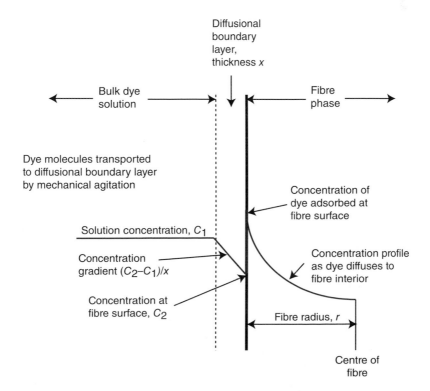

Figure 6.2   Transport of dye from bulk solution to interior of a fibre.

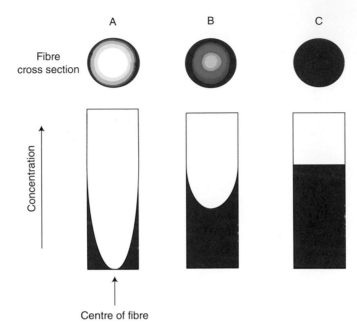

Figure 6.3 Diffusion of dye into a fibre from an infinite dyebath.

two models of the mode by which dye molecules diffuse into fibres – the pore model and the segmental-mobility model. The pore model visualises tortuous channels of water-filled pores in the amorphous regions of the fibre matrix through which dye molecules can pass. This model is particularly relevant to the dyeing of natural cellulosic fibres such as cotton and flax, which are highly crystalline in their structures but between the crystalline regions exist the amorphous regions. It is necessary for the dye molecules to be long and flat (planar) in shape so they can move along these narrow channels (Figure 6.4a and b). The segmental-mobility model envisages the displacement of the polymer molecules by the dye molecules as they enter the fibre (Figure 6.4c and d), this model applying to the diffusion of disperse dyes by hydrophobic fibres.

For some dye/fibre systems, such as anionic dyes applied to regenerated cellulosic fibres such as viscose and lyocell fibres, it is highly likely that a combination of the two processes takes place.

Another complicating issue is the fact that the dye molecules must diffuse through a medium to which they are attracted by various intermolecular forces (see Section 1.4.3). Dye molecules that are small in size are able to diffuse more quickly than large molecules, but the shape of the molecules also has an influence, with long linear-shaped molecules being able to diffuse more quickly than bulky molecules. Temperature is an important factor in governing the rate of diffusion: at higher temperatures all processes occur faster, so the rate of diffusion is higher at higher temperatures. Because diffusion, especially diffusion of the dye molecules through a solid substrate, is a relatively slow process, stage (C) of Figure 6.1 is usually the rate-determining step of the overall dyeing process. Studies of the diffusion properties of dyes in the fibre substrate are therefore important in understanding the kinetics of dyeing processes.

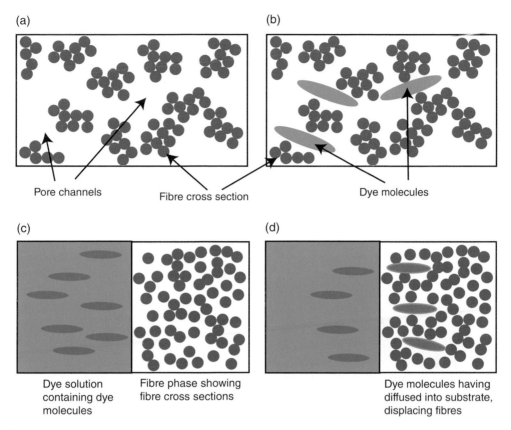

Figure 6.4 Diffusion of dye molecules into fibrous substrates. (a) and (b): pore model, (c) and (d): segmental-mobility model.

Before considering diffusion in more detail, it is useful to consider the behaviour of dye molecules in solution in water. It is well known that the molecules of many dye types aggregate in water, and this property markedly influences the ability of dyes to diffuse, both in water and in the fibre. The next section therefore considers dye aggregation in more detail.

6.3 Dye Aggregation

The ability of dye molecules to aggregate in solution is not always visually obvious, yet some dye types aggregate to such an extent that they form dispersions. The aggregation of dyes in solution leads to lower rates of diffusion of the molecules from the bulk solution to the fibre surface. Additionally, once adsorbed at the fibre surface, diffusion within the fibre is hindered, resulting in ring dyeing, as illustrated in Figure 6.3A. A lower colour yield is then obtained. Conversely, there are some advantages to aggregation of dye molecules in the fibre, as long as uniform distribution of dye throughout the fibre can be achieved. The light-fastness and wet fastness of dyes that aggregate in the fibre are very good. The improvement varies with the dye application class and the type of fibre being dyed. Table 6.1 shows the general aggregation properties of the various dye application classes.

Table 6.1 Aggregation behaviour of dye application classes.

application class	aggregation
Acid levelling	Low
Acid milling	High
Direct	High
Vat	High
Sulphur	High
Azoic	High
Reactive	Low
Disperse	High

Aggregates may form gradually over time, with the result that what was originally an optically clear solution becomes cloudy as it ages, and it is possible that the dye then precipitates out of solution. The formation of dye aggregates involves dye molecules aligning themselves closely in a regular way with, in some cases, well over 100 molecules closely bound in one aggregate. The aggregation number of a dye, which is the average number of dye molecules in an aggregate, can be determined experimentally, and it is found that different dyes all have their own individual behaviour. A number of different experimental techniques have been used to determine the aggregation number of dyes in solution. Such techniques include conductivity, light scattering, osmotic pressure and diffusion methods. Whilst the results of the various methods do not always agree with each other in absolute terms, they agree in the trends they show as the solution conditions change.

In general, the larger the size of a dye molecule, the greater is the tendency for it to form aggregates in solution. Attractive forces occur between the planar hydrophobic parts of dye molecules, and the larger the molecule, the greater the opportunity for aggregation to occur. However, two aspects of the structure of the molecule hinder aggregation:

(1) Charge repulsion
(2) Steric hindrance

Hydrophilic charged groups, such as sulphonate groups ($-SO_3^-$) present in molecules to confer water solubility, act to repel neighbouring molecules. The simple acid dye C.I. Acid Orange 10 (**6.1**) does not aggregate at all because its molecules have two negative $-SO_3^-$ (sulphonate) groups in the molecule, causing charge repulsion with nearby molecules, and this repulsion dominates over any attractive forces acting between the $-OH$ groups or between the hydrophobic aromatic rings, because the molecules are so small.

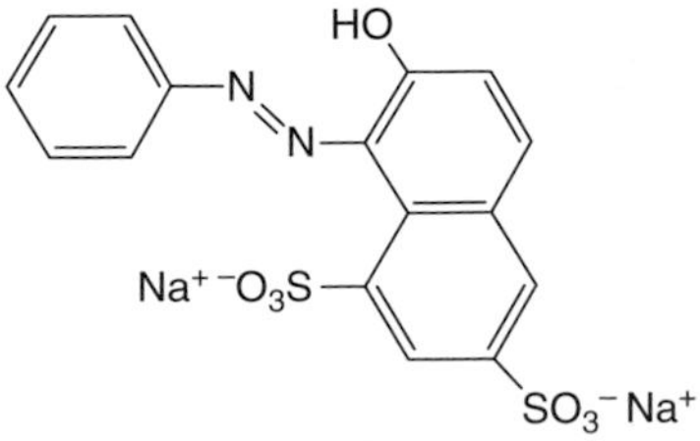

6.1

In larger dye molecules, the aggregation number is smaller if there are more $-SO_3^-$ groups and especially if they are spread apart in the molecule. The molecules of the dye C.I. Acid Red 111 (**6.2**) contain two $-SO_3^-$ groups, but they are close together at one end of the molecule, so this dye aggregates readily (**6.3**):

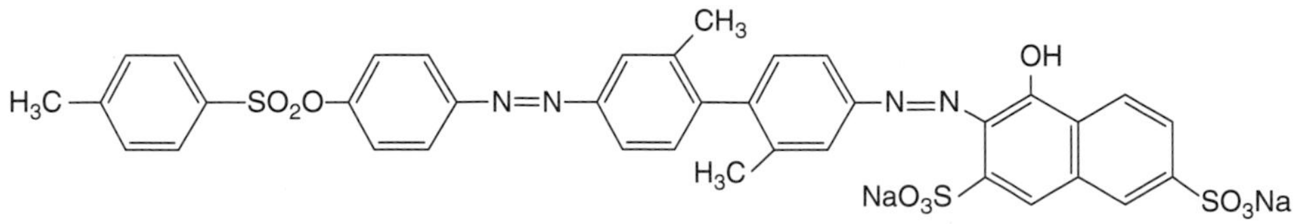

6.2

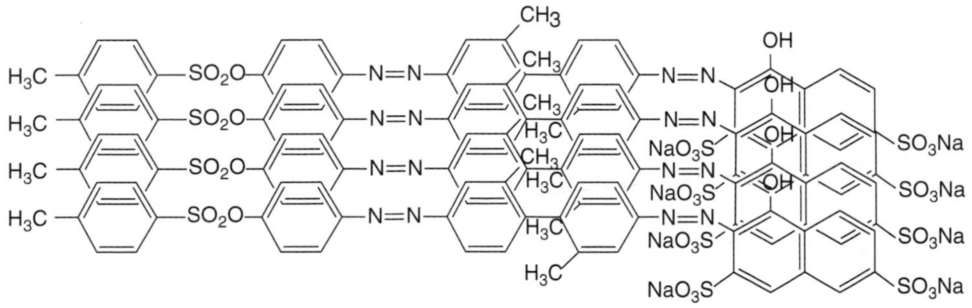

6.3

The size of a dye molecule is not the overriding factor in determining the aggregation behaviour of dye molecules; the ratio of molecular size to ionic group content is the most important factor. In comparing this ratio for the two dyes, Acid Orange 10 and Acid Red 111 (Table 6.2), it is easy to see why Acid Red 111, with its much greater ratio, aggregates so readily.

Another feature of the molecular structure of dyes that influences their aggregation behaviour is the shape of the molecule and the presence of any functional side groups in the molecule that may cause *steric hindrance*. Dyes whose molecules have flat planar structures show strong tendencies to aggregate, and typical of these types of dyes are those of the direct dye application class. A typical direct dye is C.I. Direct Red 31 (**6.4**), the molecule of which is long, thin and flat planar in shape:

Table 6.2   Size/charge ratios of dyes showing different aggregation behaviour.

dye	degree of aggregation	molecular weight	number of $-SO_3^-$ groups	ratio M.Wt/$-SO_3^-$
Acid Orange 10	Low	406	2	203
Acid Red 111	High	752	2	376

6.4

If the molecule is turned on its side, it is flat (**6.5**), and this planarity enables neighbour-ing molecules to align:

6.5

However, if bulky side groups are present, they can prevent the close alignment of dye molecules and so hinder the stacking of the molecules.

The sulphonate ($-SO_3^-$) group is bulky, having a tetrahedral shape, and if these groups are distributed across a dye molecule, they can reduce the extent to which the molecules are able to aggregate, as illustrated in Figure 6.5. In dye molecules containing even more sul-phonate groups, such as tri- and tetra-sulphonated dyes, aggregation is suppressed even more.

Aggregation tends not to occur when sulphonate groups are spread evenly across the molecule, as is the case with many reactive dyes. C.I. Reactive Red 1 (**6.6**) is a typical example of a dichlorotriazinyl reactive dye with sulphonate groups spread across the molecule:

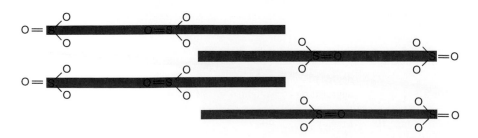

Figure 6.5 Disruption of molecular stacking due to steric hindrance by bulky sulphonate groups.

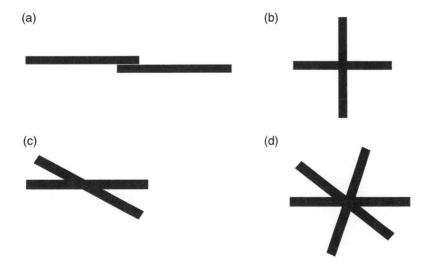

6.6

Distance is too large for meaningful intermolecular attraction

For aggregation to occur, it is not necessary for dye molecules to overlap completely, and aggregate structures such as those shown in Figure 6.6 are possible. Indeed overlapping of the type (d) has been confirmed [1] for the dye C.I. Direct Blue 1 (**6.7**). The molecules stack over each other at the central biphenyl group:

(a)

(b)

(c)

(d)

Figure 6.6 Overlapping structures of dye molecules in aggregates.

6.7

In addition to molecular size, shape and the ratio M.Wt:No. of $-SO_3^-$ groups, a number of other factors influence the extent to which a given dye aggregates in solution. These factors are:

- Concentration
- Presence of electrolyte
- Temperature
- Presence of organic solvent

Concentration

In the case of dyes that are prone to aggregation, the extent of aggregation increases with increase in concentration. This is because at higher concentrations, the repulsive forces are overcome, the influence of the charged groups being reduced by oppositely charged counter-ions. Usually at low concentrations, aggregation is minimal, and when the absorbances of dye solutions are measured, Beer's law applies (see Section 7.8). However at higher concentrations, when aggregates form, not all molecules in an aggregate are able to absorb light, so the absorbance is less than it should be. When a Beer's law graph is drawn, of absorbance versus concentration, the line deviates from linearity at higher concentrations (Figure 6.7):

Further evidence of aggregation occurring can be seen when the absorbance spectra of dye solutions at different concentrations are plotted. For example, in Figure 6.8, only one absorbance peak is observed for the dye at low concentration when the dye molecules exist as individual entities, but at a higher concentration when perhaps a dimer is formed, a second absorbance peak at a different wavelength is formed. The formation of the dimer reduces the

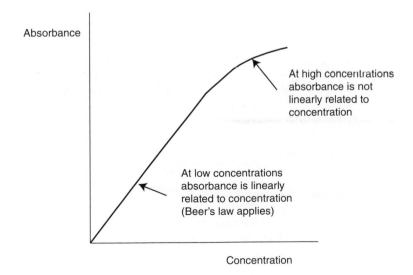

Figure 6.7 Effect of dye aggregation on Beer's law plots.

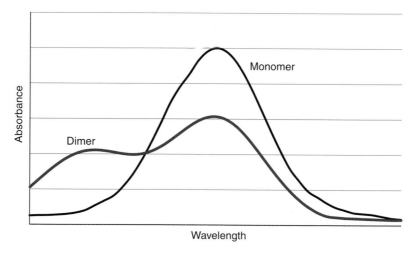

Figure 6.8 Absorbance spectrum of a dye forming a dimer at high concentrations.

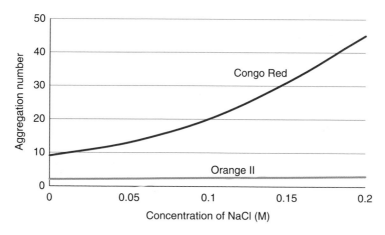

Figure 6.9 Effect of salt concentration on dye aggregation.

number of individual dye molecules in the solution, so the absorbance at the λ_{max} for the individual dye molecules decreases.

Presence of Electrolyte

In the presence of electrolyte, dyes in solution aggregate to a greater extent, because the repulsive forces are overcome, or suppressed, by the charged counter-ions. The mode of action is similar to that which occurs at the electrical double layer, and negatively charged dye ions are able to diffuse into negatively charged fibre surfaces (see Section 6.7.2). A good example of the effect of electrolyte is its significant effect on the dye C.I. Direct Red 28, known as Congo Red (Figure 6.9).

The molecules of the dye Congo Red are large and it only possesses two sulphonate groups so is prone to aggregation. The presence of electrolyte only serves to suppress charge repulsion between sulphonate groups, thereby facilitating further aggregation. In contrast, the dye C.I. Acid Orange 7 (known as Orange II) has a very simple structure (**6.8**) and does not tend to aggregate and so the salt has very little effect:

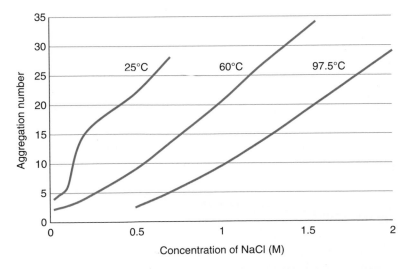

6.8

In general, dyes that have an inherent tendency to aggregate are more susceptible to the presence of electrolytes. Aggregation can cause considerable difficulties in dyeing processes by slowing both the rate of adsorption and the rate of diffusion of dye through the fibre.

Temperature

An increase in temperature causes dye aggregates to break down, so the aggregation number decreases as temperature increases. This trend is exemplified by the dyestuff C.I. Direct Blue 1 (**6.7**) as shown in Figure 6.10.

The aggregation of water-soluble dyes has consequences for the uptake of dyes by fibres during a dyeing process. Large aggregates of dye molecules diffuse more slowly through water to the fibre surface than the individual molecules, and so the rate of dyebath exhaustion is lower. As the temperature of the dyebath is raised and the aggregates break down, the rate of exhaustion gradually increases.

Figure 6.10 Effect of temperature on the aggregation of C.I. Direct Blue 1 (Source: Valko [3]. Reproduced with permission of John Wiley & Sons).

Solvent

If certain organic solvents or compounds such as urea are added to dye solutions, then the degree of aggregation (the aggregation number) reduces. The presence of a dye molecule in water, especially a dye molecule containing hydrophobic components, causes the water molecules around it to assume a more ordered structure than those further away. These water molecules therefore lose entropy. If the dye molecules aggregate, there will be fewer dye molecules to cause increased order of the water molecules, so there is greater disorder overall and the entropy of the system increases. Addition of a solvent, especially one that can preferentially form hydrogen bonds with water, such as ethanol, disturbs the water structure around dye molecules and enables dye molecules to exist more easily as individual entities. The compound urea has an especially strong action in this respect and greatly increases the solubility of dyes in water. For this reason it is often added to print pastes where the volume of water available to dissolve the dye and other print paste components is limited.

6.4   Diffusion

All naturally occurring processes react in such a way so as to achieve uniformity of conditions such as temperature, pressure or concentration in a system. If two solutions, one of higher concentration than the other, are added together, the molecules will redistribute themselves so that the concentration of the mixture is perfectly homogeneous throughout. In the absence of any external influence, such as stirring, this redistribution will occur spontaneously by diffusion. Diffusion occurs from a region of high dye concentration to one of low dye concentration, across what is called a *concentration gradient*, which is the difference between the two concentrations divided by the distance that separates them (Figure 6.11).

The steeper the concentration gradient, the faster the rate of diffusion, and this relationship is expressed by *Fick's first law of diffusion*:

$$\frac{ds}{dt} = -DA\frac{\Delta C}{x} \tag{6.1}$$

where ds is the amount of dye diffusing across an area A in a small interval of time dt, $\Delta C/x$ is the concentration gradient in which ΔC is the difference in concentration ($C_2 - C_1$ in Figure 6.11), x is the distance the two concentration regions are apart and D is the *diffusion coefficient* of the dye.

The negative sign in Equation 6.1 enables a positive value for D to be obtained, since ΔC will be negative.

The diffusion coefficient, D, is a measure of the diffusing properties of a dye in a particular fibre at a particular temperature. The higher the value of D, the greater the ability of the dye to diffuse. It is the values of D for various dyes that can be usefully compared. As might be expected, dyes of small molecular size have higher values for D than dyes of large molecular size, because they are more mobile.

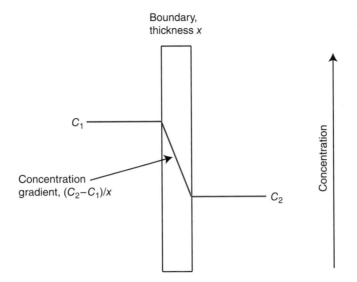

Figure 6.11 The concentration gradient.

6.4.1 Measurement of the Diffusion Coefficient of Dyes

There are different experimental techniques for determining the diffusion coefficients of dyes. Measuring D where the concentration gradient remains constant is a condition called *steady-state diffusion*. Unfortunately the method is limited experimentally because the equipment cannot be used with fibrous substrates. Instead polymer films, such as cellophane, have to be used, and although cellophane is cellulosic in nature, it does not have the complex morphological structure of cellulosic fibres such as cotton. However, to obtain D values from such experiments is easy, in that it simply involves the application of Fick's first law.

During a dyeing process the concentration gradient gradually decreases, because the dye concentration at the centre of a fibre increases, whilst that at the fibre surface decreases as the concentration of dye in the bulk solution diminishes. This type of diffusion is called *non-steady-state diffusion*, and whilst experiments of this type follow dyeing processes more realistically, actually interpreting the results to obtain D values is difficult.

Fick's second law of diffusion applies under non-steady-state conditions but is mathematically much more complex because it expresses the build-up of a dye at a given point in a medium as a function of time. It has to be remembered also that the dye diffuses towards the centre of the fibre from its surface all around its circumference. Experimentally, determining D using non-steady-state diffusion is more attractive however, and various mathematical analyses have been carried out to develop numerical equations that will enable the calculation of D from such experiments, but using what is termed an *infinite dyebath*. In this experimental method, a very small amount of fibre is dyed in a very large volume of dye liquor, so that during dyeing the change in concentration of dye in the liquor is negligible, and the concentration gradient remains fairly constant.

For these conditions an equation can be used, called *Hill's equation*, which assumes the fibres are infinitely long cylinders of radius *r*:

$$\frac{C_t}{C_\infty} = 1 - Ae^{-BK} - Fe^{-GK} - \cdots \tag{6.2}$$

where A, B, F and G are numerical constants of known value:

$$K = \frac{Dt}{r^2}$$

It is necessary to determine the value of C_∞, the concentration of dye in the fibre when dyeing has been carried out for several hours to reach equilibrium. Also dyed samples have to be prepared for short dyeing times, t, to determine C_t. Graphs relating C_t/C_∞ to K have been published, from which it is then possible to calculate D for fibres of known radius r.

There are some other mathematical approximations that work reasonably well. An equation that applies in the early stages of dyeing, that is, before the dye molecules have reached the centre of the fibre, is

$$\frac{C_t}{C_\infty} = 2\left(\frac{Dt}{\pi}\right)^{1/2} \tag{6.3}$$

To use this equation a graph is plotted of C_t/C_∞ against $t^{1/2}$. A straight line should be obtained, which has a slope proportional to $D^{1/2}$. Another simple equation, one that applies to longer dyeing times, enables D to be determined from knowledge of the *time of half-dyeing*, $t_{1/2}$ (see Section 6.5). For a fibre of known diameter, d,

$$D = 0.049\frac{d^2}{t_{1/2}} \tag{6.4}$$

These equations are useful in enabling a value for D to be obtained, and if the same techniques are used on different dyes, their diffusion properties can be compared.

However in absolute terms the values for D obtained must be treated carefully, because for many dyes, D can vary with concentration and thus with the depth of shade being dyed.

Values for diffusion coefficients vary widely, not just between the various dye/fibre systems but also within a single dye/fibre system. Also some dyes diffuse more easily through some fibres than others, as exemplified in Table 6.3 by the simple disperse dye C.I. Disperse Orange 3 and the more complex C.I. Disperse Blue 24.

It is clear that whilst C.I. Orange 3 diffuses easily through nylon, it does so much less easily through polyester, as might be expected at a temperature of only 100 °C, much below the normal dyeing temperature of 130 °C. In contrast C.I. Blue 24 diffuses much more slowly through all three fibres, though relatively easily through acetate fibres.

In general on one fibre type alone (nylon), the diffusion coefficients of dyes can vary by nearly a factor of 10, depending on their molecular complexity and the way they interact with the fibre. In the case of direct dyes, diffusion coefficients (in cellophane film) are typically about $20 \times 10^{-14}\,\text{m}^2/\text{s}$, but such is the variation that they cover a 200-fold range, depending on the size and shape of the dye molecule.

Table 6.3 Diffusion coefficients of two disperse dyes on different fibres [2].

dye structure	diffusion coefficient at 100 °C, $\times10^{15}$/m²/s		
	nylon	acetate	polyester
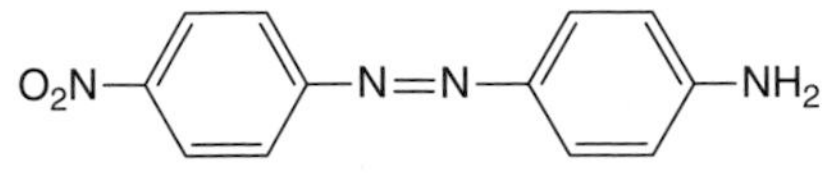 O_2N—⬡—N=N—⬡—NH_2 C.I. Disperse Orange 3	23.1	15.1	1.3
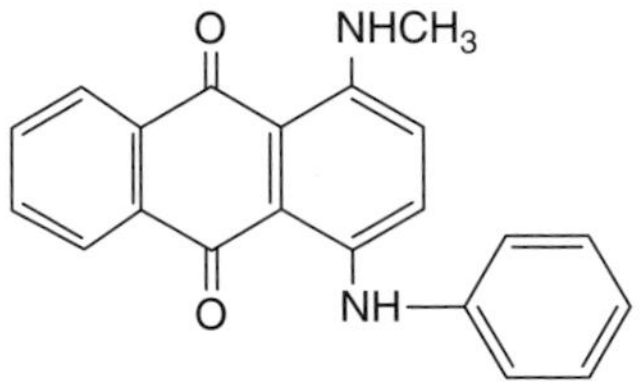 C.I. Disperse Blue 24	4.2	10.6	0.3

Source: Adapted from Renard [2].

6.4.2 Activation Energy of Diffusion

The rate of diffusion increases with temperature and a quantitative value that expresses the relationship is *activation energy of diffusion*. This concept is similar to the energy of activation of a reaction explained in Section 1.5.3. Not all dye molecules have sufficient energy to diffuse. The number of molecules with sufficient energy to break free from the surrounding polymer matrix of the fibre (i.e. 'activated molecules') is related to the total number of molecules and to the term $e^{-E/RT}$, where the activation energy of diffusion, E, is the amount by which the energy of an activated dye molecule exceeds the average energy of all the dye molecules. R is the gas constant (8.314 J/mol/K) and T is the absolute temperature.

The diffusion coefficient at any temperature T can be calculated using Equation 6.5:

$$D_T = D_O e^{-E/RT} \tag{6.5}$$

where D_O is a constant. Taking natural logarithms of Equation 6.5 gives

$$\ln D_T = \ln D_O - E/RT \tag{6.6}$$

The value of E can be obtained by plotting a graph of $\ln D_T$ against $1/T$. A straight line of slope E/R should be obtained, from which E can be easily calculated. To plot such a graph requires knowledge of the diffusion coefficients of a dye at various temperatures.

The activation energies of dyes of a particular dye/fibre system are usually very similar, though the values do vary considerably from system to system, as illustrated in Table 6.4.

Table 6.4 Activation energies of diffusion of different dye/fibre systems [4].

application class of dye	fibre	activation energy (kJ/mol)
Vat dyes	Viscose	52
Direct dyes	Viscose	59
Acid levelling dyes	Wool	92
Acid milling dyes	Wool	121
Disperse dyes	Polyester	167
Basic dyes	Acrylic	251 [5]

Source: Adapted from Bird [4].

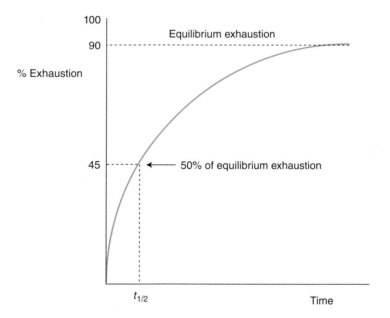

Figure 6.12 Rate of dyeing curve and time of half-dyeing.

6.5 Rate of Dyeing

The measurement of diffusion coefficients of dyes in fibres is useful because this stage of the dyeing process is usually the rate-determining stage. However it is of more use to dyers if measurements are made of dye uptake using conditions similar to those used in practice. One method for achieving this goal is to measure the percentage exhaustion (%E) of a dyebath at regular intervals during dyeing, %E being the percentage of dye originally present in the dyebath that has become adsorbed on the fibre at a given time. However, even this type of measurement is not straightforward because in many studies dyeing has been carried out at a fixed temperature, whereas in a practical dyeing process, the temperature is raised gradually as time progresses. Fixed-temperature measurements are useful though, and typically an exhaustion curve of the form is shown in Figure 6.12.

Table 6.5 Time of half-dyeing of direct dyes on viscose dyed at 90°C [6].

dye	M.Wt	$t_{1/2}$ (min)
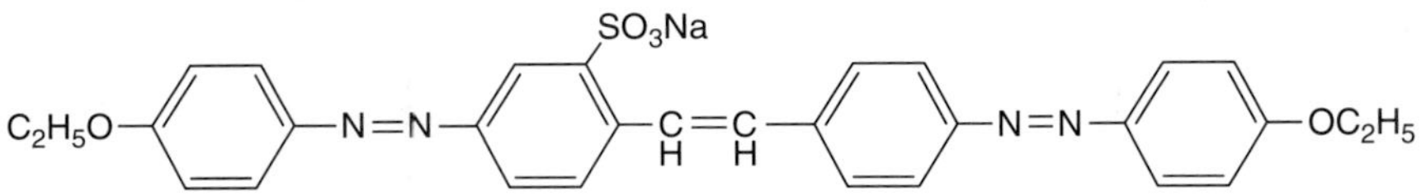 C.I. Direct Yellow 12	680	0.26
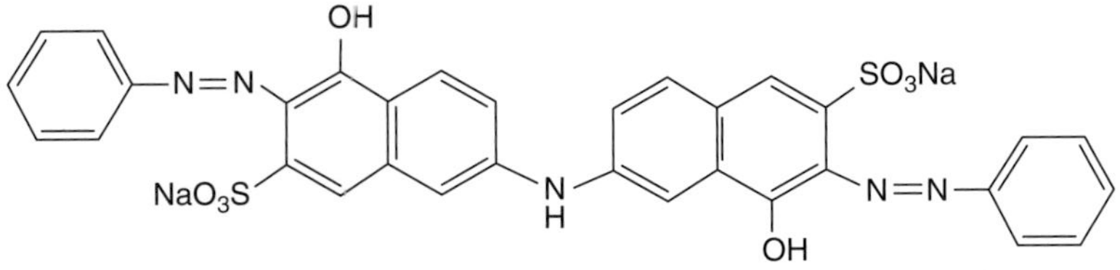 C.I. Direct Red 31	730	1.7
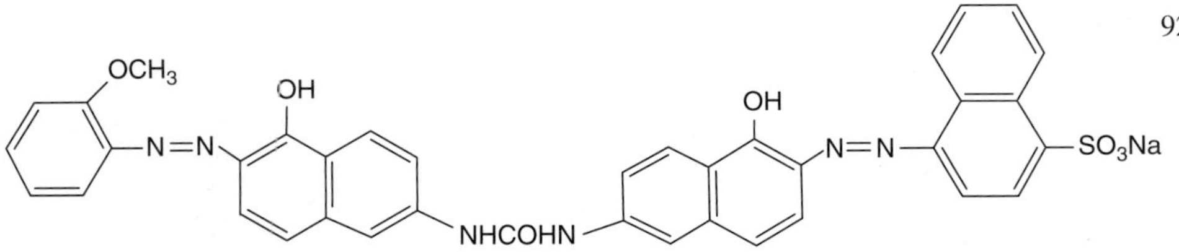 C.I. Direct Red 26	926	43.8

Source: Adapted from Boulton [6].

The rate of dyeing curve is useful in that it enables a value called the *time of half-dyeing*, $t_{1/2}$, to be obtained. This time is the time taken for the dyebath to reach 50% of its equilibrium exhaustion, the equilibrium exhaustion being the percentage dyebath exhaustion when the dyeing process has been carried out for a very long period of time. The times of half-dyeing of the three dyes referred to in Table 6.5 show how slowly the large dye molecules of C.I. Direct Red 26 diffuse into the fibre in comparison with the other two dyes of much lower molecular size. However the shape of the molecule also has a bearing on the diffusion characteristics of direct dyes in cellulosic fibres. Those dyes with long, thin-shaped molecules diffuse most easily into the fibre structure.

Some care is necessary in using $t_{1/2}$ values because some dyes have greater affinity for the fibre than others and are more readily adsorbed on the fibre surface. Such dyes are said to have a rapid 'strike', but they do not necessarily diffuse quickly into the fibre and 'ring dyeing' can result. In this case more time is required to enable the dye to diffuse from the surface to the interior of the fibre. Another factor to be considered is the equilibrium exhaustion, since if this is only low, say, 60%, then it will not take long for a dye to exhaust to half that value (30%), so $t_{1/2}$ will be low and it could be assumed wrongly the dye is building up quickly. Also it is quite possible for two dyes with very different build-up characteristics to have the same, or very similar, $t_{1/2}$ values, as demonstrated in Figure 6.13.

The rate of dyeing curves determined at different temperatures show, as might be expected, that dyebath exhaustion occurs more rapidly at higher temperatures, so the $t_{1/2}$ values are lower (Figure 6.14). However, often the equilibrium exhaustion when dyeing is carried out at higher temperatures is lower than those carried out at lower temperatures.

The reason for the lower dye uptake at higher temperatures is that the adsorption of dye by a fibre is an exothermic process so as it reaches equilibrium more dye transfers to the fibre at lower temperatures (see Section 1.5.2). This behaviour is not necessarily apparent with all dye/fibre systems but, especially with direct dyes on cotton, it is often useful to complete the dyeing process in a cooling dyebath to maximise dye uptake.

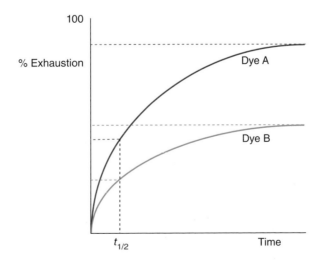

Figure 6.13 Illustration of a situation in which two different dyes of different build-up properties have the same $t_{1/2}$ values.

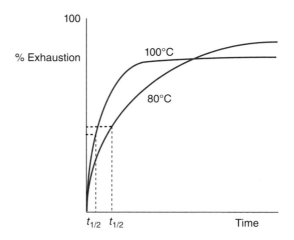

Figure 6.14 Effect of temperature on dyebath exhaustion.

6.6 Adsorption

Adsorption is a process in which a substance (the adsorbate) initially present in one phase becomes preferentially attached to another phase (the adsorbent). In a dyeing process, the adsorbate comprises the dye molecules, initially in the water phase, which are taken up by the fibre substrate (the other phase), which is the adsorbent.

There are two types of adsorption, physical adsorption and chemical adsorption (also called *chemisorption*), and both types are found in dyeing processes. The majority of dye/fibre systems involve only physical adsorption, but the application of reactive dyes to fibres involves chemical adsorption.

There are differences between physical and chemical adsorption:

- Physical adsorption only involves forces of physical attraction between the adsorbent and the adsorbate.
- Chemical adsorption involves the formation of chemical bonds, involving electron transfer, between the adsorbent and the adsorbate.

The two processes also differ in their heats of sorption, that is, the heat liberated when they occur. Heats of chemisorption are generally much higher (about 50–400 kJ/mol) than heats of physical adsorption (about 20–40 kJ/mol). Finally, the process of chemisorption requires an activation energy (see Section 1.5.3), whereas physical adsorption does not.

6.6.1 Physical Adsorption

It is useful to consider the types of forces of attraction that can exist between dyes and fibres. These are ionic bonds (Section 1.4.1) and the various types of secondary forces of attraction (Section 1.4.3), all of which involve the attraction between positive and negative charges between the dye molecules and the fibre substrate, the difference between them being the origin of the charges and their strengths.

Ionic bonds involve the attraction between oppositely charged chemical groups on the dye and fibre, where the charged groups have been formed by the loss (or gain) of an electron to form an anion (or cation). Typical of this type of bonding is that between acid dyes and wool or nylon fibres. In these cases the attraction occurs between dye ions that are negatively charged (anions) and 'sites' in the fibre that are positively charged (cations), as illustrated in Figure 6.15a. The reverse situation also exists, and the application of basic (cationic) dyes to acrylic fibres involves the attraction between the positively charged dye cations with negatively charged groups in the fibre (Figure 6.15b).

Polar forces, although they are only weak forces of attraction, provide an important contribution to the overall attraction between dyes and fibres. Figure 6.16a illustrates diagrammatically the polar forces operating between dyes and fibres in what is called *dipole–dipole interaction*, the $\delta+$ and $\delta-$ symbols representing the slight positive and negative charges, respectively. It is possible also for fully charged groups to induce a dipole in a molecule, this giving rise to the attractive force called *dipole-induced dipole interaction*. This is illustrated in Figure 6.16b, in which a charged group in a fibre creates a partial opposite charge in a dye molecule.

These types of polar force, together with *hydrogen bonding*, are responsible for the attraction between disperse dyes and hydrophobic fibres such as nylon and polyester. *π–H bonding* (see Section 1.4.3) also plays an important role in binding dyes to fibres. In this type of bonding, the negative electron clouds of a benzene ring can be attracted to an H atom of, say, an –OH group.

Finally, *van der Waals* forces, sometimes called dispersion forces, which are very weak forces of attraction between atoms and molecules, also make a contribution to the attraction between dyes and fibres. This type of force explains why hydrophobic regions in dyes and fibres, such as those exist between hydrocarbon chains, have attraction for each other.

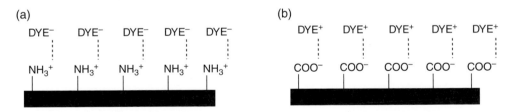

Figure 6.15 Ionic attractions between dyes and fibres. (a) Wool or nylon fibre. (b) Acrylic fibre.

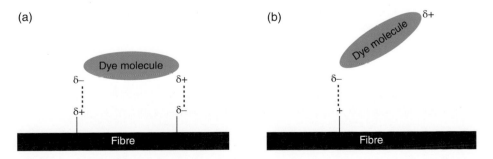

Figure 6.16 Polar forces of attraction. (a) Dipole–dipole attraction, (b) dipole-induced dipole attraction.

6.6.2   Chemical Adsorption (Chemisorption)

Chemical adsorption generally involves an attraction between the adsorbent and the adsorbate that is much stronger than that which exists through the polar and van der Waals type forces that exist in cases of physical adsorption. Certainly, when a reactive dye is applied to a fibre and becomes bound to the fibre by a *covalent bond*, the process can be referred to as chemical adsorption. Before any chemical reaction can take place, the dye has to be adsorbed by the fibre through initially a physical adsorption process. Once this has occurred, the dye and the fibre will then be in close enough proximity to each other so that chemical reaction to form a covalent bond can occur.

In a reactive dyeing process, physical adsorption is a necessary precursor to chemical adsorption, and so all of the factors that affect physical adsorption will have an influence in reactive dyeing.

6.6.3   Adsorption Isotherms

With the exception of reactive dyeing, dye adsorption occurs by physical adsorption, and at the end of the dyeing process, there is a distribution of dye molecules between the fibre phase and the solution phase. Dye adsorption isotherms are obtained by measuring how much dye is in the fibre $[D]_f$ and how much remains in solution $[D]_s$, when the dyeing has been allowed to proceed, at a constant temperature, to equilibrium. The process is repeated for different initial amounts of dye added to the dyebath, that is, for different depths of shade being applied.

Three types of adsorption isotherm can be identified, called the *Nernst, Freundlich* and *Langmuir* isotherms, and the various dye/fibre systems correspond to one of these types.

The simplest of the three types is the *Nernst adsorption isotherm* (Figure 6.17) where a graph of $[D]_f$ plotted against $[D]_s$ is a straight line. This isotherm is shown by disperse dyes on hydrophobic fibres such as polyester and nylon and is typical of the behaviour of the dye forming what is effectively a solid solution in the fibre. Both the dye and the fibre are electrically neutral, so there are no charge attraction or repulsion influences on the

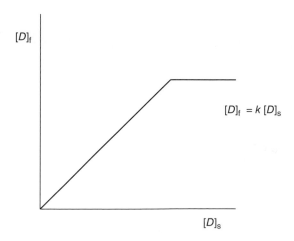

Figure 6.17   The Nernst isotherm.

adsorption of dye. The ratio $[D]_f/[D]_s$ is called the *partition coefficient*, K, and because the graph is linear, K remains constant until the saturation point of the dye in the fibre is reached. At this point, no more dye can be taken up by the fibre, no matter how much more dye is put into the dyebath.

The equation of this isotherm is quite simply stated as

$$\left[D\right]_f = K\left[D\right]_s.$$
(6.7)

If the isotherm is determined at a higher temperature, the solubility of the disperse dye in both the fibre and the aqueous phases increases. However the dye solubility increases more in the water, so the slope of the isotherm is lower and the equilibrium exhaustion of the dye is lower.

The second type of adsorption isotherm, the *Freundlich adsorption isotherm*, is shown in Figure 6.18.

Here the graph is not linear, but significantly the line keeps rising, with no apparent saturation being reached.

This type of isotherm is shown by anionic dyes, such as direct dyes, vat dyes (in their reduced form) and reactive dyes (during the exhaustion phase, before fibre reaction) applied to cellulosic fibres.

It appears that there are an infinite number of locations in the cellulosic fibre at which anionic dyes can be adsorbed. Another feature of these systems is that the anionic dyes can build up in multilayers in the fibre, and this feature contributes to the apparently unrestricted uptake of dye.

The equation of this isotherm is

$$\left[D\right]_f = k\left[D\right]_s^{x}$$
(6.8)

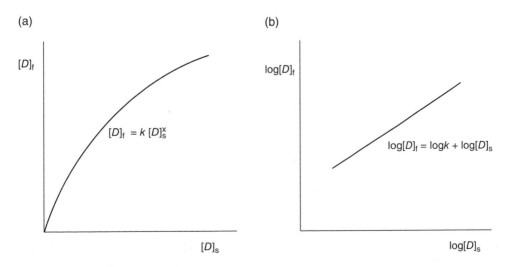

(a)

$[D]_f$

$[D]_f = k\,[D]_s^x$

$[D]_s$

(b)

$\log[D]_f$

$\log[D]_f = \log k + \log[D]_s$

$\log[D]_s$

Figure 6.18   The Freundlich isotherm.

where k and x are constants. Taking logarithms of this equation gives

$$\log[D]_f = \log k + x \log[D]_s \qquad (6.9)$$

which has the form of a simple linear equation, so if, instead of plotting $[D]_f$ against $[D]_s$, as shown in Figure 6.18a, $\log[D]_f$ is plotted against $\log[D]_s$, a straight line is obtained (Figure 6.18b).

When anionic dyes are applied to cellulosic fibres, different processes take place at the fibre–water interface than occur with disperse dye systems. Anionic dyes carry one or more negatively charged sulphonate groups, and the –OH groups in the cellulose molecule can dissociate to a certain extent, creating a negative charge at the fibre surface. This is explained later in Section 6.7.2, but the point to note here is that the adsorption of anionic dyes onto cellulosic fibres is a complex process.

The third type of isotherm is the *Langmuir adsorption isotherm*, which applies to systems in which charged dye ions are taken up at 'sites' with the opposite electrical charge in the fibre. Typical of dye/fibre systems that are adsorbed at specific sites in the fibre are acid dyes on wool and nylon, and basic (cationic) dyes on acrylic fibres (see Figure 6.15).

The shape of the adsorption isotherm exhibited by these dye/fibre systems is illustrated in Figure 6.19a. Dye builds up regularly in the fibre, but because the total number of dye sites is limited, the adsorption tails off as the sites all become occupied.

The equation of the Langmuir isotherm is

$$[D]_f = \frac{k[S]_f[D]_s}{1+k[D]_s} \qquad (6.10)$$

where $[D]_f$ is the concentration of dye in fibre and $[D]_s$ is the concentration of dye remaining in solution at equilibrium, $[S]_f$ is the concentration of dye in the fibre when all the sites are occupied. As with the Freundlich isotherm, it can be converted to a linear form, this time not by taking logarithms, but by expressing it in the form

$$\frac{1}{[D]_f} = \frac{1}{k[S]_f[D]_s} + \frac{1}{[S]_f} \qquad (6.11)$$

The point where the straight line of the graph of $1/[D]_f$ plotted against $1/[D]_s$ intercepts the $1/[D]_f$ axis gives the reciprocal of the saturation value, $1/[S]_f$, from which $[S]_f$ can be easily obtained (Figure 6.19b).

The equation of the Langmuir isotherm can be derived using the principles of reaction kinetics, explained in Section 1.5.2. The application of these principles involves some assumptions about the dye/fibre system however:

- The fibre has a fixed number of available dye sites and is uniform and homogeneous.
- All sites are equally accessible and have the same affinity for dye molecules – this means there is no preferential adsorption at particular sites.

(a)   (b)

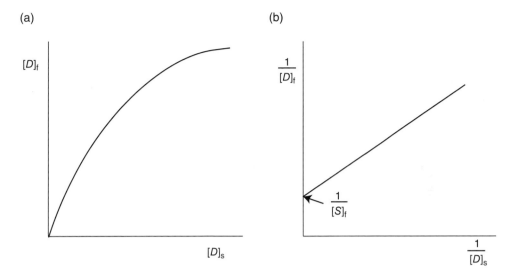

Figure 6.19 The Langmuir isotherm.

- Each site can accommodate only one dye molecule.
- There is no interaction between the adsorbed dye molecules, and the presence of a dye molecule at a site does not hinder the adsorption of another dye molecule at an adjacent site.

When a dyeing has reached equilibrium, dye is continually leaving the fibre and going back into the dye liquor, whilst simultaneously dye is continually arriving at the fibre surface and being adsorbed. By the law of mass action (see Section 1.5.2), the rate at which dye leaves the fibre (desorbs) is proportional to the concentration of dye in the fibre:

$$-\frac{d[D]_f}{dt} = k_{des}[D]_f, \qquad (6.12)$$

the negative sign indicating that concentration decreases. The rate of adsorption is proportional to both the concentration of dye in solution and the number of dye sites available (given by $[S]_f - [D]_f$), so

$$\frac{d[D]_f}{dt} = k_{ads}[D]_s \left([S]_f - [D]_f\right) \qquad (6.13)$$

Because the system is at equilibrium, these two rates, the rate of adsorption and the rate of desorption, are equal:

$$k_{des}[D]_f = k_{ads}[D]_s \left([S]_f - [D]_f\right) \qquad (6.14)$$

Rearranging Equation 6.14 gives

$$[D]_f = k_{sys}[D]_s \left([S]_f - [D]_f\right) \qquad (6.15)$$

where $k_{sys} = k_{ads}/k_{des}$. Equation 6.15 can be rearranged to give Equations 6.10 and 6.11.

As will be seen later, dyes often show uptake on the fibre to an extent that is greater than the theoretical saturation value $[S]_f$. One reason for this is that in addition to becoming adsorbed at the charged sites in the fibre by ionic bonding, dye molecules can also be adsorbed in other regions, the attraction due to van der Waals type forces – the secondary forces described in Section 6.6.1 and also Section 1.4.3.

6.7 Thermodynamic Information Derived from Equilibrium Studies of Dyeing Systems

In Section 1.5.5, the concepts of the thermodynamics of chemical reactions are explained. Just as when chemical reactions take place, so too the transfer of dye from solution into a fibre involves changes in:

- Enthalpy (ΔH) – Change in the heat energy of the system
- Entropy (ΔS) – Change in the 'disorder' or randomness of the system
- Free energy (ΔG) – The overall 'arbiter' of enthalpy and entropy changes in deciding whether a process will occur spontaneously or not

These values are useful in interpreting the driving force for dyeing processes and therefore provide a means for understanding what causes dye to transfer preferentially from the dye liquor into the fibre.

6.7.1 Standard Affinity, Standard Enthalpy and Standard Entropy of Dyeing

The driving force of a chemical reaction is given by the free energy change involved, but free energy depends on amount. For dyeing processes it is more appropriate to use what is called the *chemical potential*, which is given the symbol μ. Dye molecules have a certain chemical potential in the solution (water) phase (μ_s) and another chemical potential in the fibre phase (μ_f). Initially the chemical potential of the solution phase is greater than that of the fibre phase, but as dyeing progresses, the difference between them gradually decreases. When dyeing reaches equilibrium, the chemical potentials are equal, that is, $\mu_s = \mu_f$. The chemical attraction between a dye and a fibre, called the *affinity*, has a constant value, regardless of the process conditions under which the dyeing is carried out.

To determine the value for the affinity, the *standard chemical potentials*, μ_f° and μ_s° for the fibre and solution phases, respectively, are considered. The difference between these values $(\mu_f^{\circ} - \mu_s^{\circ})$ is a measure of the affinity – the tendency of the dye to move from the solution phase into the fibre phase. The equations that have been developed for the chemical potentials in the solution and fibre phases are

$$\mu_s = \mu_s^{\circ} + RT \ln a_s \quad \text{for the dye in the solution phase} \qquad (6.16)$$

$$\mu_f = \mu_f^\circ + RT \ln a_f \quad \text{for the dye in the fibre phase} \tag{6.17}$$

where

μ_s° and μ_f° are the standard chemical potentials
R is the gas constant (8.314 J/K/mol)
T is the absolute temperature
$\ln a_s$ and $\ln a_f$ are the natural logarithms of the *activities* a_s and a_f in the solution and fibre phases, respectively

When the dyeing process reaches equilibrium, $\mu_s = \mu_f$, so

$$-\left(\mu_f^\circ - \mu_s^\circ\right) = -\Delta\mu^\circ = RT \ln a_f - RT \ln a_s \tag{6.18}$$

and

$$-\Delta\mu^\circ = RT \ln\left(a_f / a_s\right) \tag{6.19}$$

or

$$-\Delta\mu^\circ = 2.303 RT \log\left(a_f / a_s\right) \tag{6.20}$$

In Equations 6.16–6.20, a_f and a_s are the *activities* of the dye in the fibre and in solution, respectively, at equilibrium. The 'activity' of the dye is a measure of its 'effective concentration' and is given by Equation 6.21:

$$a = fc \tag{6.21}$$

where c is the actual concentration and f is called the activity coefficient. For dilute solutions, f has a value of unity, in which case quite simply $a=c$.

The term $-\Delta\mu^\circ$ is called the *standard affinity* and is a quantitative measure of the affinity (or the 'liking') of a dye for a fibre. The higher the value of $-\Delta\mu^\circ$ (i.e. the more negative the value of $\Delta\mu^\circ$), the greater is the affinity of the dye for the fibre. It should be remembered that it is calculated from the concentrations of dye in solution and the fibre when the dyeing has reached equilibrium, which may be considerably longer than a commercial dyeing process.

Two other useful thermodynamic quantities that can be determined for a dyeing process are the changes in standard enthalpy (ΔH°) and standard entropy (ΔS°). These values can be determined from knowledge of the standard affinity at different temperatures. Starting with the equation analogous to Equation 1.12 developed in Chapter 1,

$$\Delta\mu^\circ = \Delta H^\circ - T\Delta S^\circ \tag{6.22}$$

and dividing throughout by T, gives the equation

$$\frac{\Delta\mu^\circ}{T} = \frac{\Delta H^\circ}{T} - \Delta S^\circ \tag{6.23}$$

If a graph of $\Delta\mu°/T$ is plotted against $1/T$, a straight line will be obtained, the slope of which is $\Delta H°$. Alternatively, if the standard affinity is known only at two temperatures T_1 and T_2, $\Delta H°$ can be determined using Equation 6.24:

$$\Delta H° = \left(\frac{\Delta\mu°_{T_1}}{T_1} - \frac{\Delta\mu°_{T_2}}{T_2} \right) \Bigg/ \left(\frac{1}{T_1} - \frac{1}{T_2} \right) \qquad (6.24)$$

The intercept of the graph of $\Delta\mu°/T$ against $1/T$ will give the value of $\Delta S°$.

However the value of $\Delta S°$ obtained by this method is unlikely to be reliable, because it involves extrapolating the line of the graph to the intercept on the $\Delta\mu°/T$ axis, when $1/T = 0$.

A better method is to apply known values of $\Delta\mu°$, $\Delta H°$ and T into Equation 6.22 and solve for $\Delta S°$.

The value of change in enthalpy, $\Delta H°$, indicates the loss (or gain) in heat energy of the complete system when the dyeing process has reached equilibrium. All naturally occurring processes strive to reduce their energy content, so a negative value for $\Delta H°$ will favour dye uptake by the fibre. The change in entropy, $\Delta S°$, of a process is less easy to define. It is best regarded as the degree of order of a system and naturally occurring processes are favoured if the final form has less order, that is, a greater degree of randomness or freedom. It must be remembered though that this increase in disorder applies to the system as a whole. The dye ions or molecules become attached to the fibre and have a much more ordered state, so they lose entropy, which is unfavourable to adsorption occurring. However, as dye leaves the water phase, the water molecules, which were in an ordered state around the dye ions or molecules, are given increased freedom, so they gain entropy, which is favourable. So, in entropy terms, there are winners and losers, and the overall value of $\Delta S°$ of a dyeing process is the net result of all the individual changes taking place.

Whilst values for the affinity, enthalpy and entropy of some dyes have been published, there is no official compendium of data available. Indeed those values published have usually been in research papers, and the conditions chosen by the researchers for the dyeing process (such as the temperature and the concentration of dyebath auxiliaries) vary, as does the method for determining the adsorption isotherm. The values given in the following section are taken from published articles, either in research journals or textbooks.

6.7.2 Determination of Thermodynamic Values for the Three Dye/Fibre System Types

The application of Equation 6.20 to disperse dyeing systems is relatively straightforward because the dye molecules are non-ionic and are electrically neutral. These dyes behave almost ideally, so that the activities a_s and a_f can simply be replaced by the dye concentrations in solution and in the fibre, respectively, when the dyeing has reached equilibrium. The standard affinities of these dye types can then be fairly readily determined.

For example, when applied to polyester at a temperature of 120 °C (393 K), it has been found [7] that the concentration of the dye C.I. Disperse Red 15 (**6.9**) in the fibre is 47.3 g/kg and remaining in solution is 0.147 g/l. Applying these data to Equation 6.20 gives

$$-\Delta\mu^\circ = 2.303 \times 8.314 \times 393 \times \log\left(\frac{47.3}{0.147}\right) \text{ and } -\Delta\mu^\circ = 18.87 \text{kJ / mol}$$

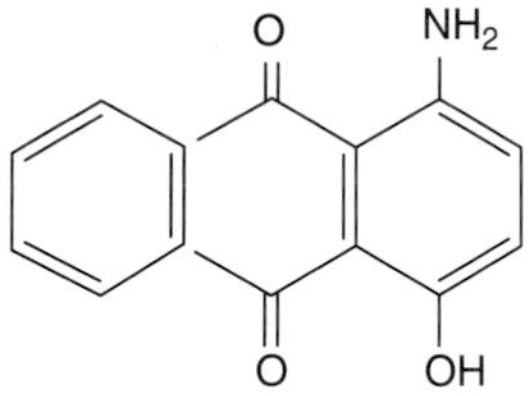

6.9

The value of affinity $-\Delta\mu^\circ$ at the lower temperature of 89 °C (362 K) is 22.15 kJ/mol, and applying these values for the affinity at the two temperatures to Equation 6.24 gives a value for the standard enthalpy change, ΔH°, of −60.45 kJ/mol. Then, application of Equation 6.22 gives a value for ΔS° of −105.8 J/mol/°C.

These two values are highly negative, showing that the enthalpy change strongly favours dye adsorption, but the entropy change does not. The system obviously loses freedom during dyeing, but this unfavourable influence is far outweighed by the significantly favourable loss in energy.

This dye when applied to cellulose acetate is found [8] to have

- An affinity $-\Delta\mu^\circ$ at 90 °C of −19.4 kJ/mol
- An enthalpy of dyeing of −46.5 kJ/mol
- An entropy of dyeing of −74.6 J/mol/°C

These values are all lower than those for applying the dye to polyester, though the same trends apply, in that the enthalpy change is the main driving force for dyeing, the entropy change being unfavourable.

In other dye/fibre systems, those involving the Freundlich and Langmuir isotherms, the application of Equation 6.20 is complicated by the fact that the dyes are ionic and do not behave ideally, either in solution or in the fibre. Table 6.6 shows the structures of some typical anionic dyes. Different treatments have been developed for assigning values for a_s and a_f for ionic dyes.

The Solution Phase

The reason why dyes do not behave ideally in solution is that they have a tendency to aggregate (see Section 6.3) and consequently their mobility is affected.

In the solution phase the activity a_s is given by the term $f[D]_s$, where f is the activity coefficient (see Equation 6.21). For example, if f has a value of 0.5, then $a = 0.5[D]_s$, meaning that the dye is behaving as though its concentration is only half of what it actually is.

Table 6.6 Structures of some typical anionic dyes.

dye	structural formula	valency of dye anion, z	abbreviated structure
C.I. Direct Yellow 50	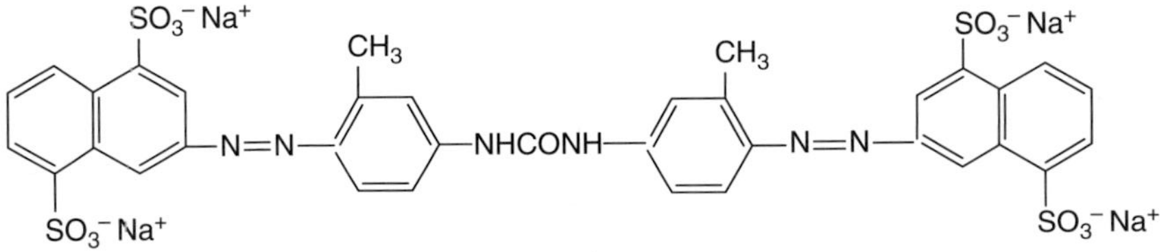 (structure with $SO_3^- Na^+$, CH_3, $N=N$, NHCONH, CH_3, $N=N$, $SO_3^- Na^+$, $SO_3^- Na^+$, $SO_3^- Na^+$)	4	Na_4D
C.I. Vat Yellow 1 (after reduction)	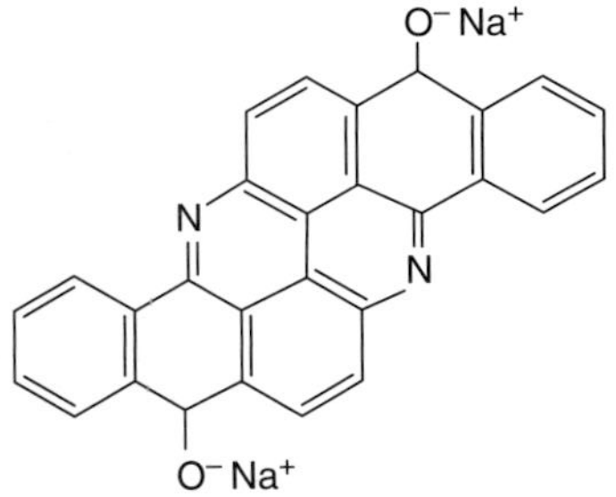 (structure with $O^- Na^+$, N, N, $O^- Na^+$)	2	Na_2D
C.I. Reactive Red 1	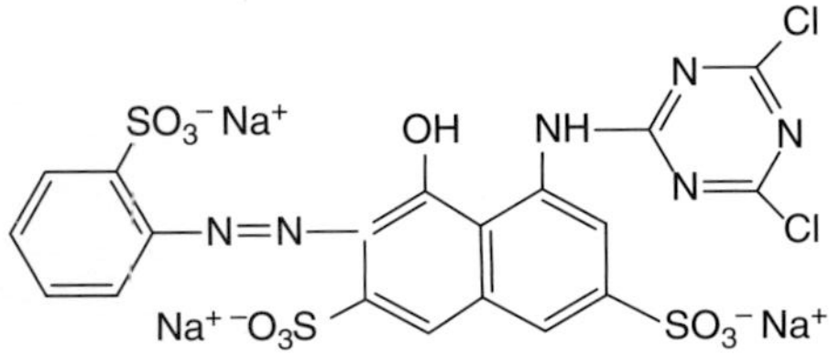 (structure with Cl, N, N, Cl, NH, OH, $SO_3^- Na^+$, $SO_3^- Na^+$, $N=N$, $Na^+ {}^-O_3S$)	3	Na_3D
C.I. Acid Red 32	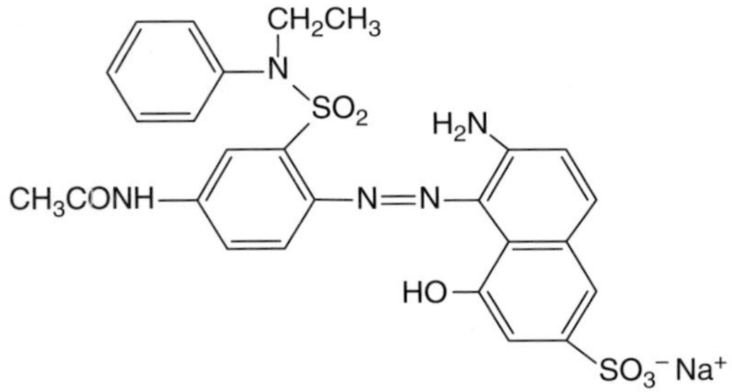 (structure with CH_2CH_3, N, SO_2, H_2N, $N=N$, HO, $SO_3^- Na^+$, CH_3CONH)	1	Na_3D

Usually, dyes behave more ideally in dilute solution – and the more dilute the dye solution becomes, the more $f \rightarrow 1$. If $f = 1$, then simply, $a_s = [D]_s$. This is assumed to be the situation in practical dyeing situations.

Anionic dyes are the sodium salts of the dye anion, having the general formula Na$_z$D, where z is the charge on the dye anion. Direct dyes, for example, contain one or more sulphonate groups –SO$_3^-$; the number they contain is called the basicity, z. So, for a dye molecule that contains two sulphonate groups and has the formula Na$_2$D, $z = 2$, the dye anion has a doubly negative charge, that is, D^{2-}.

Anionic dyes are fully dissociated into Na$^+$ and D^{z-} ions; it is necessary to write the concentration of dye as [Na]$_s^z$[D], so the equation for the *chemical potential* in solution is

$$\mu_s = \mu_s^\circ + RT \ln \left[\text{Na} \right]_s^z \left[D \right]_s \tag{6.25}$$

The Fibre Phase

Obtaining an accurate value for the activity, a_f, in the fibre phase is a much more complex matter, and different treatments are used for the two types of dye/fibre systems:

(1) Systems in which the dyes have the same electrical charge as the fibre, such as anionic dyes on cellulose, which become diffusely adsorbed on the fibre
(2) Systems in which the dyes have the opposite electrical charge as that of the fibre, such as acid dyes on wool or nylon or basic dyes on acrylic fibres, when the dye is adsorbed at specific sites in the fibre

Dye/Fibre Systems with the Same Electrical Charge   The surfaces of cellulosic fibres assume a negative charge when they are in water for two reasons.

(1) The –OH groups of the glucosidic rings of cellulose can ionise into –O$^-$ ions.
(2) Degradation, such as may be caused by over-bleaching, may convert –OH groups to –CHO and thence to weakly acid –COOH groups that can dissociate to –COO$^-$.

It is useful to consider the distribution of the various types of ions at and near to the fibre surface. The negative charges on the fibre surface create an electrical potential, causing oppositely charged positive ions (called *counter-ions*) to be attracted to it, but any negatively charged ions (called *co-ions*) to be repelled, creating an *electrical double layer*, proposed by Helmholtz and illustrated schematically in Figure 6.20a.

However this concept was revised by Gouy and Chapman who proposed a *diffuse electrical double layer*. In this model, illustrated schematically in Figure 6.20b, the concentration of the counter-ions gradually decreases, whilst that of the co-ions increases, moving away from the fibre surface.

Later studies by Stern led him to propose a combination of these two models, comprising a layer only a single counter-ion thick, at the surface, and then a diffuse second layer of co-ions and counter-ions (Figure 6.20c). This second layer is diffuse because there are competing forces of movement of ions due to thermal motion and electrostatic attractive

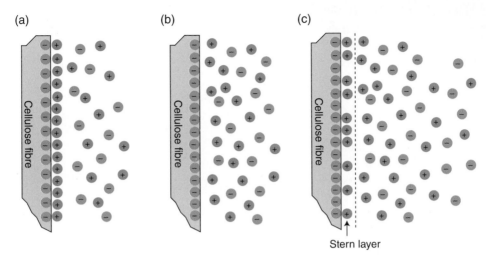

Figure 6.20   Distribution of co-ions and counter-ions at a cellulose fibre surface. (a) Helmholtz model, (b) Gouy–Chapman model, (c) Stern model.

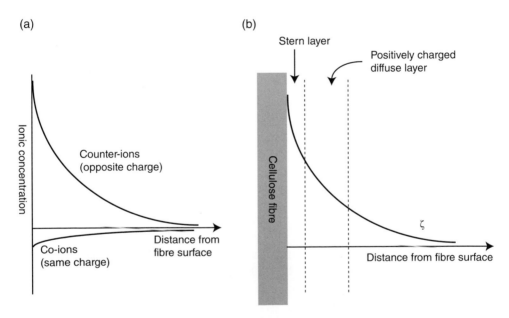

Figure 6.21   (a) Concentration of ions at a cellulosic fibre surface, (b) decrease in the zeta potential ξ on moving away from the fibre surface.

and repulsive forces. However, as in the Gouy–Chapman model, there is an excess of positively charged ions in the region of this diffuse second layer nearest to the fibre surface. The concentrations of the co-ions and the counter-ions in the region of the fibre surface is shown in Figure 6.21a. The electrical potential between the fixed layer of ions of the double layer and the diffuse part of the double layer is called the *zeta potential*, ξ, and the way it changes on moving away from the fibre surface is shown in Figure 6.21b.

If a cellulosic fibre is placed in a solution containing an anionic dye, such as a direct dye, it might be expected that a negatively charged fibre surface will repel the dye anions. However, there are opposing forces. On the one hand, there are the electrostatic repulsive forces, but on the other hand there is the inherent affinity the dye anions have for the fibre through their ability to form secondary bonds, such as hydrogen bonds, with the fibre (Figure 6.22). It is these attractive forces, giving the dye an affinity for the fibre, that dominate.

An additional factor then arises because the adsorption of the negatively charged dye anions increases the negative surface charge of the fibre. Anionic dyes are assumed to be completely dissociated into the dye anion (D^-) and the counter-ion, which is usually Na^+, in both the dyebath and the fibre. However, in order to maintain overall electrical neutrality, when the dye anions are adsorbed by the fibre, the sodium cations cannot be too far away.

In order to overcome the repulsive forces operating between the negatively charged fibre surface and the negatively charged dye anions, a neutral electrolyte, such as sodium chloride, is added to the dyebath in typical dyeing processes. The effect of adding a neutral electrolyte to the dyebath is to reduce the activation energy of adsorption. This happens because the positively charged sodium ions are attracted to the negatively charged cellulose surface and reduce the repulsive effect the surface has on dye anions. Repulsive forces operate between the negatively charged chloride ions and the fibre surface, and because the chloride ions are small and have no affinity for the fibre, their concentration at the fibre surface is zero. One model of the system is that the attractive forces between the dye and the fibre do not extend very far from the fibre surface, whilst the electrical repulsive forces act over greater distances. This results in unimolecular adsorption of dye at the fibre surface and the distribution of the various ions near the surface is as shown in Figure 6.23.

It is possible that the attractive forces between the dye ions and the fibre extend as far from the fibre as the repulsive forces. If this is the case, then the distribution of the dye ions

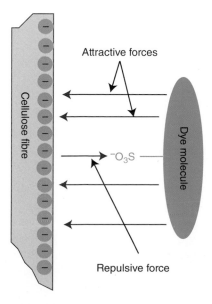

Figure 6.22 Attractive and repulsive forces between a cellulosic fibre and an anionic dye molecule.

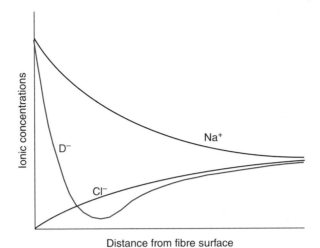

Figure 6.23 Distribution of ions near a cellulosic fibre surface, in the case of unimolecular adsorption.

will be similar to that of the Na^+ ions, giving rise to diffuse adsorption. Which of these two models is correct really depends of the charge on the fibre surface and the strength of the attractive force between the dye and the fibre. In practice the situation is likely to be a combination of both models.

In order to develop an equation to enable the affinity of a direct (or reactive) dye for cellulose, it is firstly necessary to consider the distribution of ions of an electrolyte across a *semipermeable membrane*.

A semipermeable membrane is so called because small highly mobile ions such as Na^+ and Cl^- can pass through it, but larger ions, such as dye ions D^-, cannot. In the simple case of two solutions of NaCl, of different initial concentrations, on either side of such a membrane, the transfer of one type of ion, for example, the Na^+ ion, must be accompanied by the Cl^- ion to maintain electrical neutrality. The rate of transfer of the Na^+ and accompanying Cl^- ions will depend on the product of their concentrations (see Equation 1.2 in Section 1.5.2), that is,

$$V_1 \propto \left[Na^+\right]_1 \times \left[Cl^-\right]_1 \quad \text{and} \quad V_2 \propto \left[Na^+\right]_2 \times \left[Cl^-\right]_2$$

The transfer of ions from the most concentrated solution to the weaker solution will occur, until equilibrium is reached when these two rates will be equal, so

$$\left[Na^+\right]_1 \times \left[Cl^-\right]_1 = \left[Na^+\right]_2 \times \left[Cl^-\right]_2 \tag{6.26}$$

Equation 6.26 is called the *Donnan equilibrium* and it can also be written as

$$\frac{\left[Na^+\right]_1}{\left[Na^+\right]_2} = \frac{\left[Cl^-\right]_2}{\left[Cl^-\right]_1} = \lambda \tag{6.27}$$

where λ is called the *Donnan coefficient*.

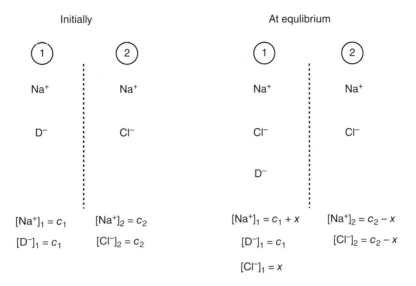

Figure 6.24 Distribution of sodium, chloride and dye ions on either side of a semipermeable membrane at equilibrium.

This equation is an important one in understanding the distribution of ions during the adsorption of direct, reactive and (leuco) vat dyes on cotton fibres.

If a solution of a dye Na^+D^- is on one side of the membrane and a solution of sodium chloride on the other, only the Na^+ and the Cl^- ions can transfer across it: the D^- ions are too large to diffuse through the membrane so they remain on their initial side, as shown in Figure 6.24. In order for Equation 6.26 to be maintained, it is necessary for some Na^+ and Cl^- ions to diffuse across the membrane. Assuming that a concentration, x, moles of Na^+ and Cl^- diffuse, then when equilibrium has established the ions will be distributed with the concentrations shown in Figure 6.24.

Substituting the concentrations of these ions into Equation 6.27 gives

$$\frac{c_1 + x}{c_2 - x} = \frac{c_2 - x}{x}$$

which on solving gives

$$x = \frac{c_2^2}{c_1 + 2c_2} \tag{6.28}$$

For electrical neutrality to be maintained on each side of the membrane,

$$\left[Na^+ \right]_1 = \left[Cl^- \right]_1 + \left[D^- \right]_1 \text{ and } \left[Na^+ \right]_2 = \left[Cl^- \right]_2$$

from which it can be inferred that, at equilibrium, $[Na^+]_1 > [Cl^-]_1$ and that $[Cl^-]_2 > [Cl^-]_1$, showing that the concentration of the NaCl is greater on the side of the membrane where there is no dye.

The diffusion of the Na^+ and Cl^- ions across the membrane cannot go to completion, and in practice total transfer is prevented by the build-up of an electrical potential difference

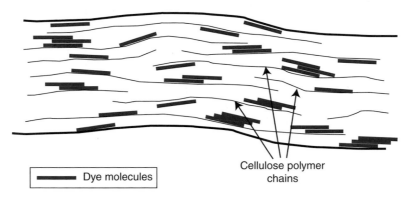

| Dye molecules | Cellulose polymer chains |

Figure 6.25   Diffuse adsorption of anionic dye ions on cellulose fibres.

across the membrane, called the *Donnan membrane potential*, which is given the symbol ψ_{Donnan}. This potential leads to the establishment of the equilibrium defined by Equation 6.27 and is related to the Donnan coefficient, λ.

The reason why a consideration of the equilibria of Na^+, Cl^- and D^- ions across a semipermeable membrane is useful in understanding the adsorption of anionic dyes by cellulosic fibres is that the dye/fibre system can be regarded as having a similar structure. As noted earlier, different theoretical treatments have been developed to account for the distribution of dye and electrolyte ions in order to determine the standard affinity of a dye for the fibre, the *unimolecular adsorption* and the *diffuse adsorption* models.

Probably the simplest of the treatments is that of diffuse adsorption in which the dye is not adsorbed on specific sites and may form multilayers within the polymer matrix, as depicted schematically in Figure 6.25.

As noted in Section 6.3, many dyes, anionic dyes such as direct dyes included, have the ability to aggregate and form multilayers. In this case, the system is divided into two phases – the surface layer (f) or internal solution and the external solution (s). The ionic concentrations of the Na^+, Cl^- and D^- ions in each of these phases are shown in Figure 6.26.

The internal solution is a volume of liquid adjacent to the fibre surface that has a volume V (l/kg). Any dye ions adsorbed onto the fibre are assumed to be dissolved in this surface liquid layer. Their concentrations are divided by V and are in the units of mol/l.

By applying Equation 6.17 the chemical potential in the fibre phase is now given by

$$\mu_f = \mu_f^\circ + RT \ln \left(\frac{\left[Na^+ \right]_f}{V} \right)^z . \frac{\left[D^{z-} \right]_f}{V} \tag{6.29}$$

The equation for the standard affinity (Equation 6.19) then becomes

$$-\Delta\mu^\circ = RT \ln \frac{\left[Na^+ \right]_f^z \left[D^{z-} \right]_f}{V^{z+1}} - RT \ln \left[Na^+ \right]_s^z \left[D^{z-} \right]_s \tag{6.30}$$

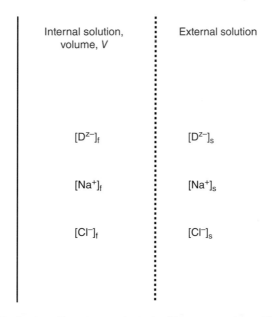

Figure 6.26 Distribution of ions in an anionic dye/fibre system giving diffuse adsorption.

All of the concentrations in Equation 6.30 can be measured, with the exception of $[Na^+]_f$. This represents a challenge, but it can be determined from experimentally measurable quantities by assuming that in the internal solution

$$\left[Na^+\right]_f = \left[Cl^-\right]_f + \left[D^-\right]_f,$$ (6.31)

which is necessary to maintain electrical neutrality, and that the Donnan equilibrium (Equation 6.24) must exist across the two phases:

$$\left[Cl^-\right]_f = \frac{\left[Na^+\right]_s\left[Cl^-\right]_s}{\left[Na^+\right]_f}$$ (6.32)

Substituting Equation 6.32 into 6.31 gives a formula from which $[Na^+]_f$ can be calculated:

$$\left[Na^+\right]_f = \left[D^{z-}\right]_f\left[\frac{z}{2}+\left(\frac{z^2}{4}+\frac{\left[Na^+\right]_s\left[Cl^-\right]_s}{\left[D^{z-}\right]_f^2}\right)^{\frac{1}{2}}\right]$$ (6.33)

One difficulty in using Equation 6.30 is assigning a satisfactory value for the volume, V, of the internal solution. It has been found [6] that a value of 0.21 l/kg cotton fibre gives the optimum fit to experimental results and it also corresponds to the amount of water taken up by cotton at 90 °C in an atmosphere of 100% relative humidity. However, for other cellulosic fibre types, such as mercerised cotton, viscose, cuprammonium rayon and lyocell, the value for V is different.

An interesting consequence of this analysis is its relationship to the Freundlich isotherm equation (Equation 6.8). Taking Equation 6.19 as the starting point and restructuring it gives

$$\frac{a_f}{a_s} = e^{-(\Delta\mu°/RT)}, \text{ or because} -\Delta\mu°, R \text{ and } T \text{ are all constant,} \quad \frac{a_f}{a_s} = \text{constant} \quad (6.34)$$

Now, if the terms derived for a_f and a_s used in Equations 6.25 and 6.29 are substituted,

$$\frac{\left[\text{Na}^+\right]_f^z \cdot \left[\text{D}^{z-}\right]_f}{\left[\text{Na}^+\right]_s^z \cdot \left[\text{D}^{z-}\right]_s} = \text{constant} \quad (6.35)$$

In the fibre, to maintain electrical neutrality, $[\text{Na}^+]_f = z[\text{D}^{z-}]_f$. Also a neutral electrolyte such as sodium chloride is usually added to the dyebath, so it can be assumed that $[\text{Na}^+]_s$ is constant. Then, Equation 6.35 can be written as

$$\frac{\left[\text{D}^{z-}\right]_f^{z+1}}{\left[\text{D}^{z-}\right]_s} = \text{constant} \quad (6.36)$$

which is the Freundlich isotherm equation.

Comparing Equation 6.36 with Equation 6.8, it follows that the slope of the graph, x, is equal to $1/(z+1)$. However, when graphs of $\log\left[\text{D}^{z-}\right]_f$ versus $\log\left[\text{D}^{z-}\right]_s$ are plotted, the slope of the line $1/(z+1)$ is not always as predicted. For example, for the dye C.I. Direct Blue 1 (**6.7**), for which the charge on the dye anion, $z=4$, the slope should be 0.2. However a slope of around 0.36 is obtained, which is probably due to the fact that the assumption $[\text{Na}^+]_f = z[\text{D}^{z-}]_f$ is not valid exactly.

Other models for the adsorption of direct dyes on cotton have been proposed, one of the more important being that in which unimolecular adsorption occurs, in which case it is necessary to consider the system as comprising the three parts illustrated in Figure 6.27. In this model it is assumed that a certain amount of the dye, $[\text{D}^{z-}]_{ad}$, is adsorbed on the fibre surface and is in equilibrium with dye in the internal solution.

Although the mathematical treatment is a little more complex than for the two-phase model, it leads to very similar equations for the standard affinity. Both of these treatments are based on relatively simple assumptions about the cotton and the components of the dyebath. The accuracy of the models can be improved by accounting for the fact that the cotton may contain carboxylic acid groups (–COOH) due to partial oxidation (e.g. from a bleaching process), so Equation 6.31 becomes

$$\left[\text{Na}^+\right]_f = \left[\text{Cl}^-\right]_f + \left[\text{D}^-\right]_f + \left[\text{COO}^-\right]_f \quad (6.37)$$

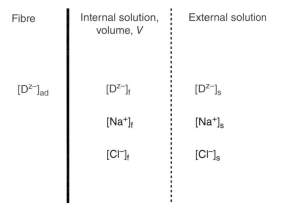

Figure 6.27 Distribution of ions in unimolecular adsorption, 3-phase model.

Table 6.7 Standard affinities of some direct dyes applied to cotton and viscose at various temperatures [9].

dye	fibre	$-\Delta\mu°$ (kJ/mol)		
		60 °C	80 °C	90 °C
C.I. Direct Yellow 12	Cotton	16.3	13.4	12.5
	Viscose	16.3	–	13.0
C.I. Direct Red 81	Cotton	15.9	13.8	13.0
	Viscose	16.3	–	13.4
C.I. Direct Red 26	Cotton	23.6	22.2	20.5
	Viscose	23.8	20.5	19.2

Additionally, for the other two application classes, reactive and vat dyes, which are applied under alkaline conditions, Equation 6.31 is further modified to account for the adsorption of $-OH^-$ ions and for the dissociation of the $-OH$ groups of the cellulose under the strongly alkaline conditions. However the model becomes more complex because knowledge of the concentrations of both $-COO^-$ and $-O^-$ needs to be established. In particular, modelling the application of vat dyes is further complicated by the fact that the dyebath contains not just sodium ions, (reduced) dye ions, hydroxide and chloride ions but also dithionite ions (the reducing agent) and its oxidation products.

The *standard affinity* of a dye, $-\Delta\mu°$, is a fundamental quantity, its value indicating the attraction of a dye for a fibre. Thus, for a given direct dye applied at a particular temperature to any cellulosic fibre, be it cotton, viscose, lyocell or cuprammonium rayon, the value of $-\Delta\mu°$ should be the same. Some values, determined many years ago, show that within reasonable limits of experimental error, this is indeed found to be true, as shown in Table 6.7. The data in Table 6.7 also show that the affinity of the dyes for the fibres decreases with temperature.

The reason for the decrease is that the enthalpy change ($\Delta H°$) for the dyeing process is negative, meaning that heat is evolved. By Le Chatelier's principle the evolution of heat is more efficient at lower temperatures; hence the absorption of dye by the fibre is favoured at lower temperature.

Another point to arise from Table 6.7 is that whilst C.I. Yellow 12 (**6.10**) and C.I. Red 81 (**6.11**) have similar standard affinity values as each other at the various temperatures, C.I. Red 26 (**6.12**) has much higher values. The molecular weight of C.I. Red 26 (938) is considerably larger than those of the C.I. Yellow 12 (680) and C.I. Red 81 (675), showing the influence of molecular size on affinity:

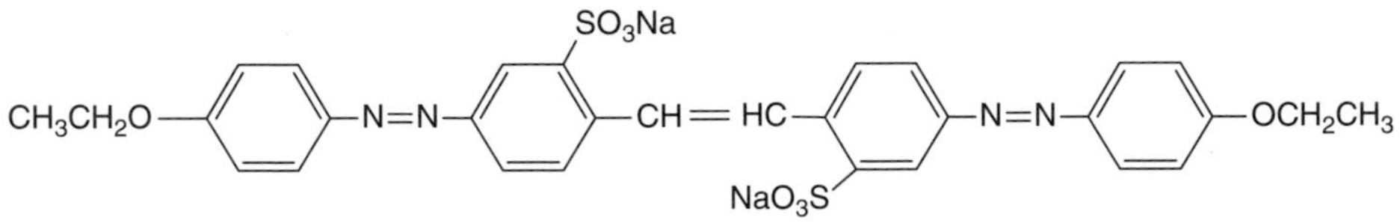

6.10

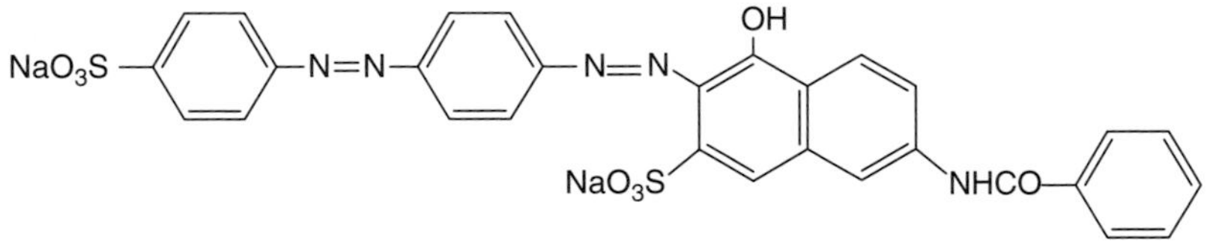

6.11

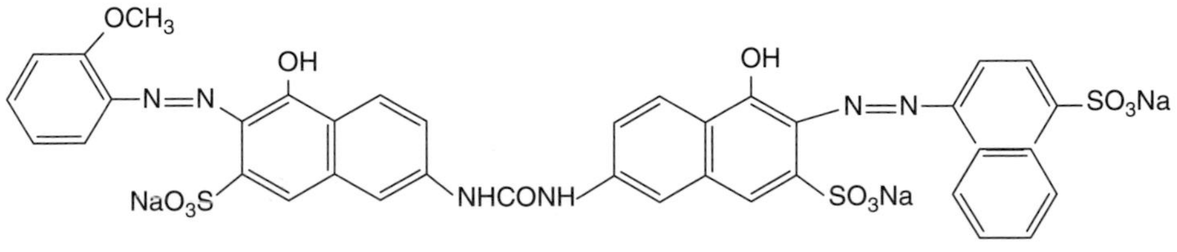

6.12

The standard affinity of C.I. Direct Blue 1 (**6.7**) applied to the cellulosic fibres viscose and lyocell (Tencel®) can be compared. The data (Table 6.8) show that there is good agreement between the values, the small difference probably due to different application conditions, especially the salt concentration.

Many vat dyes are polycyclic structures, and when applied as the leuco form to cellulose fibres, the main attractive forces are dispersion forces. It might be expected that the affinities of such dyes will be larger for dyes with larger-sized molecules, though this is not always the case, as illustrated by the dyes C.I. Vat Orange 9 and C.I. Vat Violet 10 (Table 6.9). Also the presence of other atoms, such as strongly electronegative nitrogen atoms, can have a beneficial effect on affinity, as in C.I. Vat Yellow 1.

Table 6.8 Standard affinities of C.I. Direct Blue 1 applied to viscose and Tencel at 100 °C, in the presence of 0.02 mol/l NaCl (viscose) and 0.08 mol/l NaCl (Tencel) [10, 11].

$-\Delta\mu^\circ$ (kJ/mol)	
viscose	Tencel
21.0	21.6

Source: Adapted from Marshall and Peters [10] and Carrillo et al. [11].

Table 6.9 Standard affinities of some polycyclic vat dyes applied to cotton at 40 °C [9, 12].

dye	$-\Delta\mu^\circ$ (kJ/mol)
C.I. Vat Orange 9	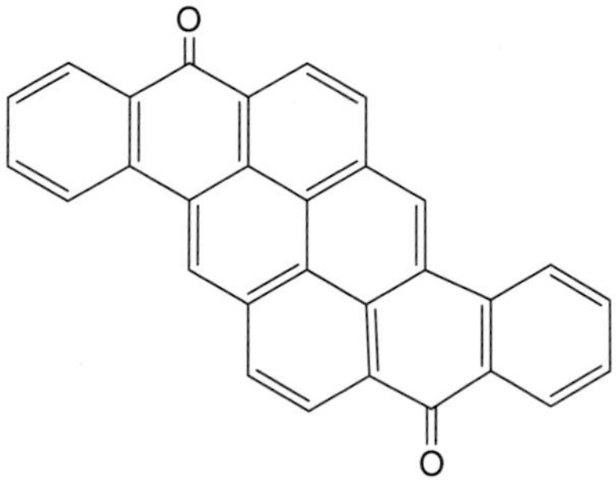 23.3

C.I. Vat Violet 10	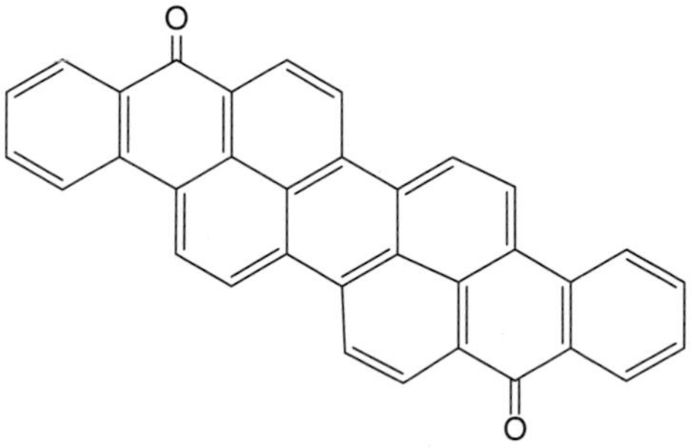 20.0

C.I. Vat Yellow 1	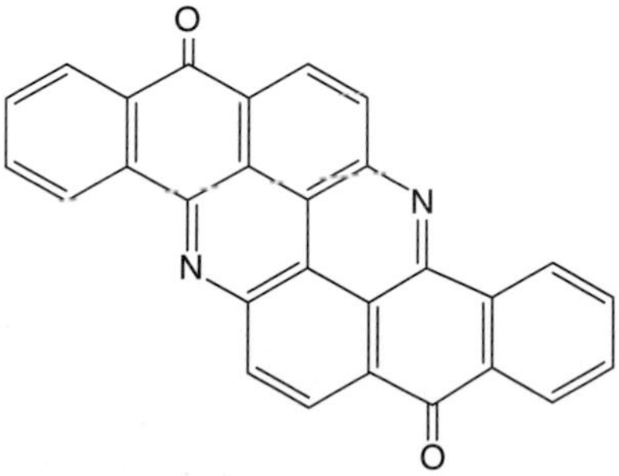 24.0

Dye/Fibre Systems with Opposite Electrical Charges   In these types of dye/fibre systems, it is assumed that the fibre contains a definite number of sites at which dye ions (of opposite charge) can be adsorbed, as represented schematically in Figure 6.28. It is also assumed that each site can accommodate only one dye ion and that the presence of a dye ion at a site does not interfere with the adsorption of dye ions at neighbouring sites.

These assumptions form the basis of the *Gilbert and Rideal* model, in which the activities of ions in the fibre (a_f) are expressed by the term

$$\frac{\phi}{1-\phi} \tag{6.38}$$

where ϕ is the fraction of sites occupied. Equation 6.38 therefore means

$$\frac{\text{Fraction of sites occupied}}{\text{Fraction of sites unoccupied}}$$

so if 20% of the sites are occupied, $\phi = 0.2$ and $\phi/(1-\phi) = 0.2/0.8$ or 0.25.

Acid dyes are typically applied to wool or nylon from an acidic dyebath when hydrogen ions (H^+) are adsorbed at either the basic groups such as $-NH_2$

$$-NH_2 + H^+ \rightarrow -NH_3^+$$

or at ionised carboxyl groups of ionic cross-links (see Section 2.6.1.1) by what is called back-titration:

$$-NH_3^+ \cdots\cdots {}^-OOC- + H^+ \rightarrow -NH_3^+ \cdots\cdots HOOC-$$

The negative dye anions are then adsorbed at the resulting positively charged $-NH_3^+$ sites. The equation for the chemical potential of a dye H_zD in the fibre phase is written as

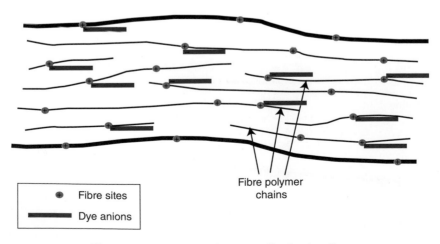

⊕	Fibre sites
▬	Dye anions

Fibre polymer
chains

Figure 6.28   Dye adsorption at specific sites in a fibre.

$$\mu_f = \mu_f^\theta + RT\ln\left[\frac{\theta_D}{1-\theta_D}\right]\left[\frac{\theta_H}{1-\theta_H}\right]^z \tag{6.39}$$

and so Equation 6.19 for the affinity of the dye for the fibre becomes

$$-\Delta\mu_f^\theta = RT\ln\left[\frac{\theta_D}{1-\theta_D}\right]\left[\frac{\theta_H}{1-\theta_H}\right]^z - RT\ln\left[D^{z-}\right]_s\left[H^+\right]_s^z \tag{6.40}$$

In the case of acid dyes applied to wool, the affinity values depend very much on the size and hydrophobic character of the molecule, together with the number of sulphonate groups in it.

For example, moving from a simple methyl group to butyl and then to a benzene ring in the series of dyes shown in Table 6.10 increases the affinity. The strength of the bonding between the dye and the fibre, as indicated by the enthalpy change, ΔH°, decreases.

Another interesting trend is the change from an unfavourable entropy loss with the dye containing the methyl group to a favourable entropy gain for the dye with the benzene ring. It is known that water molecules are highly ordered around hydrocarbon groups such as

Table 6.10 Thermodynamic values for some anthraquinone dyes applied to wool at 50 °C and pH 4.6 [13].

dye	$-\Delta\mu^\circ$ (kJ/mol)	ΔH° (kJ/mol)	ΔS° (J/mol/°C)
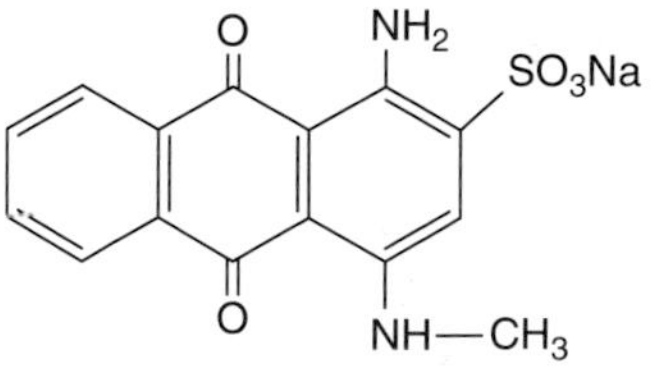 (structure: anthraquinone with NH_2, SO_3Na, $NH-CH_3$)	22.2	−31.8	−30.2
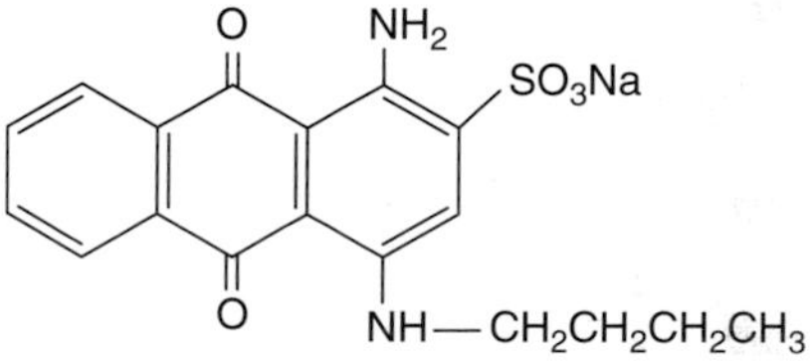 (structure: anthraquinone with NH_2, SO_3Na, $NH-CH_2CH_2CH_2CH_3$)	24.3	−29.7	−17.2
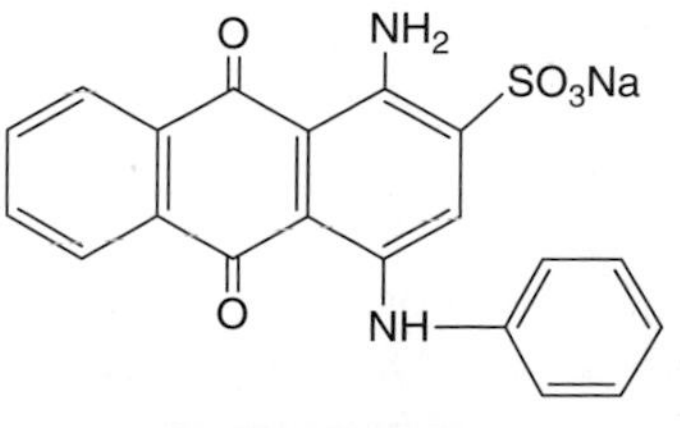 (structure: anthraquinone with NH_2, SO_3Na, NH—phenyl) C.I. Acid Blue 25	25.1	−23.5	+5.0

Source: Adapted from Iyer et al. [13].

Table 6.11   Standard affinities of a series of dyes of increasing hydrophobic character applied to nylon 6.6 at 95 °C at pH 6.1 [14].

dye	$-\Delta\mu°$ (kJ/mol)
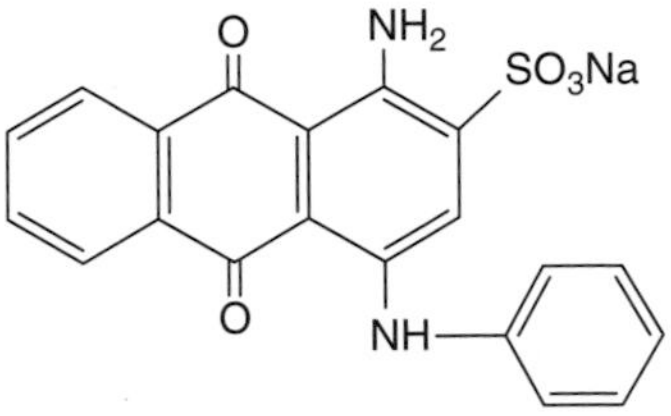 C.I. Acid Blue 25	23.9
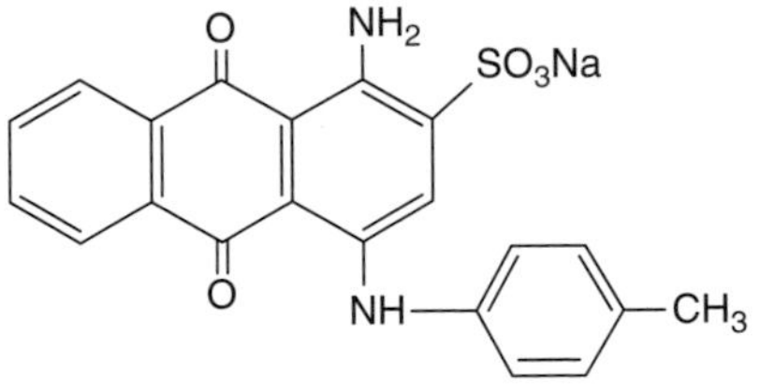	26.6
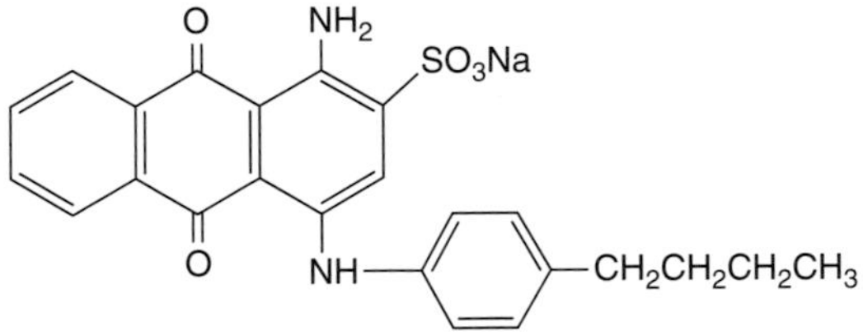	34.1
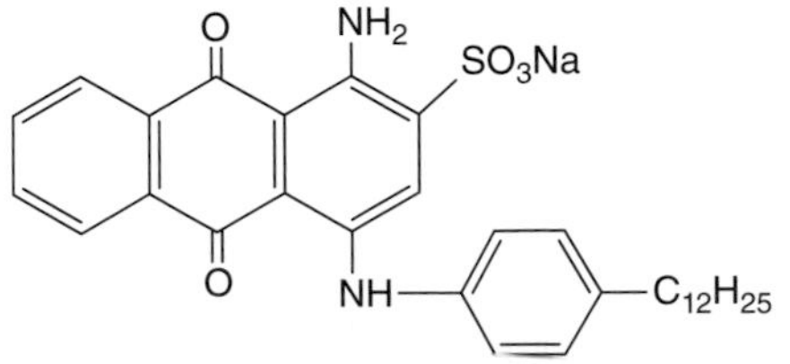	45.7

Source: Grieder [14]. Reproduced with permission of John Wiley & Sons.

aromatic rings, so when the dye molecules transfer to the fibre, the water molecules have greater freedom and therefore greater entropy.

The standard affinities of four dyes based on the C.I. Acid Blue 25 structure, when applied to nylon 6.6, are shown in Table 6.11. Whilst it is tempting to compare the affinities of C.I. Acid Blue 25 between the wool (Table 6.10) and nylon fibres, the conditions of application (pH and temperature) are different, as is the experimental technique used.

The lower value of $-\Delta\mu°$ on nylon (Table 6.11) will certainly be due in large part to the higher temperature used. Importantly, the values for $-\Delta\mu°$ shown in Table 6.11 indicate the effect of increasing the size of the hydrophobic chain on the benzene ring, as the structures change from a simple acid levelling dye (C.I. Acid Blue 25) to a supermilling acid dye type structure (that with the $-C_{12}H_{25}$ group).

Table 6.12   Effect of increasing basicity on standard affinity of acid dyes on nylon 6.6.

dye		$-\Delta\mu°$ (kJ/mol)
C.I. Acid Red 88	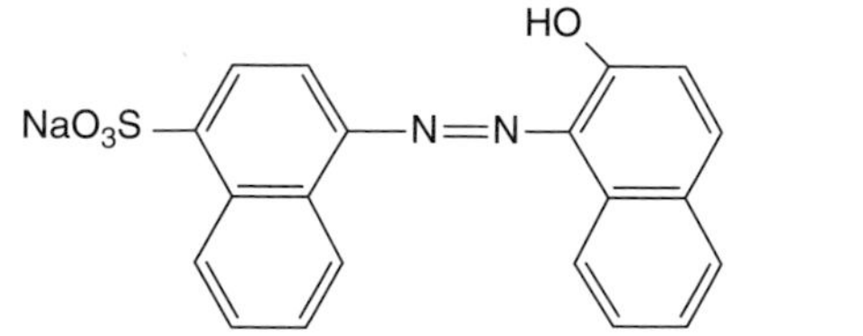	32.1
C.I. Acid Red 18	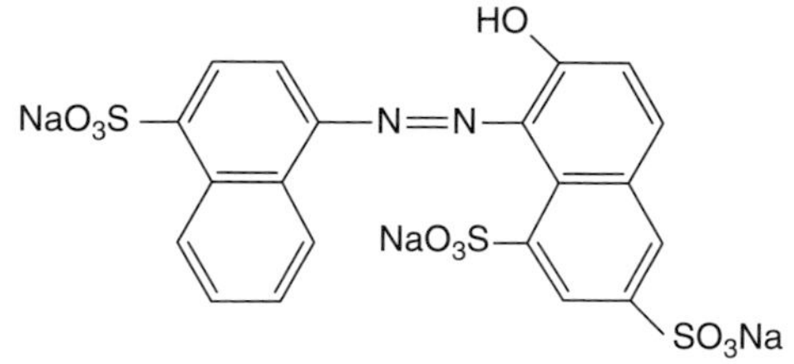	24.6

Source: Adapted from Atherton et al. [15].
Affinity determined by desorption technique at 75 °C [15].

The reduction in standard affinity caused by increasing the number of sulphonic groups is demonstrated by the two dyes: C.I. Acid Red 88, which has one sulphonate group, and C.I. Acid Red 18, which has three (Table 6.12).

An issue of importance in the application of acid dyes to nylon fibres is the adsorption of an amount of dye that is in excess of the theoretical number of sites available. This is known as *overdyeing*.

Acid dyes are adsorbed at basic groups, such as $-NH_3^+$ groups, and there are only about 30–50 mmol/kg of these groups available in nylon compared with over 800 mmol/kg in wool. In practical terms the theoretical maximum amount of an acid dye that can be adsorbed by all the sites in nylon depends on the basicity of the dye, its molecular weight and its purity.

Taking, for example, the simple monobasic dye C.I. Acid Red 88 (molecular weight 400) and assuming the commercial dye is 50% pure dye, this means that approximately a 2.4% owf shade is the maximum depth that can be applied. Indeed, some simple acid dyes are not taken up by the fibre in excess of this theoretical number of sites. However, many acid dyes, especially those with high molecular weights and/or significant hydrophobic character, such as those shown in Table 6.11, do not rely solely on basic sites in the fibre for adsorption. These dyes are capable of forming dispersion forces with the hydrophobic regions (the methylene $-(CH_2)-$ chains; see Section 2.8.1.1) of the nylon and give perfectly good wash fastness properties, even though they are not necessarily bonded to the basic sites.

A similar behaviour is known with the adsorption of basic dyes on acrylic fibres. The charge situation is the opposite of that of acid dyes applied to wool and nylon, in that basic dye ions are positively charged and the sites in the acrylic fibre are negatively charged. The negatively charged sites in acrylic fibres arise from sulphonate ($-SO_3^-$) and sulphate

($-SO_4^-$) group residues from the initiator used for the polymerisation reaction in forming the fibre and are situated at the ends of the polymer chains. The adsorption process occurs by an ion-exchange mechanism:

$$Fibre - SO_3^- H^+ + Dye^+ \leftrightarrow Fibre - SO_3^- Dye^+ + H^+$$

Overdyeing can occur when heavy depths of shade are being dyed. Again, the most probable explanation is that dispersion forces operate and the adsorption process is a combination of the Langmuir and Nernst isotherm models.

References

[1] S N Batchelor, *Color. Technol.*, **131** (2015) 15.
[2] C Renard, *Teintex*, **36** (1971) 845.
[3] E Valko, *J.S.D.C.*, **55** (1939) 173.
[4] C L Bird, *The Theory and Practice of Wool Dyeing*, 4th Edn., The Society of Dyers and Colourists, Bradford, 1972, 42.
[5] A Tokaska, *J. Soc. Text. Cellulose Ind. Japan*, **20** (1964) 100.
[6] J Boulton, *J.S.D.C.*, **60** (1944) 5.
[7] C L Bird, *J.S.D.C.*, **72** (1956) 343.
[8] C L Bird and P Harris, *J.S.D.C.*, **73** (1957) 199.
[9] T Vickerstaff, *The Physical Chemistry of Dyeing*, 2nd Edn., Oliver & Boyd, London, 1954.
[10] W J Marshall and R H Peters, *J.S.D.C.*, **63** (1947) 446.
[11] F Carrillo, M J Lis and J Valldeperas, *Dyes Pigm.*, **53** (2002) 129.
[12] J Shore, *Colorants and Auxiliaries, Vol 1 Colorants*, Chapter 3, The Society of Dyers and Colourists, Bradford, 1990.
[13] S R S Iyer, A S Ghanekar and G S Singh, *Symposium on Dye-Polymer Interactions*, Bombay (1971) 13.
[14] K Grieder, *J.S.D.C.*, **92** (1976) 8.
[15] E Atherton, D A Downey and R H Peters, *Text. Res. J.*, **25** (1955) 977.

Suggested Further Reading

A Johnson, Ed., *The Theory of Coloration of Textiles*, 2nd Edn., The Society of Dyers and Colourists, Bradford, 1989.

7

The Measurement of Colour

7.1 Introduction

With most textiles intended for the domestic market, the attention of the purchaser is first arrested by their aesthetic appeal, and at that instant colour is often a persuasive factor. If the manufactured goods are of satisfactory quality, they will conform to predetermined specifications appropriate to the intended use. The importance of quality assurance and quality control is underpinned by reliable technologies in the coloration industry to the extent that accurate, first-time dyeing is routine.

Developments in most areas of activity in the coloration industry bear witness to the effective application of modern technology, starting in the colour kitchen where dyes and chemicals are dispensed and ending with delivery of the final goods. For example, high-precision electronic balances make for more accurate recipe composition where weighing is done manually. There are also facilities for dispensing dyebath chemicals through computerised control valves that have the ability to adjust the flow of dye liquor automatically, from a large flow rate at the beginning of a dyeing cycle, down to drop-wise additions towards the end of the operation where precision is most needed. This may be accompanied by automatic recording of the consumption of dyes and chemicals for convenience in the administration of storage and purchasing.

The subject of colour measurement and colour specification has advanced to such a stage where all of the tasks relating to predicting dye recipes, carrying out the dyeing operation and assessment of the final dye goods, can be achieved instrumentally and with high accuracy, without the need for visual inspection. In this chapter the fundamental principles that enable these functions to be carried out are explained.

7.2 Describing Colour

The human perception of colour is a physiological sensation that defies absolute verbal definition. Sir Isaac Newton, during his investigations into the nature of light, came to the same conclusion, and he held the belief that light is uncoloured but that each wavelength has the power to stimulate unique sensations in the brain. Nevertheless, in the past a methodical approach to describing subjective colour assessment was the only means available for expressing opinions about the appearance of colour; this is reflected in the naming of commercial dyes and in describing colour differences.

An Introduction to Textile Coloration: Principles and Practice, First Edition. Roger H. Wardman.
© 2018 John Wiley & Sons Ltd. Published 2018 by John Wiley & Sons Ltd.

Dyers express differences in colour qualitatively through the use of the three variables hue, strength and brightness.

- *Hue* is defined as 'that attribute of colour whereby it is recognised as being predominantly red, green, blue, yellow, violet, etc.' Differences in hue between two colours therefore may result in one being described as redder, greener, bluer, yellower and so on.
- *Strength* relates to the amount of colour present, and differences in strength may be described as 'fuller' or 'thinner' to indicate higher and lower strength, respectively.
- *Brightness* refers to the greyness of a colour and is the opposite of 'dullness'. Terms such as brighter and flatter are used to describe differences in brightness.

The magnitude of a difference in any of these three attributes may also be described, using additional terms such as 'trace', 'slight', 'little' or 'much' – for example, 'much redder', 'a trace bluer' and so on. A further aid to the visualisation of a colour is given by the designatory letters that follow the names of commercial dyes and by the provision of coloured samples in the dye makers' pattern cards.

7.3 Additive and Subtractive Colour Mixing

It is useful at this stage to consider colour perception in terms of what is seen when colours are mixed. The visual effects produced are different, depending on whether coloured lights or dyes/paints/pigments are being mixed.

7.3.1 Additive Colour Mixing

Additive colour mixing refers to the visual effects produced when coloured lights are mixed. There are three *primary* coloured lights: red, green and blue. Mixing any two primary coloured lights in equal proportions gives a *secondary* colour:

$$Red + Blue = Magenta$$
$$Green + Blue = Cyan$$
$$Red + Green = Yellow$$

In this system the magenta, cyan and yellow are secondary colours. Mixing all three primary colours (in equal proportions) produces a visually acceptable white, that is,

$$Red + Green + Blue = White$$

Figure 7.1 is an illustration of the colours that are seen when the three primary coloured lights (red, green and blue) are shone on to a white screen in such a way that they overlap partially with each other.

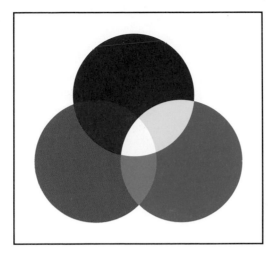

Figure 7.1   Additive colour mixing – mixing red, green and blue lights.

Using these simple relationships, it can be deduced that white can also be produced by mixing a primary colour with an appropriate secondary colour (any two colours that form white in an additive mix are called *complementary* colours):

$$Red + Cyan = White$$
$$Green + Magenta = White$$
$$Blue + Yellow = White$$

7.3.2   Subtractive Colour Mixing

This form of colour mixing refers to the mixing of dyes and pigments. When white light is incident on a dye or pigment, the wavelengths are absorbed to different extents, some more than others. The colour of the dye or pigment is due to the wavelengths not absorbed, that is, those that are reflected to the eye.

The subtractive primary colours are magenta, cyan and yellow. A yellow dye absorbs the short wavelength blue part of the spectrum, a magenta dye absorbs the middle green part and a cyan dye absorbs the long wavelength red part. If all three are mixed, then (in principle) all wavelengths are absorbed and the mixture is black. These colour effects are illustrated in Figure 7.2. In this figure, it has to be imagined that three transparent acetate sheets are placed on a glass plate, under which there is a white light source. One of the acetate sheets has been dyed with a magenta-coloured dye, another with a yellow-coloured dye and the third with a cyan-coloured dye. The three acetate sheets overlap each other partially as shown.

A yellow dye transmits green and red light, whilst a cyan dye transmits green and blue light. The colour that both these dyes transmit is green, which is why green is seen where the two filters overlap.

A magenta dye transmits blue and red light, so where the magenta-coloured filter overlaps with the yellow-coloured filter, red is transmitted, red being the colour they both transmit.

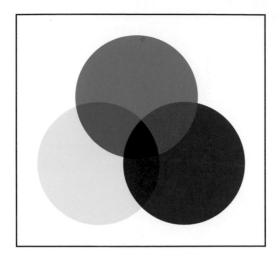

Figure 7.2   Subtractive colour mixing.

Where the magenta- and cyan-coloured filters overlap, the colour seen is blue, because blue is the colour they both transmit.

At the area where all three filters overlap, the yellow filter absorbs the blue part of the spectrum, the magenta filter absorbs the green part and the cyan filter absorbs the red part. Since all the parts of the spectrum are absorbed, no light is transmitted and hence this area appears black. In practice dyes and pigments do not absorb all wavelengths completely, so often a dark muddy brown colour is produced. For this reason ink cartridges in printers use an additional black cartridge when a black is required.

In applying dyes to textiles, the colour effects produced are also those of subtractive colour mixing. Often though, instead of using yellow, magenta and cyan dyes as the primaries, mixtures of yellow, red and blue dyes are used. When a black is required, a specialised black dye is applied to the textile.

7.4   The Colour Solid

The fact that dyers use three terms (hue, strength and brightness) to describe colours and differences between colours shows that to represent all colours a three-dimensional colour solid is required. In the objective evaluation of colour, the three axes of the colour solid are lightness (L), chroma (C) and hue (H).

- *Lightness* is defined as 'that property of a coloured object which is judged to reflect (or transmit) a greater or smaller proportion of the incident light than another object'.
- *Chroma* is 'the nearness of a colour in purity to the associated spectral colour, that is, the sensation caused by monochromatic visible light of known wavelength'.
- *Hue* has the same definition as given earlier. It is 'that attribute of colour whereby it is recognised as being predominantly red, green, blue, yellow, violet, etc.' Differences in hue between two colours therefore may result in one being described as redder, greener, bluer, yellower and so on.

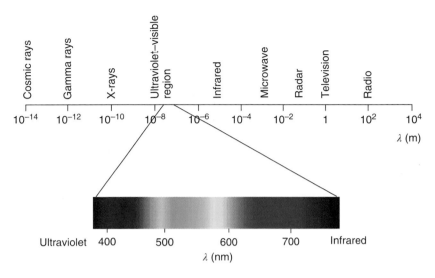

Figure 7.3 The electromagnetic spectrum.

In constructing the colour solid, it is useful firstly to consider the electromagnetic spectrum and how hues are related to each other. Visible light is just a small section of the overall electromagnetic spectrum, which ranges from rays of low wavelength but extremely high energy (cosmic rays) at one end to rays of long wavelength and low energy (radio waves) at the other end (Figure 7.3). The energy in a beam of radiation is delivered in the form of a stream of small packets known as photons. The relationship between the energy of a photon, E, and wavelength, λ, is given by Planck's equation:

$$E = \frac{hc}{\lambda} \tag{7.1}$$

where h=Planck's constant (6.626×10^{-34} J/s^{-1}) and c=speed of light (2.998×10^{8} m/s). This equation shows that the energy of a photon of light is inversely proportional to its wavelength.

The visible region of the electromagnetic spectrum, that is, the region to which our eyes are sensitive, extends from 3.8×10^{-7} to 7.8×10^{-7} m. Expressing these wavelengths in metres gives very low numbers, so for convenience, they are expressed in nanometres (nm), where $1\,\text{nm} = 1 \times 10^{-9}$ m. On this scale, the visible region lies in the wavelength region of 380–780 nm. The sensitivity of the eye is greatest to wavelengths in the middle of this range (about 555 nm) and falls away towards the two ends, for normal daytime viewing (cone vision curve in Figure 7.4). Many types of colour measuring instruments are sensitive to light with wavelengths in the range 380–730 nm. The responses of the eye to wavelengths below 400 nm and above 700 nm are so low that for practical colour measurement, the range 400–700 nm can also be used.

The curve labelled 'rod vision' in Figure 7.4 shows the sensitivity of the eye for rod vision, which is the case when viewing is carried out at very low levels of illumination (see Section 7.5.3).

It can be seen in Figure 7.3 that the hues perceived, as the wavelength increases, gradually change, and there are no sharp divisions between them. It is possible to map out a colour circle, showing how the hues relate to each other (Figure 7.5a).

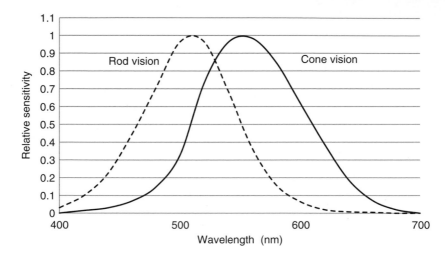

Figure 7.4   Sensitivity of the eye to visible light.

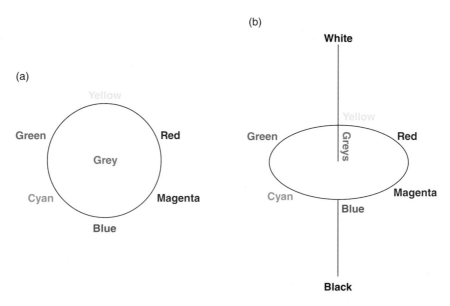

Figure 7.5   (a) The colour circle; (b) the colour solid.

Building on this concept of the hue circle, the attribute of chroma can be introduced. It can be imagined that at the outer edge, the colours are the richest that can be obtained, whilst at the centre of the circle, the colours have no hue at all; this is where the greys lie. The colour circle can therefore be used to represent the hue and chroma of colours. The third attribute of colour sensation, lightness, is then added at right angles to the plane of the colour circle to produce the three-dimensional colour solid (Figure 7.5b).

The concept of the colour solid was first used by Munsell and is illustrated in Figure 7.6. The lightness axis that runs from the top to the bottom in the centre of the solid is a visually uniform scale of greys. This scale increases to white at the top and decreases in 10 visually

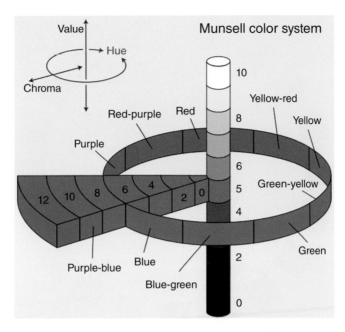

Figure 7.6   Structure of the Munsell Book of Color.

uniform steps to black at the bottom. Munsell labelled each of these greys a 'value', V. These greys have been carefully selected, so that the visual difference between one grey and the next above or the next below it is exactly the same. It is a visually uniform scale of greys between white ($V=10$) and black ($V=0$), and the midpoint of this axis (the grey of value 5) is the grey that is visually midway between black and white. The various hues radiate outwards from the grey scale, their chroma (C) increasing (like the value of the greys) in visually equal steps from the grey at the centre, where $C=0$, to the maximum full colour possible for that hue. Colours falling on the same horizontal plane (V value) in the book have the same lightness.

There are five principal hues (red R, yellow Y, green G, blue B and purple P) and five secondary hues (yellow red YR, red purple RP, purple blue PB, blue green BG, green yellow GY). These hues are illustrated in Figure 7.7. There are four pages for each hue, with the hue pages labelled 2.5, 5, 7.5 and 10. The hue page labelled 5, for example, 5R, is the purest hue (in the case of 5R, a red that is neither yellowish nor bluish). Each individual hue is represented on a page of the book, with the neutral greys at the left edge, darkest at the bottom of the page and lightest at the top.

At each lightness level (value, V), there are a series of coloured chips of increasing chroma extending outwards to the edge of the page (Figure 7.8). The maximum chroma (C) reached by each hue, at each value level, is not the same. For example, yellows reach their highest chroma at high values, whereas at low values the maximum chroma possible is only low. In the case of blues, the opposite behaviour is true.

Each coloured chip in the Munsell Book of Color is defined by a unique set of coordinates. These coordinates are the hue, value and chroma, that is, H V/C (in that order). So, for example, a dyer may match a red-coloured piece of cloth by a chip in the book with

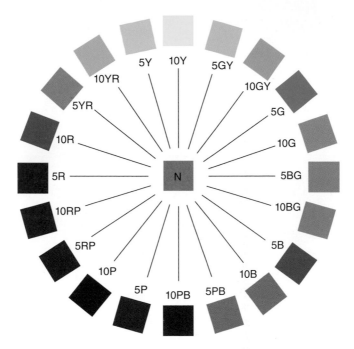

Figure 7.7 The Munsell hue circle, showing the five principle and five secondary hues of hue pages 5 and 10 (the 2.5 and 7.5 hue pages are omitted).

the coordinates 7.5R 6/4. These coordinates can be sent to someone else who also has a Munsell Book, and they can look up the colour in their book and see the colour of the dyer's cloth.

The Munsell Book of Color is just one example of many colour atlases that are available. Other atlases widely used include the Swedish Natural Color System (NCS) and the German Deutsche Institut für Normung (DIN) system, both of which have the coloured chips arranged in different ways from each other and from the Munsell Book. The main function of a colour atlas is to assist a colour user in describing, selecting and matching colours. They provide the ability of colourists to communicate colour. Colour atlases consist of a number of coloured chips mounted on the pages of a book, but the number of coloured chips is limited by cost and size. The human eye can perceive some six million different colours, and obviously only a small fraction of them can be represented in a colour atlas. Consequently the colour intervals between adjacent chips are large, but they should be as uniform as possible to aid interpolation when attempting to describe a colour.

7.5 Factors Affecting Colour Appearance

It is useful firstly to consider a typical viewing geometry – the relative positions of the light source, the surface and the observer to each other. The geometry is represented in the diagram in Figure 7.9.

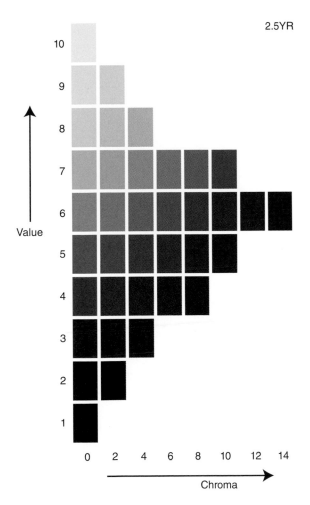

Figure 7.8   Diagram of a sample page from the Munsell Book of Color (in the Munsell book the neutral samples, where Chroma = 0, are not shown to avoid unnecessary duplication; instead they are mounted on a separate card).

The properties of each of the factors in Figure 7.9, the light source, the surface and the observer, all have an influence on the perceived colour, and the way in which these properties are specified to enable a system of colour specification to be established will now be considered.

7.5.1   Light Sources

There are three major types of source in general use: incandescent, fluorescent and light-emitting diodes (LEDs). They are characterised by their *spectral power distribution* (spd), which is the amount of light energy emitted by a source at each wavelength through the spectrum. Rather than giving the actual amounts of light being emitted, spd curves are usually given as relative amounts, that is, the amount of light emitted at each wavelength is

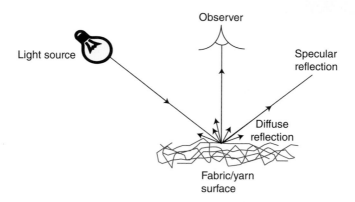

Figure 7.9 The viewing geometry.

given relative to the amount emitted at (usually) 555 nm. Showing these relative curves makes it easier to compare different light sources.

Incandescent sources emit light because of their high temperature, such as the sun or a tungsten filament lamp. The amount of light emitted at each wavelength by an incandescent light source, E_λ, can be calculated using Planck's radiation law:

$$E_\lambda = \frac{c_1}{\lambda^5 \left[e^{c_2/\lambda T} - 1 \right]} \tag{7.2}$$

where λ is the wavelength, c_1 and c_2 are constants and T is the temperature (K).

The spds of a tungsten filament lamp and average daylight are shown in Figure 7.10. The curve for tungsten light is that calculated using Planck's radiation law, whilst that for daylight is obtained from measurements. Even though both sources are incandescent, it is clear they have very different emission curves. Tungsten light emits only a relatively low amount of light energy at the blue end of the spectrum, but much larger amounts in the green, yellow and red regions. Consequently tungsten light is a fairly yellowish light. In contrast, daylight has a much more uniform energy distribution across the spectrum, with a slightly greater emission at the blue wavelengths. Daylight is therefore a much bluer light than tungsten light. The spd curve of daylight is not smooth because the light from the sun has to pass through the outer atmosphere of the Earth before reaching the surface of the Earth, which causes absorption and scattering as it does so.

Fluorescent lamps are another frequently used type of light source. Fluorescence is the process by which materials absorb light in one wavelength region and emit it at longer wavelengths. Fluorescent tubes contain mercury vapour at low pressure and are coated on the inside with special chemicals called phosphors. When an electric discharge passes through the lamp, the mercury atoms emit radiation in the ultraviolet part of the spectrum, mainly at 254, 313 and 366 nm. This radiation from the mercury atoms excites the phosphors, which in turn emit light in the visible region, notably not only at 405, 435 and 545 nm but also at other wavelengths in between (Figure 7.11). Thus the light emitted from a fluorescent source is a combination of the mercury emission wavelengths and the fluorescence emission from the phosphors.

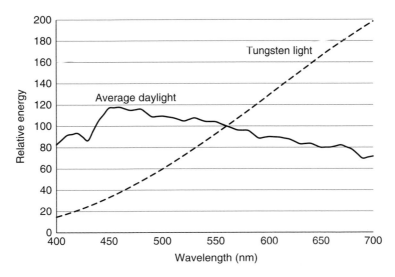

Figure 7.10 Spectral power distributions of tungsten and average daylight.

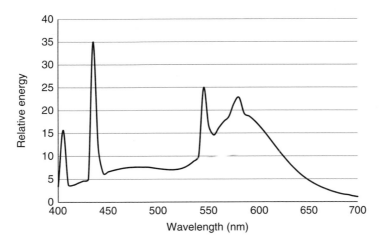

Figure 7.11 Spectral power distribution of a normal fluorescent light.

The precise spd of a fluorescent tube depends on the chemical composition of the phosphors, and variants such as 'cool white' or 'warm white' fluorescent light are produced. Another type of fluorescent tube is the '3-line' type such as TL84, which because of its high efficacy is commonly used in retail stores. The spd of this type of tube is different from other types in that it mainly comprises emission in the blue, green and red regions at the three wavelengths 435, 545 and 610 nm, respectively. Despite this the light emitted from it appears white, because a mixture of blue, green and red light appears white (see Section 7.3.1). Because the spd of 3-line sources is very different from that of daylight, some colours of materials can appear different when viewed under them than they do under daylight. This can cause considerable problems to dyehouses due to *metamerism* (see Section 7.7.3) when colour-matching shades.

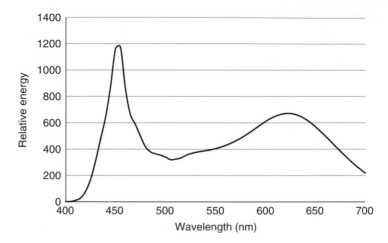

Figure 7.12   Spectral power distribution of a white LED source.

Another type of light source increasingly used is the LED type. This type of lamp emits light when an electric current is passed through a semiconductor source. The light emitted only covers a narrow range of wavelengths of about 40 nm, so they are almost monochromatic in appearance, that is, red, green or blue coloured. Their actual colour depends on the chemical composition of the semiconductor material. They are not inherently white sources, but white light can be achieved either by mixing red, green and blue LEDs or by coating a blue LED chip with a yellow phosphor. The phosphor absorbs a fraction of the blue light and fluoresces yellow light, the blue and yellow light combining to produce white light. The light produced though can be a very harsh bluish white. The spd of a typical white LED is shown in Figure 7.12.

7.5.1.1   Colour Temperature of Light Sources

Another way of describing light sources is *colour temperature*. A lump of iron raised to a high temperature will act as an incandescent source. It is a black colour at ambient temperatures, but as the temperature is raised, it glows red hot and then white hot, and then at very high temperatures, it attains a bluish hue. This is the concept of colour temperature, except that the concept in physics of an 'ideal black body radiator' is used instead of a lump of iron. The light energy it emits at each wavelength for a given temperature can be calculated using Planck's radiation law (Equation 7.2). Knowing the amount of light emitted at each wavelength, the colour properties of an incandescent light source can be calculated at every temperature. The colour temperature of a light source is the temperature of the ideal black body radiator that is nearest in colour to that of the light source. Typical colour temperatures of light sources are shown in Table 7.1.

7.5.1.2   Standard Illuminants

The Commission Internationale de l'Éclairage (CIE) is an international committee that deals with all matters relating to colour and illumination. In fact it was originally called the International Committee on Illumination but its initials (ICI) caused confusion with a major

Table 7.1 Colour temperatures of typical light sources.

type of light source	colour temperature (K)
Candle flame	1900
100 W tungsten lamp	2860
Illuminant A (represents a 100 W lamp)	2854
'Warm white' fluorescent lamp	3500
'Cool white' fluorescent lamp	4500
Overcast sky	6500
Illuminant D65 (represents average daylight)	6504

chemical manufacturer at the time, so the French equivalent name, CIE, was adopted. In 1931 this committee agreed a standard system for specifying colour and, with revisions to it in 1964, has remained in use ever since. This system sets standards for illuminants, instruments for colour measurement, and has established the standard observer, so that colour specifications can be exchanged between customers and suppliers. The committee meets regularly and makes recommendations for international standards on topics related to light and lighting.

There is an important difference between a light source and an illuminant.

A light source is a physical emitter of light, such as the sun, a light bulb or a fluorescent tube. An illuminant is a CIE specification of an spd, such as Illuminant A or Illuminant D65.

An illuminant may or may not be physically realisable as a source. For example, no light source exists, which has exactly the same spd as that defined by the CIE for Illuminant D65. Fluorescent tubes sold as 'D65' sources are a good approximation to the CIE definition. However the spd of the light emitted from a tungsten lamp corresponds closely to that of Illuminant A, as long as the lamp is not too old.

The CIE standard illuminants are numerical specifications only. Originally three were defined, called A, B and C:

- Illuminant A is light from a gas-filled tungsten filament lamp operating at a colour temperature of 2854 K.
- Illuminant B was intended to represent noon sunlight at a colour temperature of 4874 K.
- Illuminant C was intended to represent north sky light from an overcast sky at a colour temperature of 6774 K.

Illuminants B and C were later found to be inaccurate representations of daylight so are no longer in use. Instead the CIE introduced a D series of illuminants in 1964 based on measurements made on different phases of daylight. They are again defined as spds and represent daylight at different colour temperatures:

- D75: Represents cold daylight (7504 K) and is used for the critical inspection of yellowish colours
- D65: Represents average daylight (6504 K) and is widely used in colour measurement in the textile industry

- D55: Represents sunlight plus skylight (5503 K) and is used for balancing daylight films
- D50: Represents warm daylight (5003 K) and is used in the graphic arts industry

In the case of fluorescent lamps, the spds for twelve types (F1 to F12) have been compiled by the CIE, though officially they are not standard illuminants. The lamps are classified into three groups:

- F1–F6: 'Normal', or cool white, used for general illumination in offices, classrooms and other public work areas
- F7–F9: 'Broad band', or warm white, used to simulate daylight for industrial colour quality control
- F10–F12: 'Three band', for example, TL84 and Ultralume, used for lighting in retail stores

It is useful to mention at this stage the concept of the equi-energy light source, a light source that emits exactly equal amounts of light energy at each wavelength across the spectrum. No such light source exists in practice, but it is an important concept in establishing an objective method of colour specification, as detailed in Section 7.6.

7.5.2   Reflection

The diagram in Figure 7.9 shows that when viewing a matte textile surface, the diffusely reflected light is observed. For colour measurement it is therefore necessary to measure the amount of the light reflected from the surface as a fraction of the amount of light incident on it, at each wavelength through the spectrum. Such measurements are made using an instrument called a spectrophotometer.

Before considering measuring the reflectance of textiles, it is firstly necessary to define what is called the *perfect reflecting diffuser* (prd). The prd is a hypothetical surface that is the ideal white, in that it reflects 100% of the incident light, but just as importantly, it reflects it equally in all directions in the space above it, so there is no *specular* (gloss) reflectance. The prd is therefore perfectly matte. The substance that most closely has these reflection characteristics is barium sulphate powder, which has reflectance values of about 99.5% across the spectrum. Fresh snow has a whiteness that is generally acknowledged to be a good approximation to the prd.

Formally expressed, reflectance is defined as

$$R = \frac{\text{Light flux reflected by the sample}}{\text{Light flux incident on the sample}} \tag{7.3}$$

In practice it is too costly to construct spectrophotometers to measure reflectance in this way, so instead the definition is altered:

$$R = \frac{\text{Light flux reflected by the sample}}{\text{Light flux reflected by a reference white similarly illuminated}} \tag{7.4}$$

The reference white is usually a ceramic white tile, which has been calibrated accurately in terms of the prd. If R is expressed on the scale 0–1, then the term used is *reflectance factor*.

Alternatively it can be expressed on the 0–100 scale, when it is called the *percentage reflectance*. Because reflectance is measured at wavelengths across the spectrum, it is usually given the symbol R_λ. Most spectrophotometers for colour measurement provide reflectance values at every 10 nm across the visible spectrum. It is important to remember that the R_λ values are a property of the surface colour. They do not vary with the type or intensity of the incident light source.

For the measurement of reflectance, there are different designs of instrument configuration that can be used. The CIE has recommended four different illumination/viewing geometries for instrument design, illustrated in Figure 7.13. Figure 7.13b shows two geometries that include an integrating sphere, the inside of which is coated with barium sulphate or sintered polytetrafluoroethylene (PTFE). Using diffuse illumination (*d*/0) minimises the effects of surface texture because light first strikes the spherical wall of the integrating sphere and is reflected on to the sample from all directions.

With an integrating sphere, two measurements of reflectance can be made, either with the specular reflectance included or with it excluded. The user has the option of these two modes by attaching either a white specular cap or a black light trap to the wall of the integrating sphere, respectively (Figure 7.14).

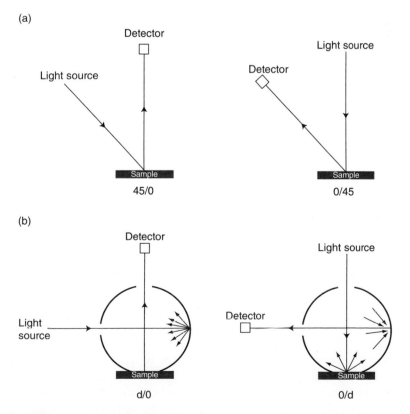

Figure 7.13 CIE recommended illuminating/viewing geometries for the measurement of reflectance by spectrophotometers.

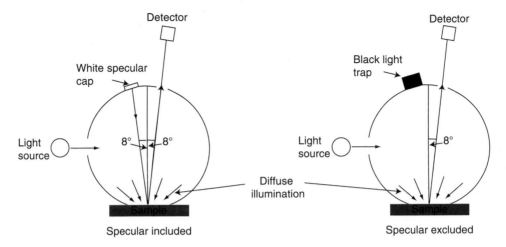

Figure 7.14 Measurement of reflectance with the specular component of reflection either included (left) or excluded (right).

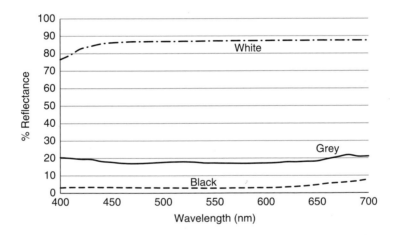

Figure 7.15 Spectral reflectance curves of typical white, grey and black colours.

Spectral reflectance curves of typical white, grey and black colours are shown in Figure 7.15. These colours are referred to as *neutral* colours. They do not contain a hue, but in practice if a set of greys (or even whites or blacks) are compared with each other, some will be perceived to be a little redder, greener or bluer than others. The perfect white does not exist, and indeed the spectral reflectance values of many 'white' textiles are much less than 100%. Fabrics with values of around 80–85% will appear to be good 'whites', though clearly the higher the overall reflectance, the better the white. Just as the perfect white has reflectance values of 100% across the spectrum, so does the perfect black with values of 0%. In other words, the perfect black absorbs all of the light that is incident on it. As with the perfect white, it is very difficult to obtain the perfect black shade. Most blacks will have spectral reflectance values close to, but slightly higher than, 0%. Also their values will not be exactly the same at every wavelength across the spectrum, giving rise to a slight hue in some blacks.

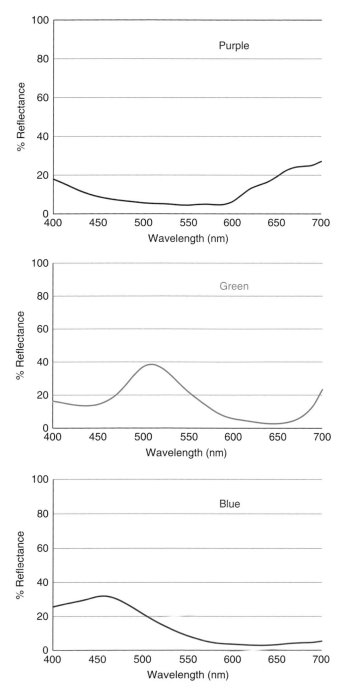

Figure 7.16   Spectral reflectance curves of typical chromatic colours.

In a similar manner to the perfect white and perfect black, pure greys should have exactly the same reflectance values at every wavelength. In practice, though, it is very difficult to dye yarns or fabrics with such consistent reflectance values. The reflectance values

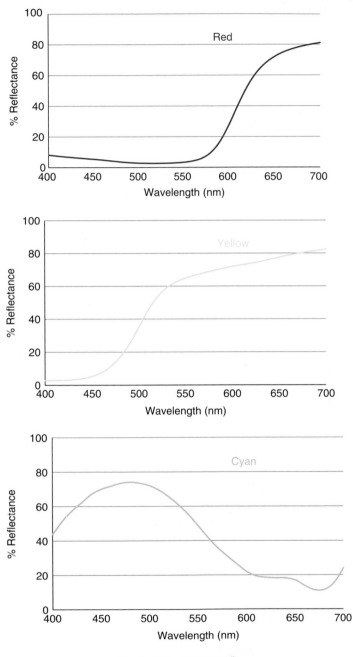

Figure 7.16   (Continued)

of most greys vary a little across the spectrum, which may cause a slight hue. The higher the overall reflectance of a grey, the lighter it will appear.

The spectral reflectance curves of typical chromatic colours, that is, those colours that possess a hue, are shown in Figure 7.16.

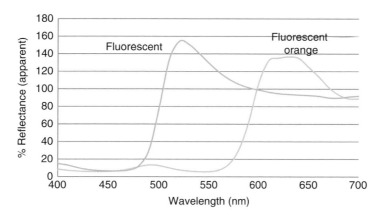

Figure 7.17   Measured spectral reflectance curves of fluorescent colours used in high visibility clothing.

The measured spectral reflectance values of fluorescent coloured materials are unusual in that they can go above 100% at certain wavelengths, which gives rise to their vivid appearance, even under dull viewing conditions. The reason for this behaviour is that the dyes (or pigments) involved absorb light at the shorter wavelengths in the visible region and then emit the absorbed energy as light at longer wavelengths. At these longer wavelengths the emitted light is viewed with the 'normal' reflected light, this outcome meaning that, effectively, more light is being viewed at these wavelengths than is being incident on the material, causing an apparent reflectance of more than 100%. The measured reflectance curves of fluorescent lime-yellow- and orange-coloured fabrics are shown in Figure 7.17.

The same mechanism of fluorescence occurs when a fluorescent whitening agent (fwa) is applied to a 'white' textile to improve its whiteness. These agents absorb radiation in the ultraviolet region of the spectrum and emit the absorbed energy at the short wavelength end (blue end) of the visible region. Because the human eye is not sensitive to ultraviolet radiation, its absorption has no influence on visual perception; it is only the emitted radiation that can be seen, this being light in the blue region of the spectrum. Even bleached 'white' goods have lower reflectance values in this region than they do at the red end, and the light emitted from the fwa overcomes the deficiency of reflection. Consequently the measured reflectance values of fabrics treated with an fwa can be well above 100% at the low wavelength end of the spectrum (Figure 7.18).

The reflectance curve of the white textile treated with an fwa shows a rapid fall towards the 400 nm. This is due to the absorption of ultraviolet radiation by the fwa, the absorption band sometimes encroaching into the visible region. A variety of fwa is available. The wavelengths of maximum emission are different so it is possible to produce whitened fabrics of slightly different hues.

7.5.3   The Eye

The final part of the viewing geometry shown in Figure 7.9 is the eye. The light from the source that is reflected from the surface is focussed by the lens to impact on the photosensitive receptors in the retina at the back of the eye (Figure 7.19). There are two types of receptors, called *rods* and *cones* because of their shapes. Under normal daytime viewing, the rods are bleached out and it is the cones that are responsible for colour vision. The rods only function at low levels of illumination but even then do not enable the perception of colour. Many rods are connected to a single nerve fibre, a feature that facilitates perception under very dark conditions, but the trade-off is that fine detail on the object being viewed cannot be perceived.

There are three types of cones: the first type being sensitive to the short wavelengths (blue sensitive), the second being sensitive to the middle region of the spectrum (green sensitive) and the third type being sensitive to the long wavelengths (red sensitive). Defective colour vision arises when one of the three types has lower than normal sensitivity, is not present or does not transmit signals to the brain. Each cone is individually linked to a nerve fibre, which enables fine detail to be perceived and also provides excellent visual

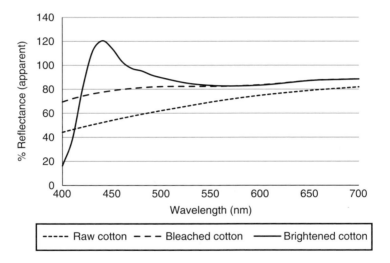

Figure 7.18 Measured spectral reflectance curves of white cotton fabrics.

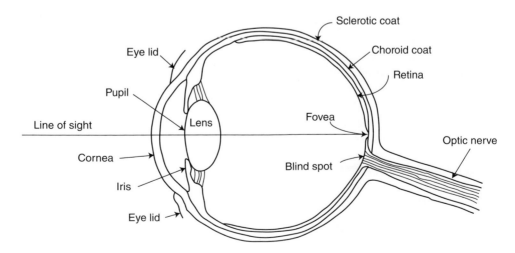

Figure 7.19 Diagram of the horizontal cross section of the human eye.

acuity. In total there are about seven million cones in an eye, about 50 000 of them are concentrated in an area of about 1 mm² called the *fovea*, which is almost directly in line of the visual axis. There are no rods in the fovea, but their concentration increases, moving outwards from it. Simultaneously the concentration of cones decreases. All of the nerve fibres exit at the back of the eye through the *optic nerve* to the brain.

7.6 The CIE System of Colour Specification

The CIE system of colour specification is based on the principle that colours can be matched using just three primary coloured lights – red, green and blue – of appropriate intensities. A colour-matching equation can be written:

$$C[C] = R[R] + G[G] + B[B] \tag{7.5}$$

which means that C units of colour [C] are matched by R units of primary [R] plus G units of primary [G] plus B units of primary [B].

7.6.1 The Standard Observer

The fundamental data for the CIE system were obtained from the classical colour-matching experiments of Wright and Guild in the late 1920s. The basis of their experiments is shown in Figure 7.20, in which observers were asked to match, using the three primary coloured lights, spectral light between 380 and 780 nm at 5 nm intervals. Guild used three filter primaries (as shown in Figure 7.20) and seven observers, whilst Wright used pure spectral

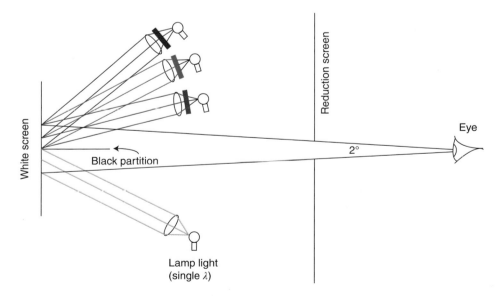

Figure 7.20 Principle of additive colour mixing to match spectral light.

primaries, at 435.8, 546.1 and 700 nm, and ten observers. The results were adjusted so that it was as though the spectral wavelengths being matched were from the equi-energy spectrum (see Section 7.5.1.2). When the slight differences in experimental design were taken into account, it was found that the results of Wright and Guild agreed very well and the average results of the total of the 17 observers who took part in the experiments define what is called the *standard observer*. These results are the amounts of the three primaries required to match unit energy of light (i.e. the same amount of light energy) at each wavelength and are called tristimulus values.

The significance of the 2° field of view shown in Figure 7.20 is that this small angle produces an image that just covers the fovea of the retina at the back of the eye. Since there are only cones and no rods in the fovea, colour perception is at its most acute, though the image seen by the observers is only tiny (it is equivalent to looking at one's thumbnail at arm's length).

The results obtained from these experiments were surprising because at all wavelengths studied across the spectrum (at 5 nm intervals), it was not possible to match a wavelength with a mixture of all three primaries. In every case (except where the wavelength being matched was the same as one of the primaries), one of the three primaries had to be mixed with the wavelength being matched, essentially to 'dilute' it, so that it could then be matched by the remaining other two primaries. What this meant was that when trying to match each wavelength, one of the primaries was required in negative amount in Equation 7.5, even though the primaries themselves were pure spectral wavelengths. For example, when matching a wavelength in the yellow region of the spectrum (colour [C] in Equation 7.5), it was necessary to add the blue primary to it, so it could then be matched by the red and green primaries. In this case of matching spectral light, Equation 7.5 could be written as

$$\left[C_\lambda \right] = \bar{r}_\lambda \left[R \right] + \bar{g}_\lambda \left[G \right] - \bar{b}_\lambda \left[B \right] \tag{7.6}$$

where $\bar{r}_\lambda, \bar{g}_\lambda, \bar{b}_\lambda$ are the amounts of the red, green and blue [R], [G], [B] primary colours required to match the wavelength λ. The 'bar' above the *r*, *g*, and *b* letters indicates that they are the average results of the observers who participated in the experiments. The subscript λ indicates that the data are available for every wavelength in the visible region of the spectrum.

All of the colours that can be matched by positive amounts of the three primaries lie in the triangle formed with the [R], [G], [B] primaries at the corners. However, the fact that a negative amount of one of the primaries was always required to match the pure spectral colours meant that these colours lay outside the triangle (Figure 7.21a).

The CIE decided that it would be better if positive amounts of all three primaries were always required, meaning that a bigger triangle was required, one that included the spectrum locus. This could only be achieved using imaginary primaries, and so new [X], [Y], [Z] primaries were established. These [X], [Y], [Z] primaries are not real primaries but instead are defined mathematically in terms of the [R], [G], [B] primaries actually used in the colour-matching experiments:

- The [X] imaginary primary can be regarded as a supersaturated red.
- The [Y] imaginary primary can be regarded as a supersaturated green.
- The [Z] imaginary primary can be regarded as a supersaturated blue.

(a)

(b)

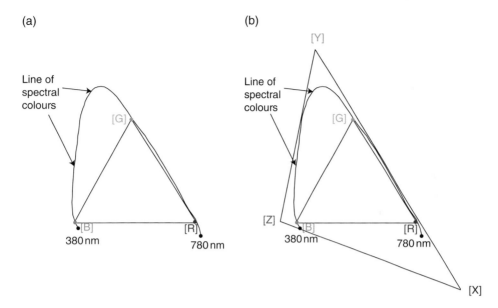

Figure 7.21 Results of colour-matching wavelengths of visible light using real [R], [G], [B] primaries and location of the [X], [Y], [Z] imaginary primaries.

These imaginary primaries were chosen so that they just enclosed the line of the spectral colours (see Figure 7.21b), and since the relationships between [X], [Y], [Z] and [R], [G], [B] were known, it was possible to convert the $\bar{r}_\lambda, \bar{g}_\lambda, \bar{b}_\lambda$ values into their corresponding $\bar{x}_\lambda, \bar{y}_\lambda, \bar{z}_\lambda$ values.

The 2° field of view does not represent normal viewing conditions, so new experiments were performed, this time using a 10° field of view. In 1964 the CIE adopted these revised $\bar{x}_\lambda, \bar{y}_\lambda, \bar{z}_\lambda$ values as the *standard observer* data, and Figure 7.22 shows how they vary with wavelength across the spectrum, for the equi-energy spectrum.

The $\bar{x}_\lambda, \bar{y}_\lambda, \bar{z}_\lambda$ values shown in Figure 7.22 are scaled relative to $\bar{y}_\lambda = 1.0$ at 557 nm. The precise relationships between [X], [Y], [Z] and [R], [G], [B] were carefully chosen so that if white light from an equi-energy light source is matched using the [X], [Y], [Z] primaries, equal amounts of them will be required. This means that the area under each of the curves in Figure 7.22 is the same. Another important point to recognise is that the shape of the $\bar{y}_\lambda$ curve closely follows that of the V_λ curve shown in Figure 7.4 (cone vision).

7.6.2   Specification of Surface Colours in the CIE XYZ System

In the CIE system colours are specified by their tristimulus values, *X*, *Y*, *Z*. These values represent the amounts of the imaginary primary colours [X], [Y], [Z] required to match the surface colour by the CIE standard observer when viewed under a CIE standard illuminant. The values indicate the responses of the red-, green- and blue-sensitive cones of the eye to the light entering it from the illuminant and reflected from the coloured surface.

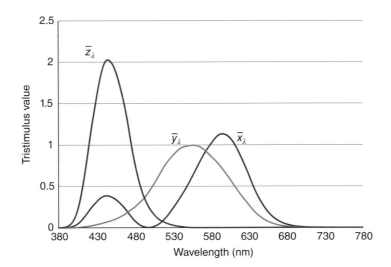

Figure 7.22   10° standard observer colour-matching values.

All three components of the viewing geometry shown in Figure 7.8, that is, the light source (specified by the spd, E_λ), the surface (specified by the reflectance factor, R_λ) and the observer (specified by the standard observer colour-matching functions, $\bar{x}_\lambda, \bar{y}_\lambda, \bar{z}_\lambda$) are used in the calculation of the tristimulus values of a coloured surface.

Consider first the tristimulus value of the coloured surface at just one wavelength. This is given by

$$X = E_\lambda \bar{x}_\lambda R_\lambda \qquad Y = E_\lambda \bar{y}_\lambda R_\lambda \qquad Z = E_\lambda \bar{z}_\lambda R_\lambda \qquad (7.7)$$

These equations allow for the fact that the light source is not equi-energy and emits different amounts of light at each wavelength, and the coloured surface only reflects a certain percentage of the light incident on it. They modify the responses of the eye (the $\bar{x}_\lambda, \bar{y}_\lambda, \bar{z}_\lambda$ values) by multiplying them by the E_λ and R_λ values.

In practice, when a coloured surface is viewed, all of the wavelengths of the spectrum are seen simultaneously, so it is necessary to add these values for all wavelengths together to get the total response of the eye. This leads to the somewhat cumbersome equations:

$$X = k \int_{380}^{780} E_\lambda \bar{x}_\lambda R_\lambda \Delta\lambda \qquad Y = k \int_{380}^{780} E_\lambda \bar{y}_\lambda R_\lambda \Delta\lambda \qquad Z = k \int_{380}^{780} E_\lambda \bar{z}_\lambda R_\lambda \Delta\lambda \qquad (7.8)$$

where $k = \dfrac{1}{\displaystyle\int_{380}^{780} E_\lambda \bar{y}_\lambda \Delta\lambda}$

These equations mean that, at every wavelength across the spectrum between 380 and 780 nm, the $E_\lambda \bar{x}_\lambda R_\lambda$, $E_\lambda \bar{y}_\lambda R_\lambda$ and $E_\lambda \bar{z}_\lambda R_\lambda$ values have to be calculated (and multiplied also by the interval between the wavelengths, $\Delta\lambda$) then added together (this is the reason for the integration, meaning summation, sign $\int$) and the total multiplied by k, which is called the *normalising constant*. Multiplying by this constant k makes $Y = 100$ for the perfect white.

In practice this method is difficult to employ for the reason that it requires R_λ values at every wavelength across the spectrum. It is too costly to build a spectrophotometer to do this – many spectrophotometers on the market only give the R_λ values at every 10 nm, some at 20 nm intervals and sometimes only between 400 and 700 nm. Fortunately Equation 7.8 can be simplified to

$$X = \sum E_\lambda \bar{x}_\lambda R_\lambda \qquad Y = \sum E_\lambda \bar{y}_\lambda R_\lambda \qquad Z = \sum E_\lambda \bar{z}_\lambda R_\lambda \qquad (7.9)$$

where the symbol $\sum$ means adding together the $E_\lambda \bar{x}_\lambda R_\lambda$, $E_\lambda \bar{y}_\lambda R_\lambda$ and $E_\lambda \bar{z}_\lambda R_\lambda$ products. To assist in the calculations, the values of $E_\lambda \bar{x}_\lambda$, $E_\lambda \bar{y}_\lambda$, $E_\lambda \bar{z}_\lambda$, 'normalised' by the k value and called *weighted ordinates*, are published by the CIE. The values are available at 10 and 20 nm intervals and for combinations of different illuminants and for the 2° and 10° standard observers (e.g. see Table 7.2). It is then only necessary to multiply them by the R_λ values at each wavelength then add them up.

To compute the tristimulus values for a surface colour, its reflectance values (as a reflectance factor R_λ, that is on the 0–1 scale) at each wavelength must be multiplied by the corresponding $E_\lambda \bar{x}_\lambda$, $E_\lambda \bar{y}_\lambda$, $E_\lambda \bar{z}_\lambda$ values and then summed.

For the perfect white, when the reflectance factor $R_\lambda = 1.0$ at every wavelength, it is just a case of multiplying every value in Table 7.2 by 1.0, which gives the same values. Therefore its tristimulus values under Illuminant D65 will be $X=94.8272$, $Y=100.0000$ and $Z=107.3906$. The perfect white will only have tristimulus values of $X=100.00$, $Y=100.00$ and $Z=100.00$ when it is viewed under the equi-energy Illuminant E. Under a different illuminant, such as Illuminant A, the weighted ordinates are different, so the perfect white

Table 7.2 Weighted ordinates for Illuminant D65, 10° observer (values are shown at 20 nm intervals).

λ (nm)	$E_\lambda \bar{x}_\lambda$	$E_\lambda \bar{y}_\lambda$	$E_\lambda \bar{z}_\lambda$
400	0.2516	0.0236	1.0906
420	3.2317	0.3301	15.3824
440	6.6805	1.1069	34.3820
460	6.0964	2.6206	35.3562
480	1.7213	4.9378	15.8979
500	0.0589	8.6695	3.9972
520	2.1845	13.8473	1.0457
540	6.8093	17.3537	0.2373
560	12.8953	17.1539	0.0025
580	16.4686	14.1481	−0.0022
600	17.2340	10.1056	0
620	12.8953	6.0212	0
640	6.2267	2.5867	0
660	2.1113	0.8268	0
680	0.5736	0.2222	0
700	0.1209	0.0460	0
Totals	94.8272	100.0000	107.3906

Table 7.3 Calculation of tristimulus values (Illuminant D65, 10° observer) for a green-coloured fabric.

λ (nm)	R_λ	$E_\lambda \bar{x}_\lambda R_\lambda$	$E_\lambda \bar{y}_\lambda R_\lambda$	$E_\lambda \bar{z}_\lambda R_\lambda$
400	0.0386	0.0097	0.0009	0.0421
420	0.0263	0.0850	0.0087	0.4046
440	0.0237	0.1583	0.0262	0.8149
460	0.0235	0.1433	0.0616	0.8309
480	0.0436	0.0750	0.2153	0.6931
500	0.1934	0.0114	1.6767	0.7731
520	0.2619	0.5721	3.6266	0.2740
540	0.2065	1.4061	3.5835	0.0490
560	0.1734	2.1090	2.9745	0.0004
580	0.1264	2.0816	1.7883	−0.0003
600	0.1214	2.0922	1.2268	0
620	0.1086	1.4004	0.6539	0
640	0.1096	0.6824	0.2835	0
660	0.1805	0.3811	0.1492	0
680	0.3754	0.2153	0.0834	0
700	0.6445	0.0779	0.0296	0
Totals		11.5010	16.3889	3.8816

The $R\lambda$ values are on 0–1 scale.

will have a different set of tristimulus values, in this case $X=111.144$, $Y=100.000$ and $Z=35.200$. For the perfect black, $R=0$, so regardless of the illuminant $X=0$, $Y=0$ and $Z=0$.

In the case of a surface colour such as a green-coloured fabric, the same procedure is carried out, giving the values shown in Table 7.3. Thus, this particular green-coloured fabric will have tristimulus values of $X=11.50$, $Y=16.39$ and $Z=3.88$ (for Illuminant D65, 10° observer).

Commercial colour measurement systems comprise a spectrophotometer linked to a dedicated computer. Once the spectrophotometer has made the measurements of reflectance and transmitted them to the computer, the calculations of tristimulus values can then be carried out. The weighted ordinates, such as those shown in Table 7.2, for all illuminant/standard observer combinations are available in the computer software, and the user only has to specify which combinations (e.g. Ill D65/10° observer, Ill A/10° observer, Ill A/2° observer) the X, Y, Z values are needed for. In normal dyehouse practice it is usually only those combinations involving the 10° observer that are used.

7.6.3 Interpretation of Tristimulus Values

The X value indicates the response of the red-sensitive cones; the Y value, the green-sensitive cones; and the Z value, the blue-sensitive cones. Thus, for example, the X value will be the highest for a red-coloured fabric, whilst the Z value will be highest for a blue-coloured fabric. The Y value not only will be highest for a green-coloured fabric but also indicates the lightness of a colour, so the higher the Y value, the lighter the colour. In the case of

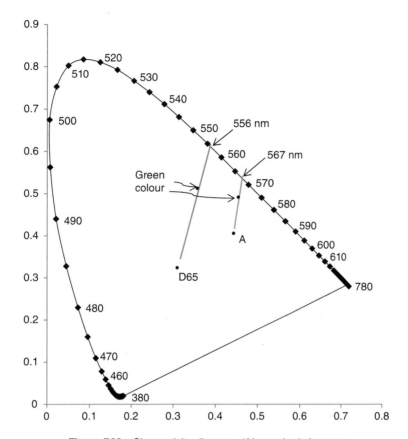

Figure 7.23 Chromaticity diagram, 10° standard observer.

neutral colours, such as greys, the values of X, Y and Z will be roughly the same. Light greys will have high X, Y, Z values overall, whilst dark greys will have low values overall.

In addition to indicating the lightness of a colour (through the Y value), it is also possible to interpret the X, Y, Z values in terms of hue and saturation. The concept of a colour space has proved to be invaluable in colour measurement. Although a three-dimensional representation is inconvenient, the practical solution has been to convert the tristimulus values into fractional quantities referred to as *chromaticity coordinates*, as shown in Equation 7.10:

$$x = \frac{X}{X+Y+Z} \qquad y = \frac{Y}{X+Y+Z} \qquad z = \frac{Z}{X+Y+Z} \tag{7.10}$$

It therefore follows that $x+y+z=1$ for all colours. Thus only two chromaticity coordinates, x and y, need to be defined, since z can always be found easily ($z=1-x-y$). Values of x and y can be plotted on a two-dimensional graph called a *chromaticity diagram*. The line joining the spectrum of colours plotted as values of x and y obtained from the tristimulus values of the standard observer is called the *spectrum locus*. The position of each wavelength on the spectrum locus is shown in Figure 7.23. The spectrum locus shows the boundary of all physically realisable colours: all colours that we can perceive in nature lie either on or within the spectrum locus and the straight line joining its two ends. This straight line shows the boundary of the non-spectral colours, such as pinks and purples.

The tristimulus values for the green fabric were obtained for Illuminant D65/10° observer. When its chromaticity coordinates ($x=0.3620$, $y=0.5159$) are plotted in the chromaticity diagram and a line from the location of Illuminant D65 ($x=0.3138$, $y=0.3310$) drawn through it and continued to the spectrum locus, it can be seen that it crosses the spectrum locus at a wavelength of approximately 556 nm. This is called the *dominant wavelength* of the colour. A feature of the chromaticity diagram is that if D65 light is mixed with monochromatic light of 556 nm, all of the greens along this line will have the same hue. They will differ only in their saturation, becoming paler as the D65 point is approached and conversely, becoming richer as the spectrum locus is approached.

Another useful feature of this diagram is that the influence of the illuminant on the colour of the fabric can be assessed. For example, if the tristimulus values of the same green colour are determined using the weighted ordinates for illuminant A/10° observer, values of $X=14.84$, $Y=15.92$ and $Z=1.59$ are obtained, giving $x=0.4587$ and $y=0.4921$. When these values are plotted in the chromaticity diagram, and a line drawn through it from the location of Illuminant A ($x=0.44758$, $y=0.40745$) towards the spectrum locus, a dominant wavelength of approximately 567 nm is obtained, showing that the fabric appears a yellower shade of green under tungsten light than it does under daylight.

7.7 Applications of the CIE System

The specification of colours by their X, Y and Z tristimulus values enables a number of very useful applications in dyeing and printing. These include

- Colorant formulation (dye recipe prediction)
- Assessment of colour difference
- Assessment of the degree of metamerism
- Assessment of colour constancy
- Colour sorting
- Assessment of the degree of whiteness

A brief outline will be given of these applications.

7.7.1 Colorant Formulation

Colorant formulation involves using the principles of colour physics and the CIE system to predict dye or pigment recipes to match standard shades. Colorant formulation involving pigments is used for pigment printing or producing paints and printing inks. In the case of dyeing textiles, the term *dye recipe prediction* is used and the equations are slightly different from those used in predicting pigment recipes. Here, only dye recipe prediction will be explained.

The aim of dye recipe prediction is to calculate the amounts of (usually) three dyes required to match a standard shade. Then, once the amounts of the three dyes are known, it is possible to calculate the amounts of the auxiliary chemicals needed for the dyeing process. The calculations can be repeated with combinations of different dyes, so a selection of dye recipes is obtained, and for each recipe, the cost and the likely degree of metamerism can be determined.

To establish a dye recipe prediction system, information about each of the dyes to be used in the recipe is required. This information takes the form of values (called *absorption coefficients*) that are measures of how strongly the dyes absorb light at each wavelength through the spectrum. These absorption coefficients are critical to the accuracy of the recipe prediction algorithm, so must be determined with great care. For each dye to be used, laboratory dyeings of each dye individually should be carried out on the same yarn or fabric that the final predicted recipe will be dyed on and using exactly the same dyeing method (time/temperature profile, dyebath auxiliaries, etc). For each dye these laboratory dyeings should be carried out at about six depths of shade, for example, 0.1, 0.25, 0.5, 0.75, 1.0 and 2.0% on weight of fibre (o.w.f.). For best accuracy and to ensure confidence, these dyeings should be duplicated.

The reflectance values of each of the dyed samples are then measured, together with those of the undyed yarn or fabric. For each dye, a 'family' of reflectance curves is obtained. These values are stored in the computer of commercial dye recipe prediction systems, which then uses them to calculate the absorption coefficients. A 'family' of reflectance curves is shown in Figure 7.24 for the dye C.I. Acid Blue 40 applied to a wool fabric.

It can be seen in this case that the lowest reflectance always occurs in the region between 600 and 650 nm. In this region absorption of light by the dye is highest. The opposite is true in the region between 450 and 500 nm. In Figure 7.25a the %R value has been plotted against concentration (% o.w.f.) at two wavelengths, 480 and 620 nm, where absorption is lowest and highest, respectively. It can be seen that in each case the lines are curved. This is a problem because it makes it very difficult to calculate accurately the %R at any other % o.w.f., which is a requirement in the recipe prediction algorithm.

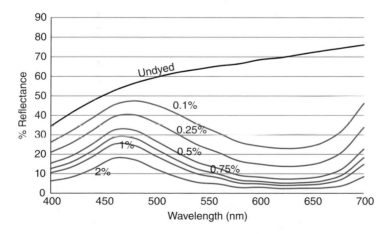

Figure 7.24   Reflectance curves of C.I. Acid Blue 40 applied to wool at different depths of shade.

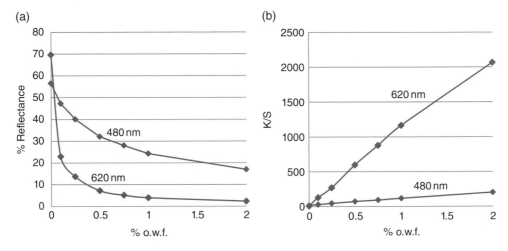

Figure 7.25   (a) % Reflectance plotted against concentration (% o.w.f.) of C.I. Acid Blue 40 on wool, at wavelengths of minimum (480 nm) and maximum (620 nm) absorption of light; (b) corresponding graphs for the Kubelka–Munk function of reflectance (K/S) plotted against concentration of dye, o.w.f.

In the late 1930s two German physicists, Kubelka and Munk, developed an equation that gives a straight-line relationship between reflectance and concentration of dye on the fibre. Their equation relates a mathematical function of reflectance to the absorption and scattering values:

$$\frac{K}{S} = \frac{(100-R)^2}{200R} \quad \text{and} \quad R = 100\left[\left(1+\frac{K}{S}\right) - \sqrt{\left(1+\frac{K}{S}\right)^2 - 1}\right] \qquad (7.11)$$

where R = % reflectance of an opaque layer of material illuminated with light of a known wavelength, K is the absorption coefficient and S is the scattering coefficient.

The K/S function is, to a good approximation, linearly related to the concentration of colorant in the substrate:

$$\frac{K}{S} = \frac{cK_{dye} + K_f}{cS_{dye} + S_f} \cong c\left(\frac{K_{dye}}{S_f}\right) + \frac{K_f}{S_f} \qquad (7.12)$$

where the subscript f refers to the undyed fibre, K_{dye} and S_{dye} are the absorption and scattering coefficients of the dye and c is the concentration of dye on the fibre, % o.w.f.

In the case of dyes on fibres, it is assumed that S_{dye} is negligible and the term K_{dye}/S_f is replaced by k, the absorption coefficient of the dye. Comparing the right hand side of Equation 7.12 with the general equation of a straight line, $y = mx + c$, a plot of K/S against concentration c should be a straight line, for which the gradient m corresponds to the absorption coefficient k. This formula has to be applied at each wavelength, so the value of k for the dye can be obtained at each wavelength across the spectrum. The plots of K/S for the two wavelengths 480 and 620 nm are shown in Figure 7.25b. It is always wise to plot these graphs, at least at the wavelength of maximum light absorption of the dye, to check for linearity and that none of the dyed samples are 'outliers'.

It is not unusual for graphs of the type illustrated in Figure 7.25b to curve, especially at higher dye concentrations. This can occur, for example, because of the scattering of light by aggregated dye particles or because the fraction of dye that fixes to the fibre changes with concentration. The computer software is able to compensate for this deviation from linearity by the application of mathematical modelling, which helps to improve the accuracy of predicted recipes, especially those involving heavy depths of shade.

The computer software automatically calculates the absorption coefficients k at each wavelength and stores them for use in the recipe prediction algorithm. The user is invited to set up databases for different dye/fibre systems, such as acid levelling dyes on wool, basic dyes on acrylics and disperse dyes on polyester. However it may be necessary to set up different databases if there are variations in the substrate, such as acid levelling dyes on loose wool, acid levelling dyes on wool yarn and acid levelling dyes on wool fabric. Different databases will be necessary especially if fabrics of different weave structures are to be dyed. All the dyes within a database will need to be compatible, in terms of rate of dyeing, reactivity (in the case of reactive dyes) and fastness properties. A typical database will probably contain the absorption coefficients of a number of dyes: 2–3 yellows, 2–3 reds, 2–3 blues and other dyes such as oranges, purples, greens and violets.

The accuracy of the dye databases should be checked before dyeing predicted recipes in bulk. This can be achieved by dyeing known recipes of trichromatic combinations of dyes. Usually it is best to dye neutral shades, say, by using equal amounts of yellow, red and blue dyes at different depth levels:

- Grey#1: $c_Y = 0.5\%$, $c_B = 0.5\%$, $c_B = 0.5\%$
- Grey#2: $c_Y = 1\%$, $c_B = 1\%$, $c_B = 1\%$
- Grey#3: $c_Y = 2\%$, $c_B = 2\%$, $c_B = 2\%$

After dyeing the three samples, they are then used as 'standards' and recipes predicted with the same dyes. The predicted recipes should be the same as those used.

An important feature of the Kubelka–Munk equation is that it is additive, so that if three dyes, such as a yellow (Y), red (R) and blue (B), are present on the fibre, Equation 7.12 can be expanded to

$$\frac{K}{S} = \frac{K_f}{S_f} + c_Y k_Y + c_R k_R + c_B k_B \tag{7.13}$$

where c_Y, c_R, c_B are the concentrations of the yellow, red and blue dyes, respectively, on the fibre and k_Y, k_R, k_B are the absorption coefficients of the three dyes.

Knowing the values of k_Y, k_R, k_B at each wavelength, the %R values at each wavelength can be calculated for any values of the dye concentrations c_Y, c_R, c_B.

To calculate a dye recipe, it is necessary to determine the concentrations of the three dyes that will give the same tristimulus values (X, Y, Z) as those of the standard being matched. So, a mathematical relationship between the three dye concentrations and the X, Y, Z values is required. This relationship has three components:

- The relationship between X, Y, Z and %R, which is given by the CIE equations:

$$X = \sum_\lambda E_\lambda \bar{x}_\lambda R_\lambda \qquad Y = \sum_\lambda E_\lambda \bar{y}_\lambda R_\lambda \qquad Z = \sum_\lambda E_\lambda \bar{z}_\lambda R_\lambda \tag{7.14}$$

- The relationship between %R and K/S, which is given by Equation 7.11
- The relationship between K/S and concentration, which is given by Equation 7.13

For known values of c_Y, c_R, c_B, X, Y, Z can be calculated:

$$c_Y, c_R, c_B \xrightarrow[\text{equation}]{\text{Kubelka–Munk}} \%R_\lambda \xrightarrow[\text{equations}]{\text{CIE}} X, Y, Z$$

However, recipe prediction requires this process in reverse, but the step $X, Y, Z \rightarrow \%R_\lambda$ cannot be done. Therefore a complex mathematical procedure called *iteration* is performed to enable the calculation of a recipe.

The information produced by a dye recipe prediction system will typically comprise:

- The concentrations of the three dyes required to match the standard shade.
- The $\%R_\lambda$ values of the standard shade and of the predicted recipe.
- The colour-difference values (ΔE) of the match under (usually) three illuminants, D65, A and F11 (TL84), but other illuminants can be specified if required, for example, F4 (WWF) or F2 (CWF).
- The amounts of auxiliary chemicals required.
- The total cost of the dye recipe.

This information will be given for various combinations of dyes, and the dyer must select the recipe that best meets the criteria for being a close match, with minimal metamerism (see Section 7.7.3) and preferably lowest cost.

Although the ultimate aim is to provide a recipe that is 'right first time', it would be optimistic to expect any system to operate with consistent perfection. Colour matching is not an exact science, and there are many deviations from the theoretical concepts upon which instrumental colour matching are based. The systems used are in a perpetual state of development and refinement. Often unexpected variables may arise for which no account has been taken in the design of the operating programme. It is easy to recognise contributory human errors in activities such as setting the operating parameters of the equipment or weighing the colour. These frequently have an easily identifiable cause, but the origin of other deviations can be more subtle and can call for careful refinement of operating procedures. For example, the dyes may behave differently when applied individually to produce the calibration dyed samples, from when they are applied in admixture, because of interaction with other dyes in the recipe. Such differences may also occur when successful small-scale trials are scaled up to full production, simply because of mechanical factors involved in dealing with a large bulk of material that cannot be foreseen from a small-scale trial.

In addition to considerations of cost and degree of metamerism in selecting the best recipe, another is how sensitive the recipes are to errors in weighing or processing deviations, for example. Often the most robust recipes are those in which the three dyes selected lie close to the shade of the standard in L^*, a^*, b^* colour space (see Section 7.7.2). If the L^*, a^*, b^* coordinates of the three dyes to be used are plotted and straight lines joined between their locations, the triangular area formed between them will represent the range of colours (the 'colour gamut') that can be obtained by mixing these dyes in different proportions. If this process is repeated for all of the dyes available, a grid of the type shown in

Figure 7.26 is obtained. This grid has to be obtained when the dyes are all at about the same depth as the target shade, so that when the L^*, a^*, b^* values of the target shade are plotted, the triangle of the three dyes that overlaps the target is the one most robust for recipe prediction in terms of resistance to variations in processing. In the example shown in Figure 7.26, the most robust recipe will be obtained if the target shade (T) is matched using the dyes C.I. Vat Orange 17, C.I. Vat Red 14 and C.I. Vat Brown 30. However these dyes may not necessarily give a match with the least metamerism or the lowest cost, so these factors have to be borne in mind when deciding which recipe to use.

Correcting bulk-scale dyeing is therefore another area in which computerised colour manipulation is of value. Instrumental measurements are able to provide values for the differences between the tristimulus values of the dyed and target patterns; these are then used to calculate a recipe suitable for bringing the dyed pattern on shade. Also computer algorithms are available that keep a history of how close predicted recipes are to their target shades. Using feedback on the concentrations of dyes predicted and the differences between corresponding tristimulus obtained when they were dyed and the target tristimulus values, the algorithms can modify the absorption coefficients of the dyes concerned so that future predicted recipes will be more accurate.

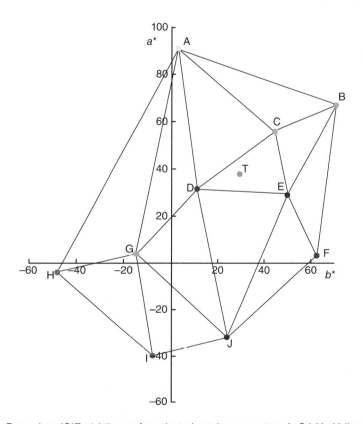

Figure 7.26 Dye colour (CIE a^*,b^*) map for selected vat dyes on cotton. A: C.I. Vat Yellow 2; B: C.I. Vat Orange 7; C: C.I. Vat Orange 17; D: C.I. Vat Brown 30; E: C.I. Vat Red 14; F: C.I. Vat Red 1; G: C.I. Vat Green 30; H: C.I. Vat Green 1; I: C.I. Vat Blue 6; J: C.I. Vat Violet 21.

7.7.2 Colour-Difference Formulae

When a batch of fabric has been dyed, it is necessary to make a decision on whether it is a close enough visual match to the standard to be acceptable. A visual check can be made but visual assessments of colour differences are not always reliable, and disagreements can often arise between the dyer and the customer. Factors such as the size of the samples being compared, the type and intensity of the lighting and the physiological health of the observer can all influence pass/fail decisions. For this reason colour-difference formulae have been developed, the function of which is to provide a number which is proportional to the size of the perceived difference between two colours.

Ever since the establishment of the CIE system of colour specification in 1931, considerable effort has been expended on developing a reliable colour-difference formula. The fundamental structure of a formula is very straightforward, involving the calculation of the distance apart in the colour space of the standard and a batch. This calculation involves the application of the Pythagoras theorem to the three dimensions of colour space (Figure 7.27).

This structure can be applied to the tristimulus values of the standard and batch (Figure 7.28), giving

$$\left(\Delta E\right)^2 = \left(\Delta X\right)^2 + \left(\Delta Y\right)^2 + \left(\Delta Z\right)^2 \tag{7.15}$$

where the Δ symbol means 'difference in'.

Equation 7.15 is a colour-difference formula, but it is found that for pairs of colours (e.g. two reds, two yellows, two purples, etc.) for which the standard and the batch are the same numeric distance apart; the visual differences are not the same. The problem is that the X, Y, Z tristimulus colour space is visually non-uniform. This defect also applies to the x, y chromaticity diagram where, again, equal changes in chromaticity coordinates do not produce equal differences in perceived colour.

The reason for the complexity of colour-difference formulae (and why so many colour-difference formulae have been published over the years) is that it has proved very difficult to modify the X, Y, Z tristimulus colour space to produce a visually uniform colour space. The greatest advance in the development of colour-difference formulae was the CIELAB formula recommended by the CIE in 1976:

$$\left(\Delta E^*\right)^2 = \left(\Delta L^*\right)^2 + \left(\Delta a^*\right)^2 + \left(\Delta b^*\right)^2 \tag{7.16}$$

or

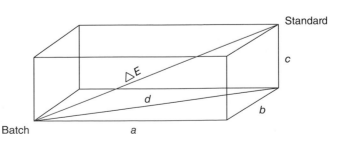

Figure 7.27 Calculation of colour difference, ΔE between a standard and a batch. $\Delta E^2 = c^2 + d^2$, and since $d^2 = a^2 + b^2$, $\Delta E^2 = c^2 + a^2 + b^2$.

$$\Delta E^* = \left[\left(\Delta L^* \right)^2 + \left(\Delta a^* \right)^2 + \left(\Delta b^* \right)^2 \right]^{1/2} \tag{7.17}$$

L^*, a^*, b^* are the three axes of a colour space that is derived from the X, Y, Z space but is much more visually uniform. The * symbols indicate that the L^*, a^*, b^* values are obtained using a cube-root approximation to a more complex set of equations originally devised. The Δ symbol again means 'difference in', so, for example, $\Delta L^* = L^*_{\text{batch}} - L^*_{\text{standard}}$.

The L^*, a^*, b^* values are obtained from the tristimulus values using the following equations:

$$L^* = 116 \left(\frac{Y}{Y_n} \right)^{1/3} - 16 \quad \text{unless} \quad \frac{Y}{Y_n} < 0.008856 \quad \text{when} \quad L^* = 903.3 \left(\frac{Y}{Y_n} \right) \tag{7.18}$$

$$a^* = 500 \left[\left(\frac{X}{X_n} \right)^{1/3} - \left(\frac{Y}{Y_n} \right)^{1/3} \right] \quad \text{and} \quad b^* = 200 \left[\left(\frac{Y}{Y_n} \right)^{1/3} - \left(\frac{Z}{Z_n} \right)^{1/3} \right] \tag{7.19}$$

Equations 7.18 and 7.19 are used for the vast majority of colours, but in the rare occasions when calculating a^* and b^*, if X/X_n, Y/Y_n and Z/Z_n are less than 0.008856, then

$$\left(\frac{X}{X_n} \right)^{1/3} \text{ is replaced by } 7.787 \left(\frac{X}{X_n} \right) + 16/116$$

$$\left(\frac{Y}{Y_n} \right)^{1/3} \text{ is replaced by } 7.787 \left(\frac{Y}{Y_n} \right) + 16/116 \tag{7.20}$$

$$\left(\frac{Z}{Z_n} \right)^{1/3} \text{ is replaced by } 7.787 \left(\frac{Z}{Z_n} \right) + 16/116$$

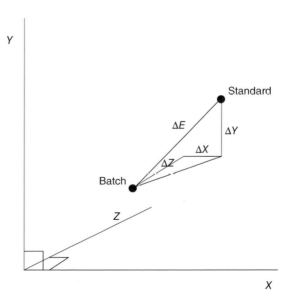

Figure 7.28 Application of Pythagoras theorem to three-dimensional X, Y, Z colour space.

Table 7.4 Tristimulus values of the perfect reflecting diffuser viewed under different illuminants (10° observer).

illuminant	X_n	Y_n	Z_n
D65	94.81	100.00	107.30
A	111.14	100.00	35.20
F11 (similar to TL84)	103.86	100.00	65.61

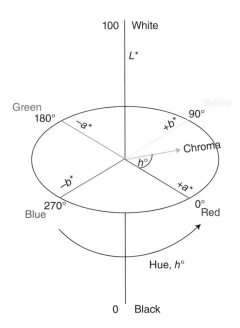

Figure 7.29 The CIELAB colour space.

In all these equations, X_n, Y_n and Z_n are the tristimulus values of the prd (see Section 7.5.2) viewed under a particular illuminant. Their values are shown in Table 7.4.

The scaling factors of 116, 500 and 200 in Equations 7.18 and 7.19 have been chosen to make the L^*, a^*, b^* colour space as visually uniform as possible and to yield ΔE values such that the tolerance limit of acceptability between a standard and a batch dyed to match it is approximately 1.0 CIELAB units.

The L^*, a^*, b^* values form the three axes of the CIELAB colour space shown in Figure 7.29. The location of a colour can also be specified by its cylindrical coordinates, L^* (lightness), C^* (chroma) and $h°$ (hue), where

$$C^* = \left(a^{*2} + b^{*2}\right)^{1/2} \quad \text{and} \quad h = \arctan\left(\frac{b^*}{a^*}\right) \tag{7.21}$$

The hue angle $h°$ is given in the degrees anticlockwise from the positive a^* axis.

The four psychological primary hues in Figure 7.29 do not quite lie at the angles shown. The purest of these hues, the hues of 5R, 5Y, 5G and 5B in the Munsell Book of Color (see Section 7.4), lie at the hue angles 27°, 95°, 162° and 260°, respectively.

A colour difference (ΔE^*) can be broken down into its components (ΔL^*, Δa^*, Δb^*), but these values are difficult to interpret in visual terms. An alternative is to express the component differences ΔL^*, ΔC^* and ΔH^* as the differences in lightness, chroma and hue, respectively. It should be noted that ΔH^* is not the difference in hue angles but has to be determined using Equation 7.22:

$$\left(\Delta E^*\right)^2 = \left(\Delta L^*\right)^2 + \left(\Delta C^*\right)^2 + \left(\Delta H^*\right)^2, \quad \text{or} \quad \Delta H^* = \left[\left(\Delta E^*\right)^2 - \left(\Delta L^*\right)^2 + \left(\Delta C^*\right)^2\right]^{1/2}$$

(7.22)

The sign (+ve or −ve) of ΔH^* is the same as the sign of the difference in hue angles ($h^{\circ}_{\text{batch}} - h^{\circ}_{\text{standard}}$).

Although the CIELAB colour-difference formula was a major achievement, it was still not wholly reliable because the L^* a^* b^* colour space is not completely visually uniform. According to the establishing principles of the CIELAB formula, the tolerance limit of acceptability should trace out a sphere in the colour space around a standard, and these spheres are all of the same size (shown as circles of constant size in the a^* b^* diagram in Figure 7.30a). However it was found that the tolerance limits of acceptability actually formed ellipses around the standards (Figure 7.30b). Not only that but also the sizes and shapes of the ellipsoids varied throughout the colour space.

In order to produce a colour-difference formula that was more accurate than the CIELAB formula, it was necessary to account for the variation in the dimensions of the ellipsoids that formed the boundary of acceptability. The formula that was developed (Equation 7.23) is called the CMC(l:c) formula, named after the Colour Measurement Committee of the Society of Dyers and Colourists, which developed it in 1984:

$$\Delta E = \left[\left(\frac{\Delta L^*}{lS_L}\right)^2 + \left(\frac{\Delta C^*}{cS_C}\right)^2 + \left(\frac{\Delta H^*}{S_H}\right)^2\right]^{1/2}$$

(7.23)

where $S_L = \dfrac{0.040975L^*_{\text{std}}}{1+0.01765L^*_{\text{std}}}$ unless $L^*_{\text{std}} < 16$, when $S_L = 0.511$

$$S_C = \frac{0.0638C^*_{\text{std}}}{1+0.0131C^*_{\text{std}}} + 0.638$$

$$S_H = S_C\left(Tf + 1 - f\right) \quad \text{where} \quad f = \left\{\frac{\left(C^*_{\text{std}}\right)^4}{\left[\left(C^*_{\text{std}}\right)^4 + 1900\right]}\right\}^{1/2}$$

and $T = 0.36 + \left|0.4\cos\left(h_{\text{std}} + 35\right)\right|$, unless h_{std} is between 164° and 345°, when

$$T = 0.56 + \left|0.2\cos\left(h_{\text{std}} + 168\right)\right|$$

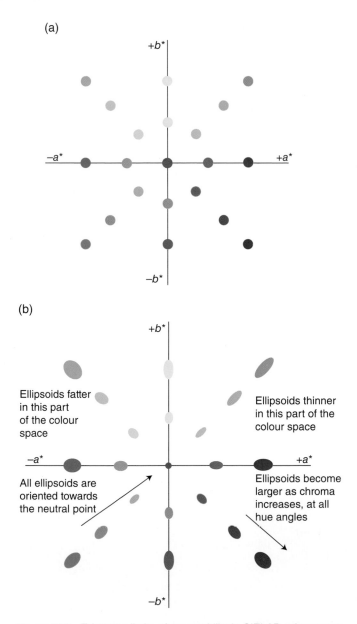

Figure 7.30 Tolerance limits of acceptability in CIELAB colour space.

The values S_L, S_C and S_H are the lengths of the semi-axes of the tolerance ellipsoid around a particular standard, as shown in Figure 7.31. These values have to be calculated for the particular point in the $L^*a^*b^*$ colour space at which the standard is located. The ellipsoid dimensions vary throughout the colour space, which is the reason for the complexity of the equations to calculate the S_L, S_C and S_H values. Always the S_H value is lower than the S_L or S_C values because differences in hue are much less tolerated than differences in lightness or chroma.

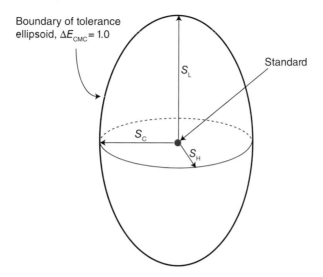

Boundary of tolerance
ellipsoid, $\Delta E_{CMC} = 1.0$

S_L

Standard

S_C

S_H

Figure 7.31 Dimensions of the tolerance ellipsoid around a standard colour.

The values of l and c in Equation 7.23 are constants that adjust the contributions of differences in lightness (ΔL^*) and chroma (ΔC^*) to the ΔE value. For quantifying the perceptibility of colour differences, l and c are both set to 1.0, in which case the differences in lightness, chroma and hue components all contribute equally to the calculated value for ΔE. For acceptability judgments $l = 2.0$ and $c = 1.0$, which has the effect of halving the contribution of differences in lightness when calculating the ΔE value. This is because differences in lightness are less critical visually when making acceptability judgements.

The CMC(l:c) formula is a British Standard (BS 6923) and an ISO standard (ISO 105, Textiles – Test for Colour Fastness, Part J03), but it was never adopted by the CIE. However the CMC(l:c) colour-difference formula is still very widely used for textiles.

In 1995 the CIE recommended a simpler version of the CMC(l:c) formula, called the CIE94 formula (Equation 7.24). This formula has the same structure as the CMC(l:c) formula but uses much simpler equations to obtain the S_L, S_C and S_H values:

$$\Delta E_{94} = \left[\left(\frac{\Delta L^*}{k_L S_L}\right)^2 + \left(\frac{\Delta C^*}{k_C S_C}\right)^2 + \left(\frac{\Delta H^*}{k_H S_H}\right)^2\right]^{1/2} \tag{7.24}$$

where $S_L = 1.0$ $S_C = 1 + 0.048 C^*_{std}$ $S_H = 1 + 0.014 C^*_{std}$ and for textiles $k_L = 2.0$, $k_C = 1.0$ and $k_H = 1.0$

In 2001 the CIE recommended another formula, called CIEDE2000 (Equation 7.25), for evaluation and it became a CIE standard in 2013. This formula is a development of the CMC(l:c) formula but takes account of the fact that some ellipsoids, especially those in the blue part of the $a* b*$ diagram, are tilted away from the neutral point. Also it modifies the $a*$ axis to improve the accuracy of the formula for neutral greys and calculates the values of S_L, S_C and S_H using different equations. The formula is

$$\Delta E_{00} = \left[\left(\frac{\Delta L^*}{k_L S_L} \right)^2 + \left(\frac{\Delta C'}{k_C S_C} \right)^2 + \left(\frac{\Delta H'}{k_H S_H} \right)^2 + R_T \left(\frac{\Delta C' \Delta H'}{S_C S_H} \right) \right]^{1/2} \qquad (7.25)$$

The equations for calculating the various parameters in Equation 7.16 are too complex to be given in this introductory book.

7.7.3   Assessment of Metamerism

At one time or another, nearly everyone has carefully chosen and purchased a garment from a store only to find that in another location the colour was not exactly what had been expected. The disappointment may have been due either to poor *colour constancy* (see Section 7.7.4) or to *metamerism*. Both phenomena are related to a change in the spectral energy distribution of the light reflected to the eye from the garment in moving from one light source to another.

Metamerism refers to the situation whereby two colours appear the same when viewed under one set of conditions but different when viewed under a different set. There are three main types of metamerism:

(1) *Illuminant metamerism*, which is the most obvious form and occurs when the light source under which the two colours are viewed is changed.
(2) *Observer metamerism*, which occurs when the observer viewing the two colours changes. In other words, two samples are a match to one observer, but not to a second observer. Another type of observer metamerism occurs when the field of view changes, so that two colours may match to the 2° observer, but not to the 10° degree observer.
(3) *Geometric metamerism*, which occurs when the angle at which the samples are viewed, or at which they are illuminated, changes.

The most common form of metamerism is illuminant metamerism, and whilst the two samples being compared may be a good visual match under one illuminant, they mismatch when viewed under a different illuminant. The reason for this behaviour is that the samples change their colour appearance to different extents when the illuminant is changed.

For metamerism to occur in cases 1 and 2, the spectral reflectance curves of the two samples must cross over at least three times (Figure 7.32). If they cross over only once or twice, the two samples will not match at all.

When viewed under one type of light, such as D65, the samples are a perfect match and their tristimulus values will be the same. Under the second type of light, when they no longer match, the tristimulus values will differ. A set of samples (Figure 7.33) marketed by the Society of Dyers and Colourists illustrate illuminant metamerism. Three samples have been prepared by careful selection of dyes so that the first matches a standard shade under

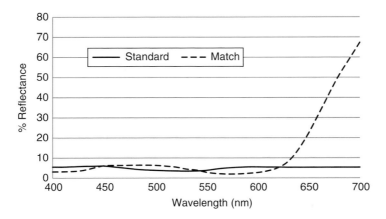

Figure 7.32   Reflectance curves of a metameric pair of grey colours.

all light sources, the second matches the standard under D65 and TL84 (Ill F11) but not under tungsten (Ill A), whilst the third matches under D65 and tungsten but not under TL84.

Both the perceived colour and the corresponding tristimulus values derived from the CIE standard observer are a function of a balance between the spectral qualities of the illuminant source and the wavelengths reflected by the coloured material. Thus, if some wavelengths critical to the appearance of the colour viewed under one illuminant are missing in a second, they will not be present in the light reflected back to the observer. The reflectance spectrum of the material will then be deficient in these wavelengths, and therefore the visual appearance of the colour will change on moving from one illuminant to another.

To obtain an indication of the size of the visual difference between the two colours a *metameric index* is required, and this can be achieved using a colour-difference formula. Figure 7.34 illustrates the principle of using a colour-difference formula as a metameric index for the two grey samples whose reflectance values are shown in Figure 7.32. In this case a ΔE value of 12.3 CMC(1 : 1) units is obtained, indicating a large colour difference between the two colours under tungsten light, mainly due to the match 'flaring' red. Considering that the two samples may well be viewed under other types of lights, it is necessary to carry out the colour-difference calculation between the two colours under different sources, such as TL84 (F11), and other fluorescent sources.

7.7.4   Assessment of Colour Constancy

Colour constancy refers to the extent to which the appearance of an individual coloured article remains the same when viewed under different light sources, irrespective of the dyes used.

Colour constancy is a term that refers to just a single sample, and it is useful to know just how colour constant a particular colour sample is, that is, how well it retains its colour appearance when the light source under which it is viewed is changed. The facility to determine this property of a colour sample was established in 2005 and is now part of ISO 105. The formula uses the CMCCON02 colour-inconstancy index, so called because it determines how much a colour sample changes when the illuminant is changed, rather than how constant it is.

Metameric samples

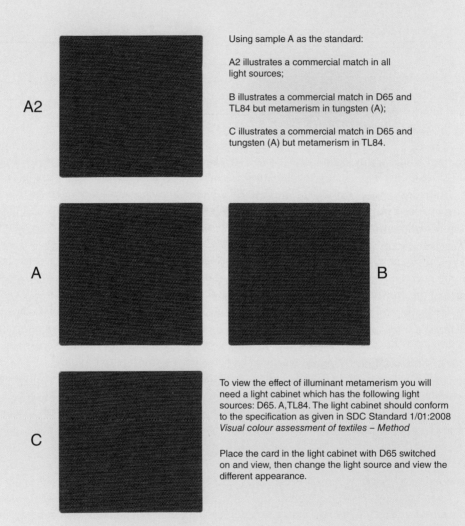

A2

A

B

C

Using sample A as the standard:

A2 illustrates a commercial match in all light sources;

B illustrates a commercial match in D65 and TL84 but metamerism in tungsten (A);

C illustrates a commercial match in D65 and tungsten (A) but metamerism in TL84.

To view the effect of illuminant metamerism you will need a light cabinet which has the following light sources: D65. A,TL84. The light cabinet should conform to the specification as given in SDC Standard 1/01:2008 *Visual colour assessment of textiles – Method*

Place the card in the light cabinet with D65 switched on and view, then change the light source and view the different appearance.

Note: for use at school or home, move to the window in good daylight then move back into the room, as you move around you will see the colours change. The colour difference will enable you to identify the light source in the room, you may have fluorescent tubes or tungsten bulbs.

Figure 7.33   SDC Metameric samples.

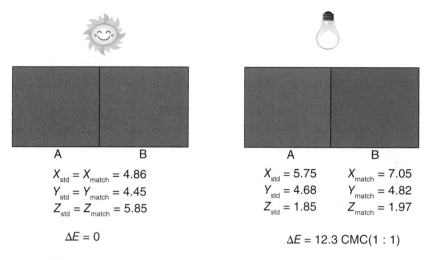

$X_{std} = X_{match} = 4.86$

$Y_{std} = Y_{match} = 4.45$

$Z_{std} = Z_{match} = 5.85$

$X_{std} = 5.75 \qquad X_{match} = 7.05$

$Y_{std} = 4.68 \qquad Y_{match} = 4.82$

$Z_{std} = 1.85 \qquad Z_{match} = 1.97$

$\Delta E = 0$

$\Delta E = 12.3\ \text{CMC}(1:1)$

Figure 7.34 Using a colour-difference formula as a metameric index.

The method involves calculating, from its reflectance spectrum, the tristimulus values (X_R, Y_R, Z_R) of the sample under Ill D65 (as the reference illuminant) and also under a second illuminant, for example, Ill A (X_A, Y_A, Z_A). A chromatic adaptation transform is used then to convert the X_A, Y_A, Z_A values into the tristimulus values of the corresponding colour in Ill D65 (X_C, Y_C, Z_C). Corresponding colours are two colours that have the same visual appearance when one is viewed under one illuminant, such as D65, and the other is viewed under a different illuminant, such as Ill A. Then, using the CMC(l:c) colour-difference formula, the ΔE value between the X_R, Y_R, Z_R and the X_C, Y_C, Z_C values is calculated. The resulting ΔE value indicates the colour inconstancy (i.e. lack of colour constancy) of the sample.

7.7.5 Colour Sorting

When it is required to dye many batches of material to the same shade, it may be that not all batches are an acceptable match to the standard, and also there is some slight variation in colour between the batches. Rather than subject the material to another dyeing process to bring them on shade, it is often acceptable to sort the batches into groups, so that within any one group, all of the batches are an acceptable match to each other, if not to the original standard. If there are sufficient batches within a group to give a meaningful production run, clothing manufacturers are often happy to take them.

Shade sorting can be done visually but it is a slow and boring task and often mistakes are made. Alternatively there are various numerical methods available that are sold in the colour management software produced by the instrument manufacturers. One of the earliest methods is the '555' system, the basis of which is illustrated in Figure 7.35. In this method, an array of cuboids is established in the $L^* a^* b^*$ colour space around the standard. Each cuboid is labelled by three digits, the first relating to lightness, the second to chroma and the third to hue. The standard is located at the centre of the block labelled '555'. If a batch is lighter than the standard, it will go into a cuboid in the plane above the 5 level, and vice versa if it is darker. The same principle applies to batches richer or paler (the second digit applies) and to batches that differ in hue (the third digit applies).

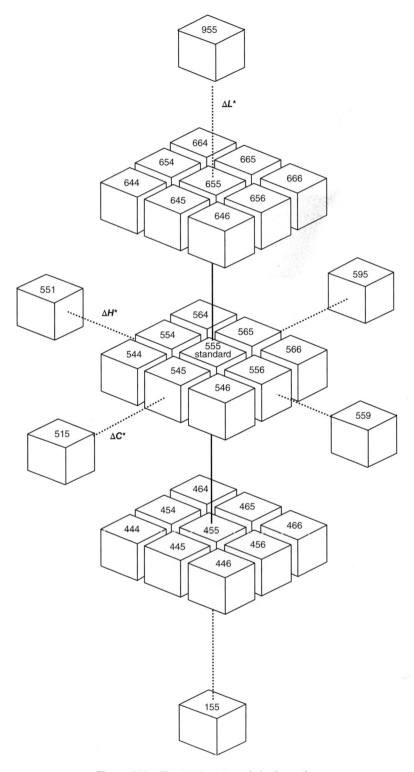

Figure 7.35   The '555' system of shade sorting.

This method of sorting has the advantage of being easy to understand and use, since the coding of each cuboid indicates the nature of its colour difference from the standard. The dimensions of the three axes of the cuboids can be adjusted to account for preferences. For example, usually the lightness dimension is much longer than the chroma or hue dimensions, because differences in lightness between samples are not as critical as differences in, say, hue.

However the method also has a significant disadvantage in that the distance from the centre of a cuboid to the corners is longer than to the centre of a side face, and batches can be sorted into more groups than is really necessary. The most appropriate shape is a sphere or, if the dimensions in the lightness, chroma and hue directions differ, an ellipsoid. However these shapes do not close pack (imagine the spaces that will exist between ping-pong balls glued together) so inevitably some batches would never be sorted into a group. Consequently other more complex shaped volumes are used, as are completely different other methods, though these are beyond the scope of this book.

7.7.6   Measurement of Whiteness

Many textile materials are sold as whites, and it is necessary to ensure that the white is of the correct quality, since any trace of yellowness is interpreted as inferior quality or lack of cleanliness. The best quality whites are those that have a high overall spectral reflectance and a low chroma. However preferred whites often have some chromaticity associated with them, as long as the hue is not a yellow. For example, if two whites have similarly high spectral reflectances, but one appears slightly bluer than the other, then the one with the bluish tint will be perceived as the better white. The presence of a yellowish tint will have quite the opposite visual impact however.

The whiteness of white textiles are commonly improved by the application of fwas (see Section 7.5.2), and at wavelengths at the blue-violet end of the spectrum, the apparent reflectance values of treated fabrics can be as high as 150%. The objective measurement and meaningful expression of the quality of white has proved technically complex. The CIE established a *whiteness index* in 1982:

$$W = Y + 800(x_n - x) + 1700(y_n - y) \tag{7.26}$$

where x, y and Y are the colorimetric values for the specimen under Illuminant D65 and x_n and y_n are the chromaticity coordinates of the Illuminant D65 (0.3137, 0.3309, respectively). Applying Equation 7.26 gives a value of $W = 100$ for the prd.

Different fwas confer slightly different hues to a white and so the whiteness index is accompanied by a tint factor:

$$T = 900(x_n - x) - 650(y_n - y) \tag{7.27}$$

If T is near zero, the specimen has a dominant wavelength of 466 nm. The more positive the value of T, the greener is the hue of the white. The more negative it is, the redder is its hue.

7.8 Solution Colour Measurement

When a dye is dissolved in a solvent to form an optically clear, coloured solution, it is possible to shine a beam of light through it and measure how much of the light is absorbed by the dye solution and how much is transmitted. This type of measurement is carried out using a spectrophotometer designed especially to measure the transmittance of dye solutions that are held in specially designed cells, shown schematically in Figure 7.36. The cells used are made of glass, and cells of different path lengths are available, for example, 0.5, 1.0, 2.0 and 4.0 cm, with the 1.0 cm size being most commonly used. Cheaper plastic cells are also available, though these should be used with care since optical imperfections in the walls may influence the transmittance readings obtained. The opposite two walls of a glass cell are made of frosted glass: these are the walls that should be used to handle the cell. The other two walls, made of clear glass through which light passes, should be kept perfectly clean.

The spectrophotometer used for the measurement of dye solutions is usually a double-beam instrument, though of a different structural design than that used for reflectance measurements. The instrument shines monochromatic light (light of a single wavelength) onto two identical cells, one of which contains the dye solution and the other the pure solvent (usually water in the case of water-soluble dyes), and records the percentage of light transmitted through the dye solution, compared with that transmitted through the pure solvent. In the case of a recording spectrophotometer, the essential structural features of which are shown in Figure 7.37, the prism slowly rotates so that gradually each wavelength passes through the narrow slit and on through the cells, so that the transmission spectrum is obtained, wavelength by wavelength. The recording devices give a continuous plot of transmission against wavelength.

Dye solutions absorb light in the visible region of the spectrum. The amount of light transmitted (the light which is not absorbed) depends on the colour of the dye and the wavelength of the incident light. The behaviour is the same as that of the light reflected by dyed fabrics, so the spectral transmission curves of dye solutions have similar shapes to those shown in Figure 7.15.

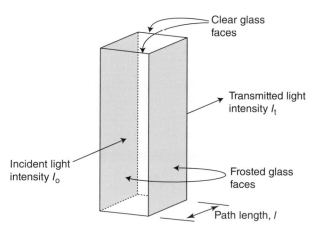

Figure 7.36 A glass cell for solution colour measurement. For most measurements a cell path length of 1 cm is used.

At any given wavelength, the amount of the incident light (I_o) that is transmitted (I_t) is given by the equation

$$I_t = I_o 10^{-\varepsilon cl} \qquad (7.28)$$

where c is the concentration of dye in solution, l is the path length of the cell and ε is the *molar absorptivity* (or molar extinction coefficient) and is a constant for a given dye at a given wavelength.

Taking logarithms of Equation 7.28 gives

$$\log\left(\frac{I_o}{I_t}\right) = \varepsilon cl \qquad (7.29)$$

If I_o is set to 100 arbitrary units, then Equation 7.29 becomes

$$\log\left(\frac{100}{\%T}\right) = \varepsilon cl \qquad (7.30)$$

where $\%T$ is the percentage transmission of the dye solution.

The term $\log(100/\%T)$ is called the *absorbance*, A, so Equation 7.30 can be simply expressed as

$$A = \varepsilon cl \qquad (7.31)$$

Equation 7.31 is the Beer–Lambert law and shows that the absorbance is directly proportional to the concentration, c, of the dye in solution (Beer's law) and also to the path length l of the cell (Lambert's law). The value of ε indicates how strongly a dye absorbs light at a particular wavelength. Dyes with high ε values are said to have high *tinctorial* strength. The higher the value of ε of a dye, the smaller the amount of dye that is necessary to apply to a fibre to obtain a certain visual depth of shade.

In addition to plotting the percentage transmission of a dye solution against wavelength, spectrophotometers also have the facility to plot the absorbance against concentration and indeed this is the most usual mode in which spectrophotometers operate.

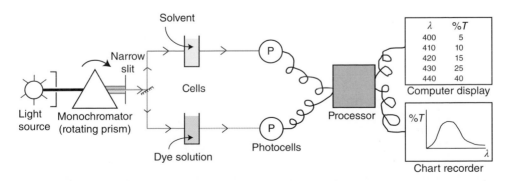

Figure 7.37 Schematic diagram of a double-beam recording spectrophotometer.

Table 7.5   Values of absorbance for different values of %*T*.

%*T*	*A*
100	0
10	1
1	2
0.1	3
0	∞

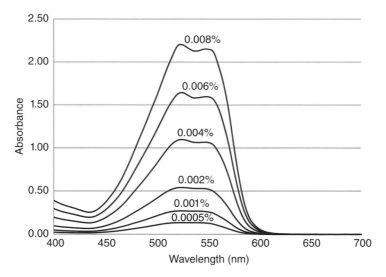

Figure 7.38   Absorbance curves for solutions of C.I. Acid Red 249, at the concentrations indicated.

Looking at the relationship between absorbance and percentage transmission $A = \log(100\,/\,\%T)$, the values for absorbance can be obtained in a range between 0 and ∞, as shown in Table 7.5.

Most spectrophotometers work most accurately in the %*T* range of 1–100, with the corresponding range for absorbance being 0–2. If it is necessary to measure a very concentrated solution of dye, then at some wavelengths its %*T* may be less than 1%, in which case the corresponding absorbance readings cannot be obtained accurately. In this situation a cell with a shorter path length than the usual 1.0 cm should be used, for example, a 0.25 or 0.1 cm cell.

Figure 7.38 shows the absorbance spectra of solutions of the dye C.I. Acid Red 249 at different concentrations. As can be seen, a 'family' of curves is obtained, with the weakest solution of the dye giving the lowest absorbance values across the spectrum and the most concentrated solution the highest absorbance values.

If one wavelength is chosen, then according to Beer's law a graph of the absorbance of the solutions plotted against their concentrations should give a straight line. In every case, the absorbance curve reaches a maximum at 525 nm, this wavelength being the $\lambda_{\max}$ for the dye – the wavelength at which it absorbs light most strongly and the value of ε is

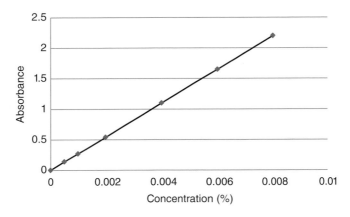

Figure 7.39   Beer's law plot for solutions of C.I. Acid Red 249 at 525 nm.

the highest. At this wavelength the absorbance of the dye varies most with concentration, and this wavelength has been chosen for the Beer's law plot in Figure 7.39.

The Beer's law plot in Figure 7.38 can be used as a 'calibration graph' and be used to determine the concentration of an unknown solution of the dye, perhaps, for example, from an exhausted dyebath. By measuring the absorbance of the unknown, its concentration can be easily read off from the calibration graph. For this type of quantitative analysis, it is always best to plot the calibration graph at λ_{max} for the dye for the following reasons:

(1)   Absorbance varies most with concentration, making the analysis as reliable as it can be.
(2)   If the spectrophotometer does not shine exactly monochromatic light (light of just one wavelength) onto the cell (and most do not), but instead a small range of wavelengths around the indicated wavelength, it does not matter too much because at λ_{max} the absorbance is fairly constant over a small range of wavelengths either side of λ_{max}.
(3)   Any absorbance due to the presence of impurities in the dye solutions will have minimal influence on the absorbance readings at λ_{max} where the absorbance due to the dye is at its highest.

Sometimes it is necessary to carry out quantitative analysis of dye solutions that contain more than one dye. It can be done but the procedures are more complex. If a solution contains a mixture of two dyes, say, a red (R) and blue (B), two wavelengths must be selected, which usually are their λ_{max} wavelengths. For a mixture of the two dyes in a solution, the Beer–Lambert law becomes

$$A_{total} = A_R + A_B \tag{7.32}$$

Expanding this equation and applying it at the two wavelengths gives

$$
\begin{aligned}
\text{at } \lambda_1 \quad & A_{total,\lambda_1} = \varepsilon_{R,\lambda_1} c_R + \varepsilon_{B,\lambda_1} c_B \\
\text{at } \lambda_2 \quad & A_{total,\lambda_2} = \varepsilon_{R,\lambda_2} c_R + \varepsilon_{B,\lambda_2} c_B
\end{aligned}
\tag{7.33}
$$

To apply Equation 7.33 it is firstly necessary to determine the values of the four extinction coefficients, ε, that is, the values for each dye at the two wavelengths. The absorbances of the dye mixture at each wavelength, A_{total, λ_1} and A_{total, λ_2}, can be measured, so all of the values in Equation 7.33 are known, with the exception of the concentrations c_R and c_B, which can be calculated by solving the two simultaneous equations.

If a dye solution is a mixture of three dyes, as is often the case when measuring exhaust dye liquors of trichromatic dye recipes, the calculation becomes more complex still. Equation 7.33 has to be applied at three wavelengths (again, usually the λ_{max} wavelengths of the three dyes) and then three simultaneous equations solved to obtain the concentrations of each of the dyes in the dye liquor. However the analyses can be inaccurate if the absorption curves of the any of the dyes overlap appreciably. It is most accurate when the λ_{max} wavelengths of the three dyes are spaced well apart in the spectrum.

Suggested Further Reading

R McDonald, Ed., *Colour Physics for Industry*, 2nd Edn., Society of Dyers and Colourists, Bradford, 1997.

C Oleari, *Standard Colorimetry: Definitions, Algorithms and Software*, John Wiley & Sons, Ltd, Chichester, 2016.

8

Fastness Testing

8.1   Introduction

The domestic consumer needs to be assured that purchased goods will be fit for their intended purpose. It has taken time for the concept of providing customer reassurance to develop: the mechanisms now adopted started to evolve during the industrial revolution through cooperation between producers of basic products and the manufacturing industry. Although it was some time later that they were adopted and rigorously applied in the textile industry, the efforts made at that time were the forerunners of current quality assurance schemes.

In the United Kingdom the British Standards Institution (BSI) began to formulate standards relevant to consumer goods during the early 1940s. At that time the natural fibres cotton, linen, silk and wool and the man-made fibres viscose, acetate and cuprammonium rayon were the only ones available. As the synthetic fibres became available throughout the 1940s and 1950s, the range of fabric types, especially when synthetic fibres were blended with natural fibres, became very extensive. It was difficult for consumers to know what properties to expect with such developments, but the use of informative labelling helped and provided a measure of quality assurance.

Many of the synthetic dyes produced during this time gave consumer dissatisfaction because of their poor resistance to both daylight and washing. This stimulated the continuing search for dyes with improved resistance to a variety of destructive agencies, and as progress was made, manufacturers began to approve various labels with indications about colour fastness. Some dye manufacturers attempted to help the consumer by approving labels for fabric dyed with particular dyes selected for their high standard of fastness. Similarly fibre manufacturers imposed the requirement for the fibre content of the fabric to appear on the label, whilst other labels have been promoted by dyers and printers. As noted in Section 2.1, it is now a legal requirement for all textile products sold in not only the EU and the United States but also throughout most of the world, to be labelled with the name and percentage content of all constituent fibres. The initial standards for textiles were concerned with constructional details. The continuing development of domestic appliances, such as washing machines, has led to the provision of labels containing more detailed information relevant to the performance of manufactured textile goods during use. Thus information is now provided on aftercare, such as washing temperature, drying and ironing.

The International Standards Organisation (ISO) was formed in 1947 and has a membership of 163 national standards bodies. It has developed standards covering all aspects of the properties of textiles. These standards provide methods for assessing properties such as the

An Introduction to Textile Coloration: Principles and Practice, First Edition. Roger H. Wardman.
© 2018 John Wiley & Sons Ltd. Published 2018 by John Wiley & Sons Ltd.

wear, strength and dimensional stability of fabrics, thermophysical comfort and flammability, as well as colour fastness. The European Committee for Standardisation (CEN) is a public standards organisation, founded in 1961. It has 33 national members that work together to develop European standards (ENs) in various sectors.

In the field of textiles, all of the ISO standards are adopted as CEN and UK standards. It is not uncommon for Europe to lead the way in the development of tests and for these tests to become adopted later by ISO. It is therefore quite common to see standard names with the letters *BS EN*, *BS EN ISO* or even *BS ISO* preceding the actual title of the standard. The *EN* prefix indicates that the standard has been adopted by all the member states of the EU. The *BS* prefix is the UK's national prefix: in other countries the *BS* prefix is replaced by the relevant national prefix (e.g. DIN in Germany, NF in France). The adoption of ISO standards is voluntary although the BSI's policy is to adopt any ISO standard except when there are substantive reasons not to do so, such as the relevant UK technical committee feeling that the ISO standard represents a potential barrier to trade or is technically unsound.

8.2   Standards Related to Coloration

The fastness of dyes to various agencies is of particular relevance and voluntary coordinated activity in this sphere began in 1927 with the Fastness Tests Committee of the Society of Dyers and Colourists (SDC).[1] Since that time the committee has been responsible for the development and continuous review of a wide range of appropriate fastness tests for dyed materials. A technical committee of ISO, the Colour Fastness Subcommittee, became responsible for bringing uniformity to the various systems adopted in different parts of the world. The internationally agreed test procedures recommended by that committee are now automatically adopted as British Standards.

The development of standards has been of great value to the colourist in both the provision of robust assessments and the transmission of relevant information between dye manufacturers and dye users. The general groupings of British Standards cover a wide range, including the following:

(a) Glossaries, which provide agreed definitions of terminology in specialist fields of activity
(b) Dimensional standards, which provide for interchangeability of manufactured components from different sources
(c) Performance standards for the specification of expected performance
(d) Standard methods of test with precise and detailed specification of operation that are of particular use in setting up quality assurance schemes
(e) Codes of practice for the design, installation, maintenance and servicing of equipment or services

---

[1] The Society, through its commercial arm SDC Enterprises, can provide a complete range of the consumables used in colour fastness testing.

These are not the only source of standard definitions. The SDC has produced a booklet 'Colour Terms and Definitions', whilst the Textile Institute has produced its book 'Textile Terms and Definitions'. In the field of colour measurement and specification, the CIE (see Section 7.6) gives definitions in its publications such as CIE Publication 15:2004, *Colorimetry*. The International Standard ISO 105-A08:2001 gives the vocabulary of terms used in colour measurement standards.

The development of standards within ISO is overseen by technical committees that comprise experts from participating countries. The ISO technical committee for textiles is TC38 and the corresponding CEN technical committee is TC248, each of which has a number of subcommittees. Within ISO the first two, TC38/SC1 (tests for coloured textiles and colorants) and TC38/SC2 (cleansing, finishing and water-resistant tests) have responsibility for the development of tests relating to the colour fastness of dyed textiles and CEN defers to the ISO-led work. The United Kingdom monitors and contributes to the development of these tests through mirror committees including TCI/81 (colour fastness and colour measurement of textiles). The SDC continues its role in this area, since writing the first BSI standards for colour fastness in 1939, by holding the secretariat for both the BSI and ISO committees. For ISO, the standards are all in the 105 series and they are coded by a letter, followed by a two-digit number and finally the year of introduction, for example, ISO 105-B02:2013. This particular example is also a British and EN standard, so is labelled BS EN ISO 105-B02:2013. The letter code (in the example given, 'B') indicates the type of agency (light and weathering), of which there are eleven as shown in Table 8.1.

Even though some of the standards were introduced many years ago, they are all reviewed every 5 years and when necessary amended or revised to ensure they remain robust and use current technology. In the United States, responsibility for the development of test methods falls to American Association of Textile Chemists and Colourists (AATCC) and to ASTM (the USA equivalent to BSI) who have developed a number of test methods and evaluation procedures. Those applicable to colour fastness are published in AATCC's technical manual. Many of these methods are similar to those of the corresponding ISO standards, although there may be subtle differences between the methods so care has to be taken when comparing tests results using data from different test methods.

Table 8.1 Colour fastness standards.

A series	General principles
B series	Colour fastness to light and weathering
C series	Colour fastness to washing and laundering
D series	Colour fastness to dry cleaning
E series	Colour fastness to aqueous agencies
F series	Specification for adjacent fabrics
G series	Colour fastness to atmospheric contaminants
N series	Colour fastness to bleaching agents
P series	Colour fastness to heat treatments
S series	Colour fastness to vulcanising
X series	Colour fastness to miscellaneous agencies
Z series	Properties of dyes

Results from the standard methods of test are widely used for the objective appraisal of the behaviour of dyes under an extensive range of circumstances. Typically a manufacturer will choose specific performance requirements applicable to the intended end use of the product, and dyes will be selected to meet those requirements.

8.3   Resistance of Coloured Fabric to Harmful Agencies

The fastness of a dyed material to various agencies may be caused either by a chemical breakdown of the dye or by removal of the dye from the fabric, or by both. In general terms, fastness properties depend on factors such as the chemical structure of the dye, the state of the dye in the fibre, the depth of shade and the presence of chemicals in the fibre.

The chemical breakdown of dye molecules is most likely to occur when the coloured material is exposed to agents capable of reacting chemically with it, such as the oxidising agents contained in detergent powders and nitrogen dioxides in the atmosphere, or even to the presence of foreign substances in the textile. In the production of man-made fibres, titanium dioxide pigment is often incorporated into the polymer mass to render the resulting fibres opaque (dull fibres), and this can reduce their fastness to light as compared with the corresponding unpigmented material (bright fibres). The yellowing of fibres with age is a chemical change that is impossible to prevent. Residual auxiliary chemicals from the dyeing process, if not thoroughly removed, may also influence fastness. The significant influence of moisture content on the fastness to light of dyed fibres is also well known. Singlet oxygen, formed by the action of light in the presence of moisture, is a highly reactive attacking species. The colour of the degradation product of the dye may well be different from that of the dye itself, and there is also likely to be a corresponding reduction in depth of shade.

Fastness assessments are also affected by the fineness of fibres, simply because a given amount of dye on a fine fibre, in contrast to coarse fibre, is spread over a larger surface area. The compactness of the yarn or fabric structure can have a bearing on how easily the dye fades or can be removed during washing treatments. The nature of the fibre itself can also markedly affect fastness to light: the fastness to light of basic dyes, for example, is very poor on cotton and wool, but excellent on acrylic fibres. Crease-resist finishes or dye-fixing agents applied to some dyed cotton also adversely affect the fastness to light of some dyes. Further influences originate from the state of the dye in the fibre. Insoluble pigments trapped mechanically inside the fibre, or dyes that have formed a strong chemical linkage with the fibre, will obviously be more resistant to removal by wet treatments than dyes that are more loosely attached. The crystallinity and state of aggregation of the dye will also influence its fastness.

In general, the deeper the shade, the better the fastness properties, although a deep dyeing of a dye with poor wash fastness will release more dye into the water than a paler dyeing. During washing this dye may well stain other fibres or fabrics. The adjacent fabrics may be part of the same garment, for example, as illustrated in Figure 8.1, or other garments being washed in the same load.

Before washing   After washing

Figure 8.1 Staining of adjacent fibres during washing.

Table 8.2 ISO 105-A series of standards.

A01	General principles of testing
A02	Grey scale for assessing change in colour
A03	Grey scale for assessing staining
A04	Method for the instrumental assessment of the degree of staining of adjacent fabrics
A05	Instrumental assessment of change in colour for determination of grey scale rating
A06	Instrumental determination of 1/1 standard depth of colour
A08	Vocabulary used in colour measurement
A11	Determination of colour fastness grades by digital imaging techniques

8.4   Principles of Colour Fastness Testing

8.4.1   The ISO Standards Outlining the General Principles

As indicated in Table 8.1, the A series of ISO standards detail the general aspects of fastness testing. The standards of this series are given in Table 8.2.

8.4.2   Grey Scales

Appraisal of the performance of a dye begins at the time of its synthesis and ends with tests designed to indicate the level of performance during its use. Obviously, consumer goods must have satisfactory resistance to domestic cleaning treatments and a reasonable resistance to fading under the action of daylight, but many other factors also need to be considered if the requirements of the textile finisher and dyer are to be met. In some cases a very high level of fastness is provided for the consumer because the dyes used are expected to withstand processing conditions that are far more severe than any likely to be encountered during normal use. In other cases special efforts are made to find new dyes that will withstand particularly intensive conditions associated with the use of a new product.

The wide ranging end uses associated with coloured textiles are accommodated by the development of realistic test methods for the fastness to wet treatments. These often involve

Figure 8.2 Grey scales: (top) the scale used to assess staining and (bottom) the scale used for the assessment of colour change in the sample (Source: Photograph courtesy of High Street Textile Testing Services Limited, Leeds, UK).

the formulation of appropriate variations in the severity of the testing conditions. The objective assessment of the effects obtained is usually made on the basis of visual comparison of the intensity of any change in the appearance of the coloured sample with standard *grey scales*. The colour fastness of the sample is rated numerically according to the contrast step in the grey scale that most closely matches the contrast between the tested and original coloured textile sample. The test is twofold: changes in both the colour of the test sample and the staining of previously undyed fabric in the presence of the test sample are examined, although different scales are used for the assessment. These scales are shown in Figure 8.2.

After the dyed material has been subjected to the test conditions, the change in colour is assessed by comparing the tested and untested dyed fabrics. A judgement of the degree of contrast between the two is then made by comparison with the relevant steps in the grey scale under the recommended conditions of illumination. The grey scale used for colour change (BS EN ISO 105-A02:1993) covers five full steps arranged in geometric progression and is a series of grey pairs in which one of each pair becomes progressively paler. Half steps in the grey scale are provided to increase the precision of the assessment, giving a nine-step scale. If there is no change in colour, a rating of 5 is assigned, and the numerical rating decreases as the colour change increases. Table 8.3 shows the colour difference (CIELAB ΔE) corresponding to each of the five steps.

The grey scale for assessing the extent of staining is used in a similar manner, by placing a sample of the unstained material alongside the stained material. A judgement of the degree of contrast between the two is then made by comparison with the relevant steps in the grey scale for staining under the recommended conditions of illumination and sample orientation, as set out in ISO 105-A01. As with the grey scale for assessing change in colour, the scale used for staining (BS EN ISO 105-A03:1993) covers five full steps arranged in geometric progression, with half steps to give a nine-step scale. Each pair comprises one white half and another half increasingly grey. When no staining occurs, a rating of 5 is assigned, but the numerical rating decreases as the staining worsens. The colour differences

Table 8.3 Colour-difference values of the grey scale for change in colour and staining.

fastness grade	ΔE (CIELAB)
5	0
(4–5)	2.2
4	4.3
(3–4)	6.0
3	8.5
(2–3)	12.0
2	16.9
(1–2)	24.0
1	34.1

Table 8.4 Terms used for the qualitative description of colour changes.

hue component	depth component	brightness component
Redder (R)	Weaker (W)	Duller (D)
Yellower (Y)	Stronger (S)	Brighter (B)
Bluer (Bl)		
Greener (G)		

(CIELAB ΔE) between the second member of each pair and the adjacent first member are the same as those shown in Table 8.3.

The rating may be supplemented by letters to indicate an accompanying change of hue or brightness. For example, the change may be indicated by a number alone to indicate the loss in depth only, but the number followed by a letter indicates other changes. Thus '3 W, Bl D' signifies that a loss in depth (weaker) corresponding to grade 3 of the grey scale is accompanied by a change of hue towards blue and that the pattern has become duller. The terms used for this purpose are shown in Table 8.4.

The qualifiers are always placed in order of magnitude. However, according to BS EN ISO 105-A02:1994, when changes occur in two or more of the attributes hue, brightness and depth, it is neither feasible nor necessary to indicate the relative magnitude of each change.

In addition to the inherent resistance of a particular dye–fibre combination to the agency in question, the assessment is also dependent upon the depth of shade. A deeper dyeing of poor fastness to wet treatments will release more dye into the water than will a paler dyeing of the same dye. Thus, both the degree of staining and the visual effect on the coloured pattern itself will be affected.

The determination of the grey scale rating involves the visual comparison of the intensity of any change in the appearance of the tested sample with the grey scales and some inaccuracy in making the comparison can be expected. To reduce such inaccuracies instrumental measurement of the colour change and the staining grey scales can be used. The CIELAB ΔE values obtained are converted to grey scale values, and the procedures for doing this are covered by the standards BS EN ISO 105-A04 and A05, respectively. A digital imaging method has been developed also, which involves scanning the original and tested samples as well as the grey scale samples, and then converts the ΔE values (CIEDE2000) to the corresponding fastness rating. This method became an ISO standard in 2012 (ISO 105-A11:2012).

8.4.3 Standard Depths

When evaluating the colour fastness properties of a dye manufacturer's dye prior to it being used commercially, the dyes are measured at *standard depths of shade*, so that dyers can compare the performances of dyes from the different manufacturers on the same basis. These standard depths of shade were developed during the 1920s by German and Swiss dye manufacturers at the time. They agreed seven sets of colours, called the Hilfstypen samples. Six of the sets comprise 18 different coloured samples, covering the hue circle, all at the same depth of shade on matt wool gabardine cloth. The six sets are labelled 2/1, 1/1, 1/3, 1/6, 1/12 and 1/25 standard depths, respectively, and the samples of each set are mounted in small booklets (Figure 8.3). The 2/1 set is dyed at twice the depth of the 1/1 set, the 1/3 set is dyed at one-third of the depth of the 1/1 set, and so on. The seventh set comprises four dark navies.

In ISO 105 (Part A01: General Principles of Fastness Testing), it is stipulated that dye manufacturers should publish fastness data obtained by testing specimens at 1/1 standard depth wherever possible, supplemented by data for one or more of the other ratios as required. The SDC has produced standard depth cards relevant to this standard. The cards have been produced according to the same principles as the Hilfstypen samples (Figure 8.4) and comply with the ISO 105 Part A01 standard.

Figure 8.3 The Hilfstypen series of standard depths.

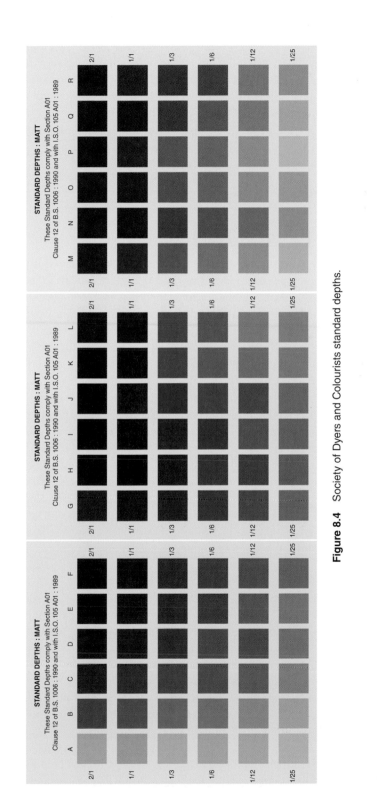

Figure 8.4 Society of Dyers and Colourists standard depths.

A problem in using the standard depth samples is that when a particular dye is to be tested, its hue may not quite match the hue of one of the 18 samples of a the set chosen, so it has to be dyed to the same depth as the sample of the set whose hue is closest to it. This involves a visual comparison, which may well have an error associated with it. To overcome this problem the standard ISO 105-A06:1995 details a numerical method of determining if a colour has the depth corresponding to a 1/1 standard depth. However, the method detailed in this standard only applies to the 1/1 standard depth.

8.5 Fastness Tests

8.5.1 Light Fastness Tests

The influence of light on the fading of dyes is a complex phenomenon influenced by many variables, making predictive tests for fastness to light difficult to establish. The depth of shade, presence of unwanted chemicals, humidity, temperature, presence of atmospheric impurities and spectral quality and intensity of the incident light all have a bearing on the end result.

When using daylight the samples are exposed outdoors but behind glass alongside *blue wool standards*, taking care to place the samples at an angle equal to the latitude of the location of the testing station, facing due south in the northern hemisphere and due north in the southern hemisphere. Adequate ventilation is essential. There are eight blue wool standards for determining fastness to light, each dyed with a different blue dye. The first standard has the lowest rating, whilst the eighth has the highest. Each successive standard requires approximately twice the exposure time of the one below it in the series to cause the same degree of fading. These blue wool standards are available from a variety of sources, such as private companies or the commercial arms of professional societies such as SDCE [1] and AATCC [2]. However, depending on the source, there may be some difference between the blue wool standards dyed in accordance with the ISO 105-B series, and those produced in the United States (L-series) or Japan, or which have been produced for use with AATCC16 methods of test.

The rate at which a sample fades is determined by partially covering both blue standards and sample with a plate and inspecting both fading standards and sample periodically. The sample is faded to the equivalent effect of a grade 4 contrast on the grey scales. At this stage the degree of contrast between the exposed and unexposed sample is compared with that for the blue standards. The number of the blue standard that exhibits the same degree of contrast (e.g. No. 6) is noted. The sample and standards are then covered with a larger plate so that only one-third is exposed and testing continued until the sample is faded to a grade 3 contrast on the grey scales. The number of the blue standard that has now faded with the same degree of contrast (e.g. No. 7) is noted. The average of the two ratings (6 and 7) gives the light fastness grade of the sample, in this case 6.5.

The time taken for testing can be inconveniently long. Furthermore, the conditions are variable from one geographical location to another. Consequently it is preferable to carry out the test in a machine using artificial light sources. For many years various light sources were used to imitate the spectral distribution of daylight, all of which have been subjected to close scrutiny. Currently the light of a xenon arc is currently favoured as the best substitute for daylight, and this source enables the time of testing to be considerably reduced.

In light fastness testing machines, the humidity at the surface of the samples has to be maintained to prevent them from drying out. The variations that occur between the results from different light sources and from operating the same light source under different conditions are mainly due to differences between the spectral composition of the lights, their intensity and the effective humidity at the surface of the sample. The effective humidity represents a combination of air temperature, the temperature of the surface of the sample and relative humidity, which governs the moisture content of the fabric. The conditions of operation are therefore specified very carefully in the standard tests and control of the effective humidity is recommended through the use of a humidity test control in the form of a fabric coloured with a particular red pigment. The relationship between the fastness to light of this fabric and the operating humidity is known with reasonable precision. Preliminary tests are then carried out to check if the humidity test control fades to the correct blue standard and adjustments in the operating conditions made if required. Once the correct conditions have been established, the testing is carried out in the normal manner. Figure 8.5 shows an example of the fading results of samples that have been tested in a light fastness testing machine.

There are many variations on the testing conditions, all of which are attempting to simulate the conditions to which dyed fabrics will be exposed during their use. Accordingly, a number of ISO standards have been developed, the 'B' series (Table 8.5), which detail the methods of exposure and assessment. The most commonly used of these standards are those using artificial daylight – B02 (indoor light) and B04 (outdoor weathering), both of which include a number of variations of test procedure.

Sometimes the colour of a dyed textile changes on exposure to light but reverts to its original state after the dyeing is kept in the dark. In such a case an additional test is carried out to indicate the extent of the change, as well as the fastness rating test. The extent of this

Figure 8.5   Example of light fastness testing of an orange coloured fabric (top) and blue wool standards for assessment of fading (bottom) (Source: Image courtesy of High Street Textile Testing Services Limited, Leeds, UK).

Table 8.5   ISO 105-B series of standard tests for light fastness.

B01	Colour fastness to light: daylight
B02	Colour fastness to artificial light: xenon arc fading lamp test
B03	Colour fastness to weathering: outdoor exposure
B04	Colour fastness to artificial weathering: xenon arc fading lamp test
B05	Detection and assessment of photochromism
B06	Colour fastness and ageing to artificial light at high temperatures. Xenon arc fading lamp test
B07	Colour fastness to light of textiles wetted with artificial perspiration
B08	Quality control of blue wool reference materials 1–7
B10	Artificial weathering – exposure to filtered xenon arc radiation

Table 8.6   ISO 105-C series of standard tests for wash fastness.

C06	Colour fastness to domestic and commercial laundering
C07	Colour fastness to wet scrubbing of pigment printed textiles
C08	Colour fastness to domestic and commercial laundering using a nonphosphate reference detergent incorporating a low-temperature bleach activator
C09	Colour fastness to domestic and commercial laundering – oxidative bleach response using a nonphosphate reference detergent incorporating a low-temperature bleach activator
C10	Colour fastness to washing with soap or soap and soda
C12	Colour fastness to industrial laundering

photochromism is expressed as a grey scale assessment given alongside the rating for fastness to light. The standard procedure for assessing photochromism is given in ISO B05. Whilst photochromism could be a problem with some early dyes, the dyes marketed for textile dyeing nowadays are rather more stable though the problem still manifests itself periodically.

8.5.2   Washing Fastness Tests

Domestic and commercial washing conditions are covered by a number of ISO wash tests, which are directed towards simulation of the conditions likely to be encountered in normal use. Table 8.6 lists the tests; the most commonly used of which is ISO 105-C06.

The test BS EN ISO 105-C06 determines the wash fastness under conditions typical of those used for normal household articles. Since there are such a wide variety of detergents available to the domestic consumer, it is necessary to specify standard detergent systems in the tests. Two detergents are specified: the AATCC 1993 Standard Reference Detergent without optical brightener (WOB) and ECE detergent with phosphates. The detergent is chosen to allow for the effects of the various components of the different brands of washing powder used in commercial and domestic laundering. The detergent composition is based on a mixture of synthetic detergents and natural soap, a phosphate, a silicate, an inorganic salt (Glauber's salt) and a compound that inhibits soil redeposition when used for cleaning off particulate matter.

The sample being tested is either stitched to a piece of *multifibre strip*, or stitched between two pieces of undyed adjacent fabric. The standard 'adjacent fabrics' are listed in Table 8.7. The specifications for the adjacent fabrics and the multifibre strip form the F series of standards within ISO 105.

Multifibre strip is a specially woven fabric strip made up of six different fibres. There are two types of multifibre strip, Multifibre DW and Multifibre TV, whose specifications are given in the standard BS ISO 105-F10. Table 8.8 shows the compositions of the two types of multifibre. In the United Kingdom, SDC Enterprises markets Multifibre DW, but since acetate fibres are of only low commercial significance for apparel, instead of Multifibre TV, it markets Multifibre LyoW® in which the acetate component has been replaced by a regenerated cellulosic component. The regenerated cellulosic component simulates the staining characteristics of lyocell, modal and viscose fibres. Another type of multifibre strip developed by SDC Enterprises is Multifibre SLW. This strip comprises silk, regenerated cellulose, cotton, polyamide, polyester, acetate and wool. However neither Multifibre LyoW nor SLW form part of the current ISO 105-C series of tests, although they may be cited by individual retailers who have developed their own versions of these tests employing these alternative multifibre fabrics.

The composite specimen and relevant soap solution composition is placed in a sealed container that is rotated in a water bath at about 40 rev/min, under the conditions of the test. There are five categories of testing conditions: A (40 °C), B (50 °C), C (60 °C), D (70 °C) and

Table 8.7 Standard adjacent fabrics for use in fastness testing.

specimen fabric	first adjacent fabric	second adjacent fabric for tests (A and B)	second adjacent fabric for tests (C, D and E)
Cotton	Cotton	Wool	Viscose
Wool	Wool	Cotton	
Silk	Silk	Cotton	
Viscose	Viscose	Wool	Cotton
Acetate/triacetate	Acetate/triacetate	Viscose	Viscose
Polyamide (nylon)	Polyamide (nylon)	Wool or cotton	Cotton
Polyester	Polyester	Wool or cotton	Cotton
Acrylic	Acrylic	Wool or cotton	Cotton

Table 8.8 Multifibre adjacent fabrics.

multifibre DW	multifibre TV
Secondary acetate	Triacetate
Bleached cotton	Bleached cotton
Polyamide	Polyamide
Polyester	Polyester
Acrylic	Acrylic
Wool	Viscose

Table 8.9 Conditions specified in BS EN ISO 105-C10 standard test for wash fastness.

test method	composition of wash liquor (g/l)	temperature (°C)	duration (hours)
1	Soap 5	40	0.5
2	Soap 5	50	0.5
3	Soap 5 + Soda 2	60	0.5
4	Soap 5 + Soda 2	95	0.5
5	Soap 5 + Soda 2	95	4.0

E (95 °C), each with variants. The variations involve the use of different numbers of steel balls and different times of test. Some variants also involve the presence of a bleaching agent, either sodium hypochlorite or sodium perborate. In all of the tests, the mechanical action is intensified by the inclusion of non-corrodable steel balls together with the composite sample in the container. Depending on the severity of the test, the number of steel balls present in the container varies between 10 and 100. At the end of the testing period, the sample is removed and rinsed. The components are separated and allowed to dry. A visual assessment of the colour change and degree of staining of the adjacent fabrics is then made using the grey scales, as described in Section 8.4.2.

The test BS EN ISO 105-C10 specifies five methods for determining the wash fastness of dyed materials to washing procedures using soap, or soap and soda, under conditions from mild to severe. These test conditions are shown in Table 8.9. The nature of the detergent to be used is stipulated in the standard, together with the concentration of sodium carbonate (soda ash) needed to make the solution alkaline in reaction. Tests 4 and 5 are of greater severity than the others. In tests 4 and 5, the mechanical action is intensified by the inclusion of ten non-corrodable steel balls together with the composite sample in the container. Again, colour changes are assessed using the grey scales.

The standard BS EN ISO 6330:2012 specifies testing procedures for domestic washing and drying. This standard specifies the different types of washing machine (e.g. front-loading or top-loading), the composition of detergents that should be used and the various modes of drying. Testing under the conditions specified can be used to assess attributes such as dimensional change, stain release and colour fastness to domestic laundering, the results of which lead to guidance on the care labelling of garments.

The tests are continually reviewed as conditions and products change. For example, in recent years there has been a trend away from phosphate-based detergents to nonphosphate carbonate-built powders and citrate-built liquid products, and it is important that test methods simulate the conditions used in domestic and commercial laundering as closely as possible.

8.5.3   Rubbing Fastness

The procedures for the determination of fastness to rubbing are specified in BS EN ISO 105-X12:2016. The tests involve rubbing the sample under test with a dry rubbing cloth and a wet rubbing cloth. For this test an instrument called a *crockmeter* is usually used, which rubs a finger, covered with cotton rubbing cloth, 10 times to and 10 times fro over the sample under test at a fixed pressure. Two fingers of different dimensions are used – for pile fabrics, a rectangular-shaped finger (19×25.4 mm) and a cylindrical-shaped finger 16 mm in diameter for other fabrics. The rubbing tests are carried out with dry and with wet cotton rubbing cloths. The degree of staining of the two cotton rubbing cloths is assessed using the grey scale for staining.

8.5.4   Other Fastness Tests

In addition to the tests described in Sections 8.5.1–8.5.3, there are other tests that form part of a common package of colour fastness tests specified by manufacturers. Tests have been developed to establish the resistance of a dyed material to a wide range of aqueous agencies relevant to different end uses and treatments in use of various types of goods. The tests most often used include fastness to water, perspiration, chlorinated water and seawater, brief descriptions of which follow.

8.5.4.1   Fastness to Water

The standard BS EN ISO 105-E01:2013 specifies testing procedures for determining the fastness to water. The sample under test, in contact with a piece of multifibre strip, is thoroughly wetted in water of composition specified by ISO 3696 then, after removing excess water it is mounted in a special holding device at a certain pressure. The device is then placed in an oven at 37 °C for 4 hours, after which time the sample and multifibre strip are removed and dried separately and the grey scale ratings for change in colour of the sample and staining of the fibres of the multifibre strip determined.

8.5.4.2   Fastness to Seawater

This test (BS EN ISO 105-E02:2013) follows a similar procedure to that of the test for fastness to water, the main difference being that the composite sample and multifibre strip is tested in a solution of 30 g/l sodium chloride solution. This concentration of salt in solution is taken as representing the salt content of typical seawater. After the test has been carried out, the grey scale ratings for change in colour of the sample and staining of the fibres of the multifibre strip are determined.

8.5.4.3   Fastness to Chlorinated Water
(Swimming Pool Water)

The main issue in swimming pool water is the action of chlorine on the dyes used in swimwear. The test BS EN ISO 105-E03:2010 assesses the change in colour (using the grey scale for colour change or instrumental colour measurement) when pieces of the sample are treated in three solutions of sodium hypochlorite of 20, 50 and 100 mg active chlorine per litre, at pH 7.5. The pieces of sample being tested are placed in sealed containers, one for each concentration of sodium hypochlorite, which are rotated in a water bath in about 40 rev/min, at 27 °C for 1 hour. The samples are then squeezed and dried before being assessed for colour change.

8.5.4.4   Fastness to Perspiration

The action of perspiration on dyed materials is replicated by treating them in two solutions containing the amino acid histidine, one at alkaline pH (8) and the other at acid pH (5.5). The sample under test is attached to a piece of multifibre strip and the composite wetted thoroughly in the alkaline histidine solution. It is then placed between two glass plates and mounted at fixed pressure in a device similar to that used for the tests for fastness to water. A second composite of the sample and multifibre strip is treated similarly in the acidic histidine solution and then mounted in the device. The device containing the composite samples is placed in an oven for 4 hours at 37 °C. Finally, after drying the samples are evaluated for change in colour and staining.

8.5.4.5   Fastness to Dry Cleaning Using Perchloroethylene Solvent

Dry cleaning involves the use of an organic solvent rather than water. The solvent most used is perchloroethylene (tetrachloroethylene) and this solvent is used in the standard test BS EN ISO 105-D01:2010. The sample under test is attached to a piece of multifibre strip and the composite placed in a cotton bag, together with twelve stainless steel discs. The bag is placed in a sealed container containing perchloroethylene solvent. The container is rotated in a water bath at 40 rev/min, at 30 °C for 30 minutes. After this treatment the composite sample is removed from the bag in the container, excess solvent removed and then the sample and multifibre strip dried. The change in colour of the sample and staining of the multifibre strip is assessed using the grey scales.

Commercial dry cleaning processes can involve a number of operations in addition to dry cleaning, such as water spotting, solvent spotting and steam pressing. These latter processes are the subject of other standard testing regimes.

8.5.5   Miscellaneous Fastness Tests

A number of tests have been established to assess the performance of dyed materials to other agencies:

E01–E16: These tests cover the fastness to a wide range of liquids with which dyed textiles come into contact. The tests for fastness to water, seawater, chlorinated water (swimming pool water) and perspiration are described briefly in Section 8.5.4. Other tests in the E series cover fastness to potting, decatizing, steaming, milling and felting as well as spotting with acid, alkali and water.

G01–G04: These tests include methods for assessing fastness to nitrogen oxides, burnt gas fumes, ozone and nitrogen oxides at high humidity. Many of these tests are relevant to the atmospheric conditions that can pertain in the atmosphere, especially in cities.

N01–N05: These tests are concerned with determining the fastness to different types of bleaching agent, commonly used in textile processing and commercial bleaching. These tests are somewhat different in that the desirable outcome of bleaching is the removal of colour.

P01–P02: These two tests are to determine colour fastness to dry heat and to steam pleating, both of which are common in textile processing and garment assembly. Colours are often changed (although just temporarily) by application of heat, particularly wet heat.

X01–X18: These are tests to miscellaneous agencies, agencies that do not fit into the other series. They include tests for fastness to rubbing (see Section 8.5.3), which is important in the assessment of colour loss through wear, as well as fastness to mercerizing, organic solvents and hot pressing.

8.6   Test Organisations for Sustainable Textile Manufacture

In view of the global environmental concern surrounding the manufacture of consumer goods, especially those that involve chemical processing, of which textile manufacture is a prime example, a number of bodies have been established with the aim of certifying products as being safe to use and produced sustainably. Reference has already been made to the Global Organic Textile Standard (GOTS) in Sections 2.5.1.5, 2.6.1.4 and 3.5.1.2, a standard of certification that relates to the sustainable manufacture of organic textile fibres and their coloration. GOTS is administered by a group of four member organisations, the Soil Association (UK), the Organic Trade Association (USA) International Verband der Naturtextilwirtschaft (Germany) and Japan Overseas Cooperative Agency (Japan).

Another system of testing and certification is the Oeko-Tex® Standard 100. Oeko-Tex is an international association of independent testing institutes, and it provides a consistent global testing methodology for textile products at all stages of manufacture. Testing for the standard is intended to identify and eliminate harmful substances in textiles, with regard to various legislative regulations such as REACH, as well as banned chemicals in textile processing. Indeed, the test criteria and permissible limit values of chemicals are more restrictive than those of many national and international standards. The Oeko-Tex label is intended to reassure end users that the textile product has been tested for harmful substances to a rigorous level, regardless of where it has been manufactured.

References

[1] SDC Enterprises Limited, www.sdcenterprises.co.uk.
[2] AATCC, www.aatcc.org.

Appendix

Some Textile Terms and Definitions

(Terms reproduced from *Textile Outlook International* with permission from the publishers *Textiles Intelligence Limited*, www.textilesintelligence.com/glo/)

Air-laid   a web or batt of staple fibres formed using the air layering process.

Air-textured yarn   a multifilament yarn that has been given increased bulk through the formation of loops, achieved by passing the yarn through air jets.

Basket weave   a textile weave consisting of double threads interlaced to produce a chequered pattern similar to that of a woven basket.

Batt   single or multiple sheets of fibre used in the production of non-woven fabric.

Braided yarn   intertwined yarn containing two or more strands.

Carding   the disentanglement, cleaning and intermixing of fibres to produce a continuous web or sliver suitable for subsequent processing. This is achieved by passing the fibres between moving pins, wires or teeth.

Circular knitting   a fabric production technique in which fabric is knitted in the form of a tube.

Commingled yarn   a yarn consisting of two or more individual yarns that have been combined, usually by means of air jets.

Condenser-spun yarn   a yarn spun from slubbing.

Core-spun yarn   a yarn comprising a central core of yarns around which a yarn of different composition is wrapped. An example is elastane fibre enclosed in cotton.

Crêpe de chine   a lightweight fabric, traditionally of silk, with a crinkly surface.

Crêpe yarn   a highly twisted yarn that may be used in the production of crêpe fabrics.

Crimp   the waviness of a fibre or filament.

Denier   a measure of linear density; the weight in grams of 9000 m of yarn.

Drafting   the process of drawing out laps, slivers, slubbings and rovings to reduce their linear density.

False-twist texturing   a process in which a single filament yarn is twisted, set and untwisted. When yarns made from thermoplastic materials are heat-set in a twisted condition, the deformation of the filaments is 'memorised' and the yarn is given a greater bulk.

Filament yarn   a yarn consisting normally of a bundle of continuous filaments.

Floats   warp float: a length of warp yarn on the surface of a woven fabric that passes over two or more weft threads; weft float: a length of weft yarn on the surface of a woven fabric that passes over two or more warp threads.

An Introduction to Textile Coloration: Principles and Practice, First Edition. Roger H. Wardman.
© 2018 John Wiley & Sons Ltd. Published 2018 by John Wiley & Sons Ltd.

Hollow spindle spinning a system of yarn formation, also known as wrap spinning, in which the feedstock (sliver or roving) is drafted and the drafted twistless strand is wrapped with a yarn as it passes through a rotating hollow spindle. The binder or wrapping yarn is mounted on the hollow spindle and is unwound and wrapped around the core by rotation of the spindle. The technique may be used for producing a range of wrap-spun yarns, or fancy yarns, by feeding different yarn and fibre feedstocks to the hollow spindle at different speeds.

Lap a sheet of fibres or fabric wrapped around a core.

Marl yarn a yarn, usually woollen spun, consisting of two or more ends of different colours twisted together.

Monofilament a yarn consisting of a single filament. Monofilament yarns can be woven, knitted or converted into non-woven structures.

Multifilament yarn a yarn made up of more than one filament or strand.

Pick a single weft of thread in a woven fabric.

Ply the number of layers in a fabric. Also used to denote the number of yarns twisted together to form a single thread or yarn.

Roving a collection of relatively fine untwisted filaments or fibrous strands that are normally used in the later or final processes of preparation for spinning.

Satin weave a warp faced weave in which the binding places are arranged with a view to producing a smooth fabric surface, free from twill.

Sett a term used to define the weft or warp density of a woven fabric, usually in terms of the number of threads per centimetre.

Shin gosen fabrics made from ultra-fine polyester filament yarns with enhanced comfort, handle, drape and aesthetics. Shin gosen fabrics are designed specifically to appeal to end users by employing a combination of sophisticated fibre and fabric processing technologies.

Staple fibres short length fibres, as distinct from continuous filaments, which are twisted together (spun) to form a coherent yarn. Most natural fibres are staple fibres, the main exception being silk, which is a filament yarn. Most man-made staple fibres are produced in this form by slicing up a tow of continuous filament.

Tenacity the strength of a fibre or yarn. Tenacity is measured by dividing the force required to break the yarn by its linear density. Usually tenacity is quoted as the applied force in centinewtons divided by the linear density in tex (cN/tex).

Tex a measure of the linear density; the weight in grams of 1000m of yarn.

Textured yarn a continuous filament yarn that has been processed to introduce durable crimps, coils, loops or other fine distortions along the length of the filaments.

Tops sliver – usually formed by combing wool to remove short fibres – which forms the starting material for worsted spinning. Man-made fibre tops can be made by the cutting or controlled breaking of continuous filament tow.

Tow the name given to an untwisted assembly of a large number of filaments; tows are cut up to produce staple fibres.

Warp yarns or threads that run in the length direction on a weaving machine (the direction in which the fabric moves during manufacture). Also, yarns incorporated along the length of a woven fabric.

Web a sheet of fibres produced by a carding machine (carded web) or combing machine (combed web). See also Batt.

Weft yarns that are incorporated across the width of a woven fabric. The threads usually run at 90° to the warp direction on a weaving machine.

Index

An Introduction to Textile Coloration: Principles and Practice, First Edition. Roger H. Wardman.
© 2018 John Wiley & Sons Ltd. Published 2018 by John Wiley & Sons Ltd.